Preface

It is axiomatic that water development projects, by their very nature, will have impacts in and around the regions they are located. The nature, magnitudes, and durations of the impacts will depend on many factors, among which are the scale and type of the projects, characteristics of the regions, development activities during and after the projects are constructed, extent to which the projects can act as catalysts to bring new development to the region, social and environmental implications of the projects, governmental policies that are linked to water and associated development sectors, and efficiencies of management of the projects.

Water projects affect regional development through a variety of pathways, known and unknown. The impacts can be direct, indirect, or tertiary. They could vary over time and space, both within and outside the region. What is certain is that large water development projects will invariably have numerous impacts on the regions, some of which can be managed with appropriate policy interventions but others may be difficult to control. The question thus is not whether water management projects can affect regional development, because invariably they will, but rather how a water development project can be planned, implemented, and managed right from the very beginning so that its positive impacts are maximized and negative impacts minimized. In other words, the focus should be on how to maximize the net benefits and the number of beneficiaries from a project and also on how the people who may have to bear the direct costs of a project can be transformed into beneficiaries.

What is somewhat surprising, however, is that while no sane individual will argue with the fact that large water development projects are bound to affect the regions concerned in a variety of ways, neither water resources professionals nor development professionals have comprehensively studied how such projects can affect the regions from the pre-planning phase (that is once a decision is announced on the construction of a project) to its post-operational phase. Unless these impacts are carefully and objectively monitored and evaluated at periodic intervals, it simply is not feasible to formulate and implement policy options which can maximize the benefits and minimize the costs, thus ensuring maximum net benefit to society as a whole. The number of water projects anywhere in the world, which have had an impact on regional development and have been seriously monitored and evaluated at periodic intervals during their operational phase, can be counted on one's fingers with most of the fingers left over!

The paucity of reliable studies has contributed to several negative developments, among which are the following:

a) The validity of many of the hypotheses that are now almost universally accepted are not known. Anecdotal evidences indicate that many of the hypotheses that are currently being used need to be carefully reassessed. Some hypotheses may have to be modified in the light of the experiences and findings in the real world, and a few may have to be jettisoned completely.

b) Currently, limited knowledge is available on what works and what does not work, where and why, and under what conditions. Unless the existing knowledge base in this area is expanded, net benefit maximization from individual projects cannot be assured. It is now known that what works in one country may not work in another for a variety of physical, technical, economic, social, cultural, environmental, legal, and institutional conditions. Thus, what could contribute to a successful project in China may not produce similar results in Brazil or India, and vice versa.

c) There is increasing opposition to large-scale development projects, irrespective of their overall benefits, from certain non-governmental organizations. To the extent such oppositions can encourage rethinking of planning, construction, and operation of projects with a view to improve their efficiencies and net benefits, these should be welcomed. What is unwelcome is the blanket condemnation of all large-scale water development projects, irrespective of their benefits, because of dogmatic reasons, hidden agendas, and/or vested interests of certain groups. It is not generally known that globally NGOs have become a thousand-billion-dollar industry, many of which use media manipulation, and high-power marketing to impose their views and agendas on people and policy makers around the world. Passionate dedication of some of their members to the cause (in this case anti-water development projects), self-righteous belief in their own moral superiority (corollary, anyone who does not share their views is totally wrong), and slick organization to reach their goals have meant that they have made little effort to understand the overall benefits and costs of properly planned and implemented water activities. Given their current influence, this ignorance has already proved to be especially harmful to most developing countries.

The main objective of this book is to review the experiences from different parts of the world to see how specific water development projects have influenced regional development so that appropriate lessons (both good and bad) can be learnt. By synthesizing these experiences, and learning from these findings, it should be possible to plan, design, implement, and operate future water projects better, so that they can contribute more to regional income distribution, poverty alleviation, and environmental protection, than what have been witnessed in the past. It is also hoped that the facts, figures, and analyses contained in the case studies of this book will encourage all parties to discuss objectively and constructively how best water development projects can be used in developing countries to accelerate regional development, contribute directly to poverty alleviation, and improve the quality of life of people in the regions concerned.

Finally, on behalf of the Third World Centre for Water Management, I would like to express our most sincere appreciation to the Nippon Foundation for supporting this study on impact evaluation of large water development projects. The present book is the first major publication from this project. Several other books will follow during the next three to four years, which will be the direct outputs from this project. We are especially grateful to Reizo Utagawa, managing director of the Foundation, Shuichi Ohno, director, and Masanori Tamazawa of the International Affair Division of the Nippon Foundation. Without their support, encouragement, and advice, the present book simply could not have been prepared.

Asit K. Biswas
President
Third World Centre for Water Management, Mexico

Contributors

ZAHIR UDDIN AHMAD, Development Advisor, The Royal Netherlands Embassy, Dhaka, Bangladesh.

FELIPE B. ALBERTANI, Consultant, Environment Division, Sustainable Development Department, Inter-American Development Bank, Washington, DC, USA.

NASER JAMES BATENI, Chief, Division of Planning and Local Assistance, California Department of Water Resources, Sacramentos, California, USA.

ASIT K. BISWAS, President, Third World Centre for Water Management, Mexico.

LILIAN DEL CASTILLO DE LABORDE, Professor of Law, University of Buenos Aires, Law School, Argentina.

LUIS E. GARCIA, Consultant, Inter-American Development Bank, Virginia, USA.

RAJIV K. GUPTA, Commissioner for Higher Education, Gujarat, India.

TSUYOSHI HASHIMOTO, Executive Director, RECS International Incorporated, Tokyo, Japan.

JAMES E. NICKUM, Professor of International Relations, Tokyo Jogakkan College, Tokyo, Japan.

DIEGO J. RODRÍGUEZ, Economist, Environment Division, Sustainable Development Department, Inter-American Development Bank, Washington, DC, USA.

YUTAKA TAKAHASI, Professor Emeritus, University of Tokyo, Tokyo, Japan.

CECILIA TORTAJADA, Vice-President, Third World Centre for Water Management, Mexico.

ANTHONY R. TURTON, Head, African Water Issues Research Unit, Rant en Dal, South Africa.

OLCAY ÜNVER, Former President, GAP Regional Development Administration, Ankara, Turkey.

Contents

Tables and Figures   xi
Abbreviations   xvi

1. Water and Regional Development   1
   Asit K. Biswas
2. Asian Experiences in Water and Regional Development   14
   Tsuyoshi Hashimoto
3. Water Resources, Impoverishment, and Regional   54
   Imbalances: Some Reflections from the State of Gujarat,
   India
   Rajiv K. Gupta
4. Water Development Potential Within a Basin-wide   83
   Approach: Ganges–Brahmaputra–Meghna (GBM) Issues
   Zahir Uddin Ahmad
5. Water and Regional Development in the Yellow River   114
   Basin
   James E. Nickum
6. Current Situation and Future Prospects of Water   137
   and River Management in Japan
   Yutaka Takahasi
7. Participative Water-based Regional Development in   154
   the South-Eastern Anatolia Project (GAP): A Pioneering
   Model
   Olcay Ünver • Rajiv K. Gupta
8. South-Eastern Anatolia Project: Impacts of the   190
   Atatürk Dam
   Cecilia Tortajada
9. Evolution of Water Management Institutions in Select   251
   Southern African International River Basins
   Anthony R. Turton

10.   Water Resources Development in California   290
      Naser James Bateni
11.   Contribution of Water Resources Development to   309
      Regional Development: Case Studies from Latin America
      Luis E. Garcia • Diego J. Rodríguez • Felipe B. Albertani
12.   Salto Grande: A Binational Dam on the Uruguay River   331
      Lilian del Castillo de Laborde

      Index   368

Tables and Figures

Tables

2.1 Changes in per capita GRDP in different regions of   18
     Thailand showing increasing disparities
2.2 Changes in household income and other socio-   20
     economic indices in Bangkok and north-east Thailand
2.3 Integrated regional development planning: Cases   25
     supported by JICA
2.4 Evaluation of water resources potential for north-east 32
     Thailand
2.5 Estimate of available groundwater potential in   33
     north-east Thailand
3.1 Water availability in dams in Gujarat   56
3.2 Villages with no water sources and facilities   57
     throughout the year (as on 1 April 1998), Gujarat
3.3 Relationship between sources of drinking water and   60
     selected HDIs
3.4 Gujarat—Census of families below the poverty line   61
     (BPL) in rural areas (as of 1 April 2000)
3.5 Gujarat—Comparison of families below the poverty   63
     line in rural areas in the Eighth and Ninth Five-
     Year Plans
3.6 Absolute change in utilizable groundwater   67
3.7 Fluoride situation in selected areas of Gujarat   75
3.8 Fluorosis amongst children in Gujarat   75
4.1 Electricity consumption, 1996   100
4.2 India—Installed power capacity, 1998   101
4.3 Electricity consumption in the countries of the GBM   103
     region, 1996
4.4 Commercial energy use (oil equivalent)   104

5.1	Three reaches of the Yellow River	120
5.2	Areas of Yellow River provinces in the Huang–Huai–Hai basins	121
5.3	Average farm household income, 1990 and 1997 (Shandong = 100)	124
5.4	Yellow River diversions in Shandong province	126
5.5	Total economic losses attributed to flood, drought, and urban shortage, 1950–90 (in million Yuan Renminbi at 1990 prices)	129
5.6	Allocation of consumptive use of water (in million m^3 per annum) from the Yellow River prior to the completion of the south to north transfer project(s)	131
6.1	History of river management in Japan	151
7.1	Decadal population growth rates in 2000 (percentage)	156
7.2	Vital statistics and age groups in 2000	156
7.3	Distribution of population in 2000 by three major age groups (per cent)	157
7.4	Share of households with ten or more members (1995), GAP region	157
7.5	Average household income (1994)	158
7.6	Per capita GDP in GAP provinces according to DPT (State Planning Organization) and DIE (State Institute of Statistics) sources (at 1998 prices)	159
7.7	Daily poverty lines and poor households in terms of per capita and per household minimum food expenditure and cost of basic needs	161
7.8	Ranking of GAP provinces in terms of their socio-economic development status	162
7.9	Labour force status in the GAP region, 2000	163
7.10	Turkey—GAP region school enrolment	164
7.11	Rank of the provinces in the GAP region in gender empowerment measurement (GEM) index	166
7.12	Breakdown of drinking water facilities in villages of the GAP region	167
7.13	Land use in 1985	169
7.14	Present irrigation in the GAP region	169
7.15	Electricity production in the GAP region of Turkey	170
7.16	Labour force need in the GAP region with full Irrigation	172

7.17 Gross agricultural output value (GAOV) and value added from irrigation in the Harran Plain as a sample area — 173

7.18 GAP-GIDEM services (15 September 1997–31 May 2000) — 175

7.19 Projections as per the development plan alternatives — 184

7.20 Investment and areas to be irrigated under the development alternative plans — 185

8.1 Water and land resources development projects in the GAP region — 192

8.2 Fishery activities conducted by the DSI — 210

8.3 Existing fish species in the Atatürk Dam Lake — 211

8.4 Fishing activities in the Atatürk reservoir — 212

8.5 Freshwater fish yield in Adıyaman, 1993–7 — 215

8.6 Freshwater fish yield in Şanlıurfa, 1993–7 — 218

8.7 Resettlement of population due to the construction of the Atatürk Dam — 225

8.8 Resettlement of population due to the construction of the Atatürk Dam — 229

8.9 Number of families resettled due to the construction of large dams in Turkey, 1970–97 — 237

9.1 International river basins in southern Africa (after Pallett, 1997: 71) — 253

10.1 Historic development of reservoirs in California — 300

11.1 Water-related projects approved by the Inter-American Development Bank, 1961–2001 (in 1995 US$ million) — 314

11.2 Paradigm shift in water resources approach — 316

Figures

2.1 Procedure of integrated regional development planning — 28

2.2 Water supply and demand balance in Central Luzon — 36

2.3 Projected water deficits and identified water resources development projects in Central Luzon in the Philipines — 37

2.4 Water balance in 2015 (without project) with maximum potential lowland development and proposed cropping patterns — 39

2.5   Uma Oya multipurpose project with trans-basin water   40
      diversion and contour canals for effective soil and
      water conservation

2.6   Complementary development of economic spheres   49
      with water diversion from the Salween to Chao Phraya

2.7   Complementary development of border regions   50
      through water diversion and technical cooperation
      between Laos and Thailand

3.1   A comparison of surface water availability and the   55
      population of Gujarat with India as a whole

3.2   Per capita availability of water in Gujarat   56

3.3   Utilizable surface water (31,500 mcm) in Gujarat   58

3.4   Sources of irrigation in Gujarat   66

3.5   Literacy rate in Gujarat   69

3.6   Main workers in Gujarat (as a percentage of the   71
      total population)

3.7   Non-workers (females) in Gujarat (as a percentage of   71
      the state's population)

3.8   Small-scale industrial units in Gujarat, 1999   72

3.9   Industrial water requirements   73

3.10 Density of forests in Gujarat   74

3.11 Area affected by salinity   76

4.1   Map of the Ganges–Brahmaputra–Meghna region   87

5.1   Estimated indirect loss due to droughts in the   128
      Yellow River Basin, 1950–90

7.1   Map of the GAP region in Turkey   155

7.2   DIE, education ratios in the GAP region according   164
      to cities and gender (%)

7.3   Region-wise cotton production in Turkey, 2000   173

7.4   Manufacturing firms (with more than ten employees)   175
      in Şanlıurfa, Adıyaman, Diyarbakır, and Mardin

8.1   Number of people employed at the Atatürk Dam   203
      site, 1983–96

8.2   Annual gross salaries for personnel working for the   205
      Akpinar Construction Company, 1985–97

8.3   Monthly gross salaries of unskilled workers at the   205
      Atatürk Dam, ATA Construction Company, 1984–96

8.4   Freshwater fish yield in Adıyaman, 1993–7   215

8.5   Freshwater fish yield in Şanlıurfa, 1993–7   218

10.1	Relief map of California	291
10.2	Annual variability of the flow of the Sacramento River	293
10.3	California's major water projects	295
10.4	The Sacramento and San Joaquin rivers converge to form the delta, the hub of California's surface water delivery system	297
10.5	Regional imports and exports of water	303
12.1	The Uruguay River Basin	332
12.2	The Salto Grande Dam site	353
12.3	The Salto Grande lake	354
12.4	The Salto Grande power network	359

Abbreviations

AHI	average household income
AMS	aggregate measurement of support
ANC	African National Congress
bcm	billion cubic metres
BOO	build-own-operate
BPL	below poverty line
COPEL	Companhia Paranaese de Energia
CVP	Central Valley Project
DIDP	Davos Integrated Development Programme
DRC	Democratic Republic of Congo
EIA	environmental impact assessment
FAO	Food and Agriculture Organization
FDI	foreign direct investment
GAP	Güneydoğu Anadolu Projesi
GDI	gender development index
GIDB	Gujarat Infrastructure Development Board
GIS	geographic information system
GRDP	gross regional domestic product
GRP	gross regional product
GNP	gross national product
GWh	gigawatt-hours
IBT	inter-basin transfer
IDB	Inter-American Development Bank
IIPS	International Institute for Population Sciences
IRDP	integrated regional development planning
IUIDP	integrated urban infrastructure development project
IWRM	integrated water resources management
JBIC	Japan Bank for International Cooperation
JIA	Joint Irrigation Authority
JICA	Japan International Cooperation Agency

JPTC	Joint Permanent Technical Commission
kWh	kilowatt-hours
LBPTC	Limpopo Basin Permanent Technical Committee
LHDA	Lesotho Highlands Development Authority
LHWC	Lesotho Highlands Water Commission
LHWP	Lesotho Highlands Water Project
maf	million acre feet
MAR	mean average runoff
NEPA	National Environmental Policy Act
NGO	non-governmental organization
NSDP	net state domestic product
ODA	Official Development Assistance
OECF	Overseas Economic Cooperation Fund
OKACOM	Permanent Okavango River Basin Water Commission
ORASECOM	Orange–Senqu River Commission
PAPs	project affected persons
PPM	parts per million
PRODURSA	Global Programme for Urban Development and Sanitation
SAARC	South Asian Association for Regional Cooperation
SADC	South African Development Community
SADCC	South African Development Coordination Conference
SARCCUS	South African Regional Commission for the Conservation and Utilization of the Soil
SDP	state domestic product
SOIWPD	Southern Okavango Integrated Water Development Project
SWP	State Water Project
TCTA	Trans-Caledon Tunnel Authority
TPTC	Tripartite Permanent Technical Committee
TWh	Terawatt hours
UTE	Usinas y Transmisiones Electricas
VAT	value added tax
VNJIS	Vioolsdrift and Noordoewer Joint Irrigation Schemes
WHO	World Health Organization
WTO	World Trade Organization

1

Water and Regional Development

Asit K. Biswas

Introduction

Water resources management, like the management of any other natural resource, has been an evolving process. As our knowledge base increases, technology improves, and societal needs, views, and aspirations change, management practices change as well. Over the past fifty years, water management practices and processes have changed, mostly incrementally. The rate of these changes has accelerated during the 1990s. It is highly likely that the world of water management will probably change more during the next twenty years, compared to the past 2000 years. Thus, it is absolutely essential to consider how best to cope with these changes, pre-emptively, correctly, and in a cost-effective manner.

If the twentieth century is considered, historically, the main objective of water resources development has been the concept of economic efficiency, and the technique used for its evaluation has traditionally been benefit–cost analysis. Thus, the Federal Reclamation Act of the United States (US) of 1902 required economic analyses of projects, and the 1936 US Flood Control Act stipulated that benefits to whomsoever they may accrue should exceed project costs. Such laws set the stage for water resources planning and decision making for decades to come. Even a century after the US Federal Reclamation Act was passed, economic analyses of water projects continue to be an essential requirement for decision-making purposes.

As the twentieth century progressed, new objectives for water resources management were added. For example, during the 1930s, water development projects were used in the US as a means to reduce unemployment, and promote regional development. By the 1950s, use of water development projects to promote regional income redistribution was considered to be a valid objective. However, in spite of the merit of this objective, and also its conceptual attractiveness, it has not received the type of wide acceptance or permanence that has been accorded to the economic efficiency objective. During the post-1980 era, explicit use of regional income redistribution objective had already taken a back seat for planning and managing new water projects.

As the 1970s and 1980s progressed, another new objective was gradually added: it was that of environmental quality. It should be noted that this new objective was added not only for water planning, but also for all other types of development projects. During the 1990s, environmental quality considerations became an increasingly important objective, a development that should be welcomed. Unfortunately, however, the concept of using water projects as an engine for regional development and regional income redistribution lost ground steadily as an objective.

At the beginning of the twenty-first century, it is now becoming increasingly apparent that, at least for the developing countries, the role water projects can play, and should play, to promote regional development, thus improving the lifestyles of the people, especially the poor and the landless, needs to be urgently re-examined. Prima facie, it appears that this objective should receive as high a priority as those of economic efficiency and environmental quality.

Water and Regional Development

Mustafa Kamal Atatürk said: 'I am leaving no sermon or dogma, nor am I leaving as my legacy any commandment that is frozen in time and cast in stone. What I leave behind is reverence for scientific knowledge and decent judgement.' The water profession will do well to take note of Atatürk's recommendation in terms of 'reverence' for 'scientific knowledge and decent judgement' to meet the changing needs and aspirations of the society. It appears that the profession threw away the baby with the bathwater, when it allowed the objective of regional income redistribution to gradually fade away during the post-1980 period.

During the 1980s and 1990s, water development projects have come under concerted attacks from social and/or environmental activists, often because of wrong or dogmatic reasons, and/or due to the vested interests of the people concerned. Unfortunately, these attacks were neither taken seriously nor properly countered by water resources development professionals for at least twenty years. Furthermore, just as the activists were well organized, the water experts were equally unorganized, thinking that these attacks on water projects were short-term developments. Thus, it was felt that the best strategy was to ignore these attacks and they would disappear, or at least decline in intensity, within a reasonably short period of time. Unfortunately, exactly the reverse happened: their silence and apathy encouraged the activists to attack water projects in even more strident terms. In addition, while the opponents of the water projects were media-savvy, the proponents were not. As a result, public perception in many places, especially in developed countries, gradually became that water projects did more harm than good to the society and the environment.

Unquestionably, these unwarranted and often misguided developments have seriously hampered the construction of many good water projects in developing countries, or at the very least delayed their implementation by years, if not decades. One serious adverse impact of the current state of affairs has been that these delays have seriously affected the poverty alleviation efforts through regional development, using water as the engine for economic growth.

Studies from different parts of the world indicate that properly planned, constructed, and managed water development projects can make significant and lasting contributions to the social and economic development of the project areas. The impacts can be primary, secondary, and tertiary, and they often vary in the project areas in terms of time, spatial distributions, and intensities. Many impacts may extend well beyond the project area, to the country as a whole, or even beyond national boundaries.

While there is no question that the overall impact of properly planned water projects are overwhelmingly positive, it is equally certain that they contribute to some negative developments as well. This is not a special case for water development projects, but is also equally valid for any infrastructural development project, or, for that matter any development policy. For any development project, or for any policy, it is axiomatic that some people will benefit and equally others will pay the cost, at least in terms of higher taxes. It is absolutely impossible to consider any project or formulate a development policy that can have only beneficiaries, and not a single person will have to pay the cost.

Since all projects and all policies result in both benefits and costs, it is necessary for decision-making purposes to determine the following:

a) nature of the beneficiaries, that is, who benefits and who pays the costs because of the construction of a project;

b) how to make the people who are likely to pay the costs of the projects become its beneficiaries, or at the very least reduce their cost burden to the minimum possible and also to compensate them properly for the costs (both financial and intangible) they have to bear;

c) take appropriate measures to maximize the benefits and minimize the costs; and

d) ensure that the projects result in overwhelmingly positive benefits for the society as a whole, including the environment.

The corollary of these considerations is that if the total costs for any project or policy exceed the benefits to the society as a whole, that particular project or policy should not be implemented. This conclusion is valid not only for water policies and projects but for any type of development policy or project.

Water Development as an Engine for Regional Growth

Water resources development professionals have not seriously considered the fact that water projects can significantly improve regional development processes. As such, the various interrelationships and interlinkages between the two are not known accurately and comprehensively at present, except for a very few specific projects, and that too in a somewhat anecdotal fashion. There are at present many anecdotal evidences of positive interlinkages between the two, which also have never been studied either systematically or in-depth. In addition, an important policy issue that has been mostly ignored by both water resources and development professionals is how a water project should be planned, designed, constructed, and operated from the very beginning so that it can have the maximum possible positive impact on the development of the region concerned, and also contribute to improving the standards of living of the maximum number of people, especially the poor.

Many regional development implications, which stem from the construction and operation of water projects, can easily be identified, both conceptually and also by some available facts and figures.

Provision of reliable and clean water supply to domestic, commercial, and industrial consumers is an important issue since the developing world is becoming rapidly more and more urbanized. As the urban population increases, the catchment areas in which these urban centres are located experience concurrently both water scarcities and increased water pollution. At a certain stage, industrial growth and the attendant employment opportunities may become constrained, unless the consumers receive the appropriate quantity and quality of water they need. If adequate quantity and quality of water is not available to the domestic consumers, they will face many problems, most important of which could be the following:

a) Health: Incidences and frequencies of occurrence of water-related diseases may increase.

b) Economic: People, especially the poor, are often forced to buy poor-quality water from vendors at high costs, which adversely affects their finances as well as their health. Poor health also means additional economic burden on families in terms of obtaining medical help and medicines, as well as loss of income due to sicknesses. This is an especially important issue for daily labourers, who receive no income whatsoever for the period they are unable to work because of illness.

c) Social: In many urban areas, and in most rural areas, people (mainly women and children) are forced to spend considerable time in collecting water for domestic consumption and use. The time wasted in making journeys to collect water each day is time which could have been used for more productive activities.

The three types of problems noted here are closely interrelated: one affects the other, and, in turn, is affected by the others. These problems can be seen for both urban as well as rural families.

These three issues can be considered to be traditional, at least in the sense that they are well known to the water resources and development professionals. However, once the scope of water planning is broadened, and water development projects are specifically planned to promote regional development and to improve the standards of living of the people, especially the poor, the implications increase manifold. Among the new implications are the following issues.

a) Food availability: Introduction of irrigation increases the total agricultural production of the area in two main ways, by increasing the yields per unit of land and also by expanding the total area cultivated.

The adverse impacts of floods and droughts are very significantly reduced by water projects.

b) Nutrition and food security: Overall, both nutrition and food security improves in the irrigated areas because of crop diversification and more reliable food production. Per capita food production in the area increases, and families grow more food in their land, as well as more varieties of crops.

c) Employment generation: Employment opportunities for skilled and unskilled labour increases significantly in the project area because of: (i) construction-related activities for dams, canals, drainage systems, other hydraulic structures, roads, and other related infrastructures; (ii) irrigated agriculture which significantly increases the need for labour, and thus employment opportunities; and (iii) increased to agricultural production which brings new agro-industrial developments to the region, thus further increasing its employment needs. All these, in turn, contribute to the growth of all types of service facilities, and thus further increase the employment potential of the area.

d) Transportation: Transportation facilities increase very substantially during the construction of a project and immediately thereafter. Dams are normally constructed in the upper catchment areas, which often suffer from the lack of adequate transportation and communication facilities while the lower catchment areas which are generally more populous, have greater employment opportunities and better standards of living. One of the very first activities of water development projects is to link the dam sites in the upper catchment areas to the towns and cities in the lower regions. This immediately links the highlands with the lowlands in terms of trade, commerce, access to educational and medical facilities, introduction of new ideas and technology, and numerous other direct and indirect linkages.

Road systems are generally built at least on one side of the embankments of the canals. This links up the villages in a way that did not exist before. The development of new transportation networks open up new opportunities for increased commercial activities and social interactions. As these activities increase, more and better transportation networks have to be built, which further improves the transportation facilities within and between towns and villages of the region. Thus, as construction begins, transportation facilities of the region start to improve, often dramatically. In other words, construction of a water project acts as a catalyst to open up areas that were somewhat inaccessible before by improving very substantially the transportation facilities that existed earlier.

e) Energy availability: Water development projects increase the energy availability of the region in three distinct ways. First is the generation of hydroelectric power. Electricity is an essential requirement for improving the lifestyles of rural and urban people, and also increasing the opportunities for commercial and industrial development. At present, the electricity needs for most developing countries are increasing exponentially. In recent years, even for populous countries like Brazil, China, India, and Turkey, total electricity consumption has increased at an annual rate of 5–12 per cent compounded.

Water projects not only generate electricity, but since hydropower is a non-consumptive use, water released after power generation is subsequently used for agricultural, commercial, and industrial purposes, as well as for domestic consumption. Furthermore, hydropower is a renewable source of energy, which means that non-renewable resources like oil, gas, or coal can be saved. This is an especially important issue for most developing countries, which do not have adequate hydrocarbon resources nationally for power generation. Foreign exchange in hard currencies is thus saved by not importing petroleum and natural gas from other countries.

For most large dams, electricity generated is significantly more than what can be efficiently used in the region. Thus, the electricity generated is transferred through the national grid, which means much of the country benefits from such development activities.

There are other ways through which the energy situation improves the social and economic conditions of the region. One is through increases in crop production and higher cropping intensities. Changing rain-fed agriculture to irrigated agriculture means that one crop a year is replaced by two or three crops per year. Both the higher cropping intensities and increased agricultural production result in the generation of significantly higher quantities of agricultural residues. Experiences from India, China, and Thailand indicate that poorer households burn these agricultural residues for cooking purposes, thus reducing the need for fuelwood.

Another indirect energy-related issue is that the use of commercial fuel increases with increase in the incomes of the people due to the new economic opportunities available. Poorer people move from exclusive use of fuelwood for cooking to commercial energy like kerosene and bottled gas. All these developments contribute to reduction in deforestation rates. These are very positive changes which occur frequently due to water projects. Regrettably these have has been mostly ignored by professionals in the field of water resources development and environment.

f) Availability of better social facilities: Social facilities like better healthcare and education invariably become an important by-product of water development projects, even when these projects are not planned as an integrated component of regional development. When the Atatürk Dam in Turkey was constructed, the first pharmacy of the region was established near the construction site. This meant that people of the area did not have to travel for hours, using poor transportation facilities, to get the required medicines. With improved transportation and communication facilities and higher levels of economic activities, educational levels of children improve for a variety of reasons. Among these reasons are: (i) more primary and secondary schools are opened because of increased economic and commercial activities, and immigration of people; (ii) access to schools is improved. For example, before the Atatürk Dam was construed, many young people had to walk 4–5 km each way to go to schools, but as soon as proper roads were constructed, students started to travel by buses. These improved facilities have also meant that more girls are going to school than even before; (iii) quality of teaching improves. Before the projects are constructed, teachers from outside the project area often refuse to come because of the isolated nature of the region. However, once the area opens up, more and more teachers decide to move to the region, thus improving the quality of teaching in schools.

g) Gender-related issues: The indirect and tertiary impacts on women from water and regional development has been a totally neglected issue. The evaluation of the Bhima Command Area Development Project in India indicates that it had a very high impact on the education of children, especially girls. Before the project was completed, many families had to continuously travel from one place to another, looking for employment opportunities. Girls invariably moved with their families, and thus they seldom, if ever, attended schools. After the Bhima irrigation project was completed, employment opportunities within the region expanded exponentially (Biswas, 1987). As a result, it was no longer necessary for families to move from one location to another in search of jobs. These nomadic labourers have now become sedentary. One direct impact of this permanency has been that they are now sending their daughters to schools, without any encouragement from the government or any donor agency. Similar progress can be noted due to the constructions of the Teesta Barrage in Bangladesh.

h) Environmental impacts: Much has been written on the environmental impacts of water development projects. There are few issues that are worth noting.

Very few large water development projects have been scientifically and objectively assessed for their environmental impacts. In contrast, thousands of environmental impact assessments (EIA) are now available. These EIAs are forecasts. Regrettably, no one checks if these forecasted impacts are correct. For example, an objective review of the environmental impacts of the Aswan High Dam in Egypt, after some thirty years of operation, indicates that most of the forecasts were hopelessly wrong (Biswas, 2002). In the absence of such serious ex-post impact studies, errors in forecasting the environmental impacts have continued to be perpetuated.

Environmental impacts, in the context of water development projects are always considered to be negative. The positive impacts are never considered. This mindset of the analysts needs to be changed. What is needed in the future is identification and estimation of both positive and negative impacts in an objective and unbiased manner.

The negative impact of water development projects, both real and imaginary, have received extensive coverages. Accordingly, in this chapter only the positive impacts are outlined here based on studies carried out by the Third World Centre for Water Management.

i) Reduction of deforestation: Assessment of the Bhima Project indicates that it has reduced the rate of deforestation very significantly (Biswas, 1987). Need for fuelwood for cooking has declined because of the reasons mentioned earlier. In addition, as the incomes of the people have increased, and more fodder is now available, livestock holdings have exploded compared to the pre-project conditions. Dried cow dung has now become an important fuel for cooking. Thus, numerous families who depended upon fuelwood for cooking, now do not use them, or have reduced their use, because of: (a) increased availability of agricultural residues and cow dung; and (b) higher income, which has allowed them to move up the economic ladder, and thus use commercial fuels like kerosene or cooking gas. These two unexpected developments have reduced the rates of deforestation in the area significantly.

ii) Increase in biodiversity: For the Bhima Project and the Indira Gandhi Nahar Project in Rajasthan, India, both constructed in highly arid areas, all the available anecdotal evidences indicate that they have significantly increased the biodiversity of the area. What is now needed is serious and thorough scientific studies which can compare the levels and the natures of biodiversity before and after the projects were constructed. Regrettably, not even one such authoritative study of biodiversity has ever been carried out anywhere in the world, to enable comparision of pre- and post-project conditions.

Water-Based Regional Development: The Case of Bhutan

One of the best examples of using water development as an engine for economic growth of a region is the case of Bhutan.

Bhutan, often known as the Hermit Kingdom, was basically inaccessible to the world until 1960. In 1961, it initiated its first five-year development plan. In 1960, it had one of the lowest per capita income for any country, not only in south Asia but also in the rest of the world.

Bhutan is a landlocked country in the Himalayas. Because of the topography, its agricultural potential is very limited. It has a hydropower potential of about 20,000 megawatts (MW), which is slightly less than one-quarter of the potential of its western neighbour, Nepal. The country realized long time ago that water development is not an end in itself but a means to an end, where the end is to improve the lifestyles of the people of the region through a variety of pathways. Poverty alleviation and income distribution are two important objectives of such water-based regional development.

Bhutan and Nepal have followed different paths to develop the water resources of the international rivers that flow through them. Short on capital and expertise, Bhutan initiated a plan around 1980 to develop the hydropower potential of the Wangchu cascade of Chukha in collaboration with its southern neighbour, India. The Chukha is a 336 MW project constructed by India on the basis of 60 per cent grant and 40 per cent loan. The cost was estimated at Rs 2,450 million. A 220 kilovolt (kV) transmission line links the Bhutanese capital Thimpu and the city of Phuntsholing on the Indian border, from where electricity is supplied to four Indian states. The electricity generated is used first to satisfy Bhutan's own needs, which earlier was supplied by diesel and mini-hydro stations. The country's total electricity requirement has more than doubled since its initial use of 12 MW. In addition, the unit cost of electricity has steadily declined because of this higher and more economic scale of production. Gradual expansion of the electric network to different parts of Bhutan has meant reduced use of fuelwood than what would otherwise have been the case, as well as of diesel which had to be imported. Reduced fuelwood use has had a beneficial impact on the forests and the environment.

The electricity produced in excess of the requirement of Bhutan is purchased and used by India as peak power through its eastern electricity grid. Initially, the two countries agreed to have different pricing patterns for firm and secondary powers. Later, the two tariffs became the same. Subsequently, the tariff was increased. The revenues that Bhutan has been receiving from its electricity sales to India not only has serviced its debt load without any difficulty, but also left enough surplus to finance other development projects and support some social services, including, increasing the salary scales of its civil servants.

The Bhutanese development policy in recent decades has been to ensure that national economic development proceeds parallel to the country's social, economic, and institutional absorptive capacities. Environmental conservation has been considered to be another important national objective. The King of Bhutan wishes to ensure that there is no excessive tourism and industrial development, which could adversely impact on its culture, general lifestyles of its people, and the environment. The main focus is thus on developing the hydropower resources of the country, which could enhance the economic and social development processes of the country. The excess electricity generated will continue to be sold to India, which faces serious energy shortages.

India has further assisted Bhutan with the construction and funding of a 45 MW run-of-the-river power station at Kuri Chu. Similar collaborative efforts have taken place, or are under active consideration for Chukha II (1000 MW) and Chukha III (900 MW with a storage dam). An agreement was signed with India in 1993 to study the feasibility of a large storage dam on the Sunkosh River. If all these projects are completed, and assuming that unit price paid for electricity will continue to increase periodically, Bhutan can easily earn over $100 million per year from the export of its hydropower alone during the post-2010 period. For a country with a small population, this is indeed a very significant income which will accrue regularly year after year.

The 'win–win' approach adopted by Bhutan and India is a good example of how water resources can be used by a region (two countries share this region) to improve the lifestyles of the people, contribute to poverty alleviation, and simultaneously protect the environment. The developments were not narrowly focused on water only: it encompassed other primary water-related activities like hydropower, agricultural and industrial development, intermodal transportation system, within which navigation could be an important component, and environmental conservation. These, in turn, could directly lead to secondary impacts like employment generation, regional development, capacity building of the nation, and poverty alleviation. Viewed from any direction, the collaboration between Bhutan and India has been most beneficial to the two countries, including ensuring regional peace and stability. All these developments have resulted in Bhutan's per capita income increasing from one of the lowest in south Asia to almost one of the highest within a very brief time span of only two decades. The future looks even brighter if the focus on water-based regional development continues with associated safeguards.

Conclusion

If properly planned and implemented, water projects can act as the engine for regional development, especially in the arid areas of developing countries. If water development projects are to be successful in improving the lifestyle of the poor, and in ensuring environmental protection, they need to be carefully planned and executed within an overall framework of integrated regional development. This can be done, and this needs to be done, but this cannot be done until and unless water resources development professionals, regional planners, and development experts concurrently accept this thesis. Fortunately, current indications are that all the three groups are proceeding in this general direction. There is thus a convergence of ideas, and the time has come to use water development as the engine for economic growth for the regions concerned.

Even without direct and deliberate integration of water and regional development activities in the past, current indications are that water development projects have an impact on regional economic growth. For many underdeveloped areas in arid regions of developing countries, water development projects should be designed in such a way from the very beginning so that they become an integral component of an overall regional development framework. This will not only accelerate development activities of such areas, but will also improve the standard of living and the quality of life of millions of people.

References

Biswas, A.K., 'Aswan Dam Revisited: The Benefits of a Much-Maligned Dam', *Development and Cooperation*, No. 6, 2002, pp. 25–27.
Biswas, A.K., 'Monitoring and Evaluation of Irrigated Agriculture: A Case Study of Bhima Project, India', *Food Policy*, February 1987, pp. 47–61.

2

Asian Experiences in Water and Regional Development

Tsuyoshi Hashimoto

Introduction

The basic idea behind regional development is the utilization of indigenous resources in a region by and for the benefit of the local community in that region. The most inherent indigenous resources are water, land, and human resources. The development of human resources is the ultimate goal of development in any region, and other resources are utilized to attain that goal. These resources include forest resources, tourism resources, mineral resources, and, more broadly, capital stock, social habits, and institutions. Secondary resources may be defined based on these primary resources. These include education, the management of human resources, and agricultural products based on land and water resources.

Tourism which relies primarily on domestic tourists can make a modest contribution to regional development. Tourism-based regional development aimed at attracting international travellers can be quite volatile given that it will have to compete against other tourist destinations. Regional development relying on mineral resources can be risky as evaluating the mineral resources potential is generally difficult.

Sound regional development should be based primarily on water and related land resources in combination with other resources. Most resources can be introduced from outside under proper management for sound regional development. Water resources may be introduced by trans-basin water diversion, and even human beings may be introduced by voluntary or planned transmigration. These possibilities involve the classic argument between 'water to humans' and 'humans to water'. In principle, it is more natural and desirable to guide development and move population to a region rich in water resources rather than to bring water over long distances in order to fulfil the water requirements of people in a metropolitan region. In reality, however, trans-basin water transfer could become an important means to fully realize development potentials in regions rich in other resources and which already have a large population.

There exist a wide range of opportunities for water and regional development in Asia. These range from more authentic regional development based primarily on indigenous resources to opportunities for trans-basin water diversion and complementary development of neighbouring regions with shared water resources. The latter includes cases which involve the sharing of the waters of international Asian rivers such as the Mekong, the Ganges, the Brahmaputra, and the Salween among the neighbouring countries.

In this chapter, the Asian experiences in water and regional development in recent years are examined, and reference is made to representative cases of regional development of various kinds. In the next section, characteristics of water resources and regional development in Asia are analysed and implications of the former to the regional development in Asia are clarified. In the following section, how the water sector is treated in regional development planning is explained using Asian case studies. Subsequently, new issues and opportunities for water and regional development are discussed in general and also in the Asian context. Finally, conclusions are drawn to see if the Asian models may be developed to meet the challenges presented by new issues and opportunities.

Characteristics of Water Resources and Regional Development in Asia

Characteristics of Regional Development of Developing Countries in Asia

The basic idea behind regional development is essentially the same in both developing and developed countries as long as the local people are the focal point of the issue. It is concerned with how the local people utilize various available resources so as to satisfy their own development needs or how they improve various conditions both inside and outside the region for utilizing their resources most effectively to satisfy their development needs better. Initial conditions for regional development, however, vary very widely between developing and developed countries. The most serious difference is in the availability of development funds and human resources of high quality.

In most developing countries in Asia, the following conditions related to regional development are commonly observed:

a) large populations and high population growth and density;
b) large inter-regional disparities;
c) insufficient integration of national socio-economy;
d) unclear division of responsibilities between administrations at different levels under the general trend of localization; and
e) insufficient provision for basic infrastructure and services, including those for water and sanitation.

While these conditions are shared by most developing countries in other parts of the world, in the case of many Asian countries successful economic growth in recent years makes the combination of these conditions unique to these countries. High economic growth has created some of the largest population concentrations in these Asian countries, widening the inter-regional disparities in economic wealth, social services delivery, and provision of basic infrastructure. In some cases this has undermined socio-political stability. Localization of development administration may be more advanced in some developing countries in Asia, but this tends to highlight the issue of proper division of responsibilities between administrations at different levels. Specific conditions are discussed further.

Urbanization and Inter-regional Disparities

There exist more than ten major cities in developing countries in Asia which have a population over 10 million as of 2000. Mumbai, in India, leads the list with a population of 18.1 million while Shanghai in China has a population of 17.4 million. These are centres of economic wealth and power, and also centres of varying and increasing problems including water-related ones such as water shortages and pollution.

It is difficult to measure changes in inter-regional disparities in most countries in the developing world, but a recent study for Thailand indicates that disparities between the Bangkok metropolitan region and most other regions have actually widened in the past two decades of high economic growth (Table 2.1). In particular, the income disparity as measured by the per capita gross regional domestic product (GRDP) between the Bangkok metropolitan region and the poorest region of the north-east increased from 5.3 times in 1975 to 9.8 times in 1993. The disparity decreased thereafter to 8.2 times in 1998. A similar trend is observed for household income, although the disparity is smaller. It peaked at 3.5 times in 1992 and reduced to 2.9 times in 1998. Disparities in social services and basic infrastructure have generally decreased in the past decade (Table 2.2).

Development Administration

Localization and decentralization of development administration have been discussed for decades in some developing countries in Asia. In Sri Lanka, the devolution process was pursued by Bandaranaike, a government minister, as early as in the 1940s, to achieve three objectives: democratic decentralization, a responsive government administration, and people's participation in development. It was not until the Bandaranaike–Chelvanayagam pact was signed in 1957 that Sri Lanka started to progress towards the devolution process. The main features of the pact were: (i) creation of regional councils; (ii) delegation of administrative powers to the councils by an Act of Parliament; and (iii) block grants, taxation, and borrowing as the basis for finances of the councils. The process of devolution is still going on, and functional division between the Central government, provincial councils, and local governments are still not clear.

The Thai government has been taking a series of regulations to effect partial decentralization of development administration since 1932 when the completely centralized system under the monarchy was replaced by a constitutional democracy. Thailand's development administration, however, is still highly centralized. Of the three administrative levels: Central, regional, and local, regional administration represents de-concentration of Central administration rather than decentralization. The local administration represents some official duties assigned to local governments, where all or part of the officials are elected by the people, but their performance is constrained by limited financial and administrative resources.

Development administration in the Philippines consists of the Central government and three tiers of local government. Recently, many functions of government agencies have been delegated to their regional

Table 2.1: Changes in per capita GRDP in different regions of Thailand showing increasing disparities

	Bangkok Metropolitan	Central	Eastern	Western	Northern	North-eastern	Southern	Thailand
1975	18,827	6,291	9,541	9,344	5,388	3,527	5,899	7,220
1980	42,155	11,333	19,220	15,832	9,585	6,257	12,710	14,703
1985	59,003	16,749	25,603	21,047	13,353	8,124	15,358	20,220
1986	59,338	19,324	29,118	12,968	14,361	9,193	17,519	21,051
1987	69,065	21,359	31,165	21,333	14,361	9,193	17,519	23,355
1988	82,241	26,032	36,620	23,513	17,097	10,698	20,329	27,621
1989	96,329	30,587	45,751	28,434	18,833	11,981	21,956	32,001
1991	118,494	31,709	51,837	28,825	19,093	12,664	25,950	37,329
1992	125,209	35,382	56,882	30,888	20,491	13,253	26,806	39,840
1993	134,454	38,256	63,789	32,176	20,925	13,670	28,152	42,647
1994	141,452	43,960	71,236	34,020	22,264	14,940	30,521	45,908
1995	149,635	48,009	78,925	37,120	23,634	16,410	32,475	49,416
1996	152,531	52,200	89,952	37,630	25,089	17,122	33,900	51,820
1997	144,681	51,344	94,728	36,527	24,164	16,458	33,361	50,444
1998p	123,175	44,562	86,990	32,872	22,731	15,043	31,561	44,871

Per capita GDP/GRDP (Baht)

(Contd.)

18

Table 2.1 Contd.

	Bangkok Metropolitan	Central	Eastern	Western	Northern	North -eastern	Southern	Thailand
1975	5.3	1.8	2.7	2.6	1.5	1.0	1.7	2.0
1980	6.7	1.8	3.1	2.5	1.5	1.0	2.0	2.3
1985	7.3	2.1	3.2	2.6	1.6	1.0	1.9	2.5
1986	6.5	2.1	3.2	1.4	1.6	1.0	1.9	2.3
1987	7.5	2.3	3.4	2.3	1.6	1.0	1.9	2.5
1988	7.7	2.4	3.4	2.2	1.6	1.0	1.9	2.6
1989	8.0	2.6	3.8	2.4	1.6	1.0	1.8	2.7
1991	9.4	2.5	4.1	2.3	1.5	1.0	2.0	2.9
1992	9.4	2.7	4.3	2.3	1.5	1.0	2.0	3.0
1993	9.8	2.8	4.7	2.4	1.5	1.0	2.1	3.1
1994	9.5	2.9	4.8	2.3	1.5	1.0	2.0	3.1
1995	9.1	2.9	4.8	2.3	1.4	1.0	2.0	3.0
1996	8.9	3.0	5.3	2.2	1.5	1.0	2.0	3.0
1997	8.8	3.1	5.8	2.2	1.5	1.0	2.0	3.1
1998p	8.2	3.0	5.8	2.2	1.5	1.0	2.1	3.0

Index of per capita GDP/GRDP compared with the north-east as a base

Notes: 1. Actual data for 1991–7 at 1988 prices.
2. Provisional data for 1998.
Source: JSID, March 2001.

19

Table 2.2: Changes in household income and other socio-economic indices in Bangkok and north-east Thailand

	Household income (B×10³/year)			Per capita energy sales (kWh equivalent)			Per capita bank deposit (B×10³)			Average wage (B/month)			Engel's coefficient (%)		
	BKK	NE	Gap	BKK	NE	Gap	BKK	NE	Gap	BKK	NE	Gap	BKK	NE	Gap
1987															
1988	94.5	36.8	2.6									32.7	37.8		1.2
1989															
1990	140.7	42.3	3.3	2,915			113.8	4.2	27.0				29.8	40.9	1.4
1991				3,162			131.2	4.9	26.9						
1992	191.4	54.3	3.5	3,411	211	16.2	149.0	5.9	25.4				29.6	37.6	1.3
1993				3,646	241	15.1	176.8	6.8	26.2						
1994	197.0	67.2	2.9	3,847	272	14.1	192.7	7.8	24.8				29.8	36.6	1.2
1995					304		216.3	9.3	23.2						
1996	263.4	88.7	3.0		321		242.5	10.6	22.9	8,761	4,152	2.1	29.2	34.1	1.2
1997								11.1		8,833	4,881	1.8			
1998	299.1	102.6	2.9					11.2		9,550	4,950	1.9	29.5	40.4	1.4
1999	320.9	97.7	3.3										28.5	38.3	1.3

(Contd.)

20

Table 2.2 Contd.

Year	Population per doctor			Population per nurse			No. of public telephones per 1000 population			No. of telephone lines per 1000 population			Population per private vehicle			Population per private and commercial vehicle		
	BKK	NE	Gap	BKK	NE	Gap	BKK	NE	Gap	BKK	NE	Gap	BKK	NE	Gap	BKK	NE	Gap
1987							1.75	0.07	25.0	3.25	0.32	10.2						
1988	519	12,128	23.4	211	3,920	18.6	1.72	0.08	21.5	8.24	0.34	24.2	9.8	355.9	36.3	4.0	20.7	5.2
1989	436	11,691	26.8	174	3,631	20.9	1.68	0.09	18.7	9.30	0.38	24.5	11.4	431.8	37.9	4.4	19.9	4.5
1990	443	11,175	25.2	173	3,280	19.0	1.74	0.11	15.8	10.83	0.44	24.6	8.9	397.6	44.7	3.6	18.0	5.0
1991	455	10,690	23.5	188	2,888	15.4	1.90	0.14	13.6	12.30	0.55	22.4	8.9	355.2	39.9	3.6	15.2	4.2
1992	440	10,526	23.9	178	2,748	15.4	2.11	0.19	11.1	13.74	0.69	19.9	8.1	327.9	40.5	3.1	13.3	4.3
1993	448	10,751	24.0	178	2,597	14.6	2.32	0.24	9.7	16.52	0.83	19.9	7.4	220.6	29.8	2.8	10.6	3.8
1994	450	10,655	23.7	165	2,443	14.8	2.50	0.27	9.3	17.52	0.93	18.8	6.7	205.9	30.7	2.6	9.1	3.5
1995		10,746			2,233		2.89	0.31	9.3	17.66	0.95	18.6	6.5	169.6	26.1	2.4	8.1	3.4
1996		10,183			2,232		3.20	0.36	8.9	28.25	1.52	18.6	6.2	143.5	23.1	2.2	6.6	3.0
1997		9,754			2,090		3.50	0.73	4.8	46.52	1.83	25.4	5.8	120.7	20.8	2.4	6.0	2.5
1998							4.07	0.97	4.2	30.98	2.04	15.2	5.4	116.1	21.5	1.9	5.6	2.9
1999							5.57	1.29	4.3	31.60	2.16	14.6	5.3	103.2	19.5	1.8	4.9	2.7

Source: JSID, March 2001.

21

offices (de-concentration). Regional development councils, created in 1972, have been strengthened to improve development planning and coordinating functions at the regional level. The local government consist of provinces, cities, municipalities, and *barangays*. Each local government has executive officials and the local development council as the legislative body with elected members.

In addition to the hierarchical system of development administration, a separate institution is created sometimes to take charge of a specific region that is considered strategically important for the country as a whole. Among them are the Southern Development Authority in Sri Lanka, and the Metro Manila Development Authority and the Subic Bay Metropolitan Authority in the Philippines. Such a regional authority has not been established in Thailand, and the most successful case of the Eastern Seaboard (ESB) regional development has been managed by its committee established in 1992 under the chairmanship of the deputy prime minister. In Indonesia, the Brantas River development office established in 1967 worked effectively as a regional authority with clear mandates and autonomy to exercise its budget for this highly successful project. An innovative institutional set-up in the Philippines is the alliance of local governments established for the Davao Integrated Development Programme (DIDP). The main thrust of the DIDP is to pursue more integrated and effective development of combined jurisdictions of local governments by mustering various resources, including not only natural resources but also administrative, financial, and human resources.

Approach to Regional Development in Asian Developing Countries

The high economic growth in the recent past attained by south-east and east Asian developing countries was realized through the so-called state command economic development or top-down approach to development in regional development terminology. Although the period of high economic growth varied among the countries, regional development by state initiative was basically efficiency oriented, focusing mainly on the capital region of each country with less emphasis on equity including inter-regional balance. The exceptions were China and Indonesia due to their large territories. In China, the state command economic development was directed earlier to provinces in coastal areas and the north-eastern provinces as well as the capital region. In Indonesia, East Java was a focus of development since the late 1950s centring on the Brantas river basin. Also, the transmigration policy has been consistently taken since the first National Development Plan (BAPENAS, 1969–70) to develop the outer islands and to reduce the population pressure on the island of Java, although it was not very successful.

In other countries, emphasis of regional development has shifted slowly but steadily from the capital region to other regions. This reflects generally increasing urban problems due to over-concentration of population in the capital region and the impoverishment of rural areas. Another shift has been that from the dominance of engineering solutions to regional development to increasing emphasis on institutional measures. Typical examples of engineering solutions are highways, power plants, ports, and other infrastructure for the Eastern Seaboard of Thailand and coastal areas in China, and multipurpose dams for East Java in Indonesia. A popular approach adopted subsequently was the establishment of various free zones such as special economic zones and export processing zones, which provided a combination of improved infrastructure and incentive measures to attract private sector investments. Other institutional measures have been introduced in relation to designated areas under regional development policies.

Throughout this period of shifting emphasis, regional development in developing countries Asia has effectively utilized official development assistance, particularly technical cooperation and financial support from the Japanese Government. Development measures initiated by the Brantas river basin in East Java, Indonesia, presents a classic case and a success story of continued support of Japan's Overseas Development Assistance (ODA) as discussed elsewhere. A more recent case is the development of the Eastern Seaboard in Thailand, for which technical cooperation from the Japan International Cooperation Agency (JICA) and concessional loans from the Overseas Economic Cooperation Fund (OECF, presently the Japan Bank for International Cooperation—JBIC) were continuously provided.

Another aspect of the shift in emphasis with regard to regional development in Asian countries may be seen from the JICA's technical cooperation facility of integrated regional development planning (IRDP). This cooperation seeks to prepare a master plan for the long-term development of a region selected in principle by a respective country. Selection of a region reflects, to a large extent, the regional development policy of the respective government. The IRDP projects undertaken by

JICA in the past decades are summarized in Table 2.3. In Indonesia, regions in Java were selected during the early days, and the emphasis has now shifted to regions in the outer islands such as Sumatra and Kalimantan. In China, regions of strategic importance far from the capital region have been consistently selected. Regions in Thailand were selected apparently in relation to the capital region until recently when the border region with Laos was selected. In the Philippines, two regions neighbouring on the capital region of Metro Manila were selected first. The emphasis has now shifted to more remote regions of high potentials and strategic importance.

As the regional development policy is adopted, a bottom–up or participatory approach to regional development has been slowly effected, but to different degrees in developing countries in Asia. This process is also reflected in the regional development planning by the JICA's IRDP. The Philippines may be the most advanced in this respect of all the countries included in Table 2.3. For the Calabarzon region, the development plan was initially prepared basically by a top-down approach, but the results were made open to the public and extensive discussions took place during a series of workshops. The final plan was prepared incorporating the discussions. For the subsequent Central Luzon case, an alliance of non-governmental organizations (NGOs) was mobilized throughout the planning period to effect the bottom-up approach within the framework of top-down planning. The participatory planning approach was more fully implemented for the DIDP in Mindanao. An alliance of local governments served as a counterpart agency for this technical cooperation, and over 150 consultative workshops and meetings were conducted at different levels throughout the planning: provincial, city/municipality, and community levels. A similar participatory approach was taken for cases in Thailand, Indonesia, and Sri Lanka to a lesser extent, but not for cases in China and Vietnam.

Water Resources for Regional Development in Asia

Asia is considered relatively rich in water resources. The average annual precipitation generally ranges between 1,000–2,500 mm in most part of east, south-east and south Asia, while the world average is around 800 mm. This is however offset by the large population in Asia, which the rainfall has to support. Consequently, the per capita water endowment is not so favourable.

Most developing countries in Asia have a tropical to subtropical climate with high temperature and heavy rainfalls. This makes land

Table 2.3: Integrated regional development planning: Cases
supported by JICA

Country	Region	Year	Population (in thousands)	Area (km²)
Indonesia	East Java	1975	26,383	47,922
Indonesia	Central Java	1976	24,000	34,206
Tanzania	Kilimanjaro	1976		13,200
Indonesia	Southern coast of East Java	1978	5,600	8,310
Jordan	Northern region	1978	3,000	600
Philippines	Bohol	1978	760	4,120
Egypt	Southern region	1978	200	36,000
Brazil	Three provinces	1978	5,220	500,000
Malaysia	Trengganu south	1982	160	5,370
Thailand	Upper southern region	1982	1,100	22,300
Brazil	Greater Carajas	1982		
Kenya	Victoria lake region	1985	8,100	47,709
China	Hainan island	1985	5,980	33,900
Jordan	Karak	1985	270	8,100
Thailand	Upper central region	1988	2,700	16,600
Indonesia	Northern Sumatra	1988	20,000	264,000
Philippines	Calabarzon	1989	6,349	16,229
Indonesia	Southern Sumatra	1991	13,960	201,100
Thailand	Lower north-east and upper east	1992	10,300	89,000
China	Jiujian	1992	430	699
Philippines	Central Luzon	1993	6,200	18,230
Philippines	Cebu	1993	2,330	5,088
Sri Lanka	Southern area	1994	2,610	10,951
Indonesia	West Kalimantan	1995	4,840	29,900
Thailand	Western Seaboard	1996	2,900	43,700
Poland	Konin	1996	480	5,000
China	Jilin	1996	7,800	45,806
Vietnam	Central region	1996		
Philippines	Davao	1997	3,290	19,671
Turkey	East Black Sea	1998	2,870	39,203
China	Haicheng	1998	1,070	2,734
Thailand and Laos	Border regions	1999	3,940	63,000
Mozambique	Angonia	2000	761	40,000
Cambodia	Phnom Penh-Sihanoukville corridor	2001		

productivity inherently higher and organic materials decompose more easily. These are considered generally favourable conditions for development but it is not so. Human and other wastes are sometimes more difficult to treat in tropical regions especially under humid conditions. Shallow groundwater is abundant and easily available, but vulnerable to contamination by human wastes through improper treatment such as the use of unsealed septic tanks. In tropical and subtropical Asia, water and sanitation are, inevitably, more closely related.

Large cities tend to develop along major rivers, particularly at their mouths. Such a location has favourable conditions for the development of cities due to the flat topography of such regions, ease of collection and distribution of goods and movement of people by river, ease of access to markets, and convenience of water supply, and disposal of wastes and wastewater. Consequently, most capital regions have developed along major rivers. This means, conversely, that development of other regions is generally less favourable with respect to water availability and access to major rivers. Development and management of water resources, therefore, constitute an important aspect of regional development.

As the world is dominated increasingly by free trade and open economy, economic interactions with neighbouring countries have become increasingly important for the development of a country. This situation presents new opportunities for frontier regions of any country, which tend to be economically backward due to distance and poor access from the capital region. Such regions may turn the disadvantage into an advantage by utilizing their closeness to neighbouring countries. Often, international rivers constitute boundaries between neighbouring countries. Joint development and management of international rivers hold the key of developing some frontier regions.

Water Sector in Regional Development Planning—Asian Case Studies

Scope of the Water Sector and Regional Development Planning

Water resources constitute an integral part of regional development planning in both developed and developing countries. The water sector comprises the use of water resources for various purposes including water supply for domestic and industrial uses, irrigated agriculture, hydropower generation, recreation, and others. Flood control and low flow augmentation by storage reservoirs and other measures for downstream uses are also important part of the sector for some regions. Planning for water resources encompasses both infrastructure facilities such as dams, irrigation systems, water reticulation networks, hydropower plants, and river training works, and institutional aspects such as water rights, user charges, water users' associations, and cost allocation/sharing.

In this section, the main body of works involved in the water sector for regional development planning is described. It is a quantitative analysis of water supply and demand in a region and formulation of water resources development projects based on such an analysis. Institutional aspects are mentioned only incidentally, and water quality aspects are not dealt with.

A general procedure of integrated regional development is illustrated in Figure 2.1. A simplified procedure consists of four main steps: (i) analysis of existing conditions; (ii) evaluation of potentials and constraints; (iii) establishment of development objectives and strategy; and (iv) formulation of measures—projects, programmes, and institutional measures. First, existing conditions are analysed covering all the sectors such as the socio-economy, infrastructure, and resources. Existing development policies and measures are also analysed as related to regional development. Second, development potentials are evaluated and constraints identified. Evaluation of land capability and water resources potentials is the most essential part of this step. Third, development objectives are set and basic strategy established for development of the region as a whole. Usually, multiple regional development objectives are defined covering economic, social, and environmental aspects. More specific sector-wise objectives and strategies are established in such a way as to support the attainment of regional development objectives under the basic strategy. Fourth, specific measures are formulated under the strategies including development projects, institutional measures, and combinations of both programmes.

In the water sector, water resources potentials are evaluated and combined with the analysis of existing conditions and the evaluation of other development potentials to clarify the need for water resources development and management and to formulate specific projects with supporting institutional measures. As a region aims at structural

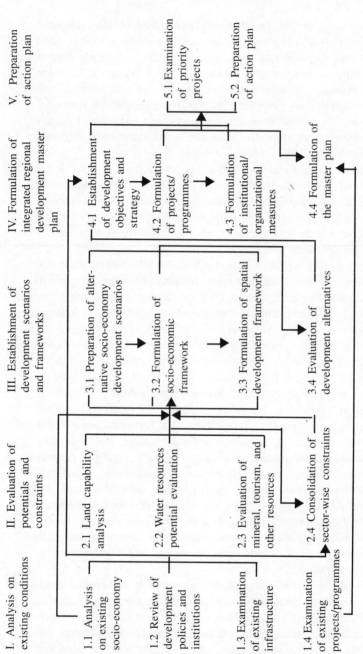

Figure 2.1: Procedure of integrated regional development planning

28

changes in the medium to long term through planned development efforts, water resources development often proves to be instrumental. For instance, inter-basin transfer and diversion of water may promote the development of a water deficient but otherwise high potential area, resulting in changes in the distribution of economic activities and population, leading even to fundamental changes in the nation's social and geopolitical structure.

Planning for water resources, as for any other resource, is formally treated as an optimization problem under a set of constraints. Planning aims basically to maximize—under the given constraints—the net benefit, defined as the total benefit derived from various water uses less the total cost involved in water resources development and management. Constraints include those related to water rights and maintenance flow requirements, environmental and other regulations, and priority by use as well as water resources endowment and budget constraints. Unlike other resources, water resources face two unconditional constraints. One is the absolute need to secure the minimum amount of water necessary for survival, and the other is the protection of human lives from major floods. These aspects are usually treated separately, despite some attempts to quantify the value of human lives for inclusion in the benefit–cost analysis framework.

Water Resources Potential Evaluation

Procedure and Methods for Evaluation of Water Resources Potential

Evaluation of water resources potential for regional development planning is conducted quantitatively based on the best available data. Primary data would not be generated through hydrological measurements or groundwater tests for planning purposes with limited manpower and time. The minimum requirement for quantitative evaluation is the availability of isohyet curves. Combined with run-off coefficients which are determined on the basis of topography, soil, vegetation, and other observable conditions, annual average discharges may be calculated by basin/sub-basin. Hydrological data should be used if available. It is desirable to have hydrological records for at least fifteen years or so, but the best available data should allow quantitative analysis in any case. Any quantitative analysis, even a crude one, should support better planning judgements and decisions. For regional development planning purposes, monthly flow data would suffice.

In some regions, dry year discharges need to be quantified, even if rudimentary, in addition to the annual average discharge. A dry year may be taken as the year when once-in-five-year low flow occurs (the probability of this is 80 per cent). If monthly flow data are available for twenty years, they are arranged in decreasing order, and the forty-eighth flow from the bottom is taken as the once-in-five-year low flow.

Potential evaluation for groundwater is more difficult as reliable data is largely lacking in most developing countries. To quantify the potential, the amount of evaporation and surface water flow is subtracted from the precipitation to set an upper bound of groundwater endowment. This may be broadly broken down into shallow groundwater and deep groundwater. Unless reliable data for deep groundwater are available, only shallow groundwater should be taken for local uses for conservative planning. Endowment of shallow groundwater is observable to some extent through existing dug wells. For some regions, groundwater potential may be assessed roughly as a percentage of the total precipitation. In tropical regions in Asia some 10 per cent of the precipitation may be taken as the groundwater potential.

Endowment of groundwater depends on topography, soil, vegetation, and geology as well as precipitation. For surface water and groundwater, relationship with topography, soil, and vegetation are mutually complementary. As for geology, sedimentary rocks are considered generally more favourable for groundwater endowment, while volcanic rocks have low possibilities. For the latter, groundwater may endow only in faults and fissures so that the evaluation should exclude deep groundwater except springs.

Water Resources Potential Evaluation for Central Luzon

Surface water potential in Central Luzon in the Philippines was estimated by an isohyet map and run-off coefficient. The available isohyet was examined in the light of recent observation data and used with modifications. Run-off coefficient was derived from observed relationship between the annual rainfalls and river discharges, and determined to be 0.50 for the Pampanga river basin (which is the largest river in the region) and 0.65 for other river basins. To reflect different discharge patterns for tributaries, the variable run-off coefficient was determined by land use to check the macro relationships determined earlier. The observed relationships were reconstructed by setting run-off coefficient at 0.45 for paddy fields and cultivated areas with upland crops, maize, and vegetables. For forests, shrubland, grassland, and urbanized areas this is 0.65. The total surface water potential was estimated at 25 billion m³ annually, of which the Pampanga River accounts for 11 billion m³, and the once-in-five-year low flow at 15 billion m³ or 60 per cent of the average annual discharge.

Groundwater potential was evaluated on the basis of rapid assessment by the National Water Council. This assessment identified shallow groundwater areas, deep groundwater areas, and areas where it was difficult to utilize groundwater based on geology, topography, elevation, precipitation, and groundwater table. Groundwater cultivation was taken to be 10 per cent of the precipitation, which was assumed to be available in identified shallow and deep groundwater areas. The total groundwater potential was estimated at 3.2 billion m³ , consisting of 2.2 billion m³ from shallow reserves and 1.0 billion m³ from deep aquifers annually.

Evaluation of Water Resources Potential for North-east Thailand

In the lower north-east and upper east region of Thailand, surface water potentials were estimated by dividing the regions into sub-basins and applying the area, average annual rainfall, and run-off to each sub-basin. Results were compiled by province, and the total annual discharge was calculated to be 17.4 million m³. For selected rivers, probability distribution of discharges was examined to determine 80 per cent of the probable flow at 4.6 billion m³ or some 25 per cent of the average. As the regions are generally arid, soil salinity is high in some areas. Distribution of areas affected by soil salinity was determined by province, and surface run-offs from such areas were excluded from the surface water potentials available for drinking and irrigation purposes. The calculation procedure and results are summarized in Table 2.4.

The groundwater potential of the regions was examined based on hydro-geological maps produced by the Mineral Resources Directorate of the ministry of industry. These maps subdivide the regions into seven classes, of which only four are promising for groundwater endowment. For each class of areas, the number of wells that may be dug were determined, and the groundwater potential was calculated by assuming daily pumping for ten hours over 300 days annually. As summarized in Table 2.5, the total groundwater potential was calculated to be 120 million m³, which is much less than the conservative estimate of the surface water potential of this region.

Table 2.4: Evaluation of water resources potential for north-east Thailand

(1) Calculation of total discharge

Province	Catchment area (km^2)	Annual precipitation (mm)	Run-off coefficient	Estimated annual discharge (million m^3)
Nakhon Nayok	448	1677	0.424	347
Prachin Buri	11,108	1626	0.225	4064
Nakhon Ratchasima	15,965	1086	0.135	2341
Buri Ram	9835	1165	0.270	3094
Surin	10,739	1224	0.186	2445
Si Sa Ket	7723	1226	0.083	786
Ubon Ratchathani	19,309	1520	0.126	3698
Mukdahan	3498	1467	0.177	908
Yasothon	2095	1369	0.186	533
Total	80,760	–	–	18,216

(2) Estimation of available surface water potential

Province	80% probable discharge	Ratio of land not affected by salinity	Available surface water potential (million m^3)
Nakhon Nayok	87	50	44
Prachin Buri	1016	25	254
Nakhon Ratchasima	585	50	293
Buri Ram	773	80	618
Surin	612	50	306
Si Sa Ket	197	70	138
Ubon Ratchathani	924	80	739
Mukdahan	227	40	91
Yasothon	133	80	106
Total	4554	–	2589

Water Supply and Demand Balance Analysis and Project Formulation

Procedure and Methods of Water Supply and Demand Balance Analysis

The water supply and demand balance analysis compares water demand in the region over the planning period with the water resources potentials evaluated. Water demand consists of water for domestic,

Table 2.5: Estimate of available groundwater potential in
north-east Thailand

Province	Area with groundwater endowment by class (km^2)				Available groundwater potential (million m^3/year)
	Highest	High	Medium	Low	
Nakhon Nayok	877	–	130	–	9.7
Prachin Buri	491	–	3,200	–	12.6
Nakhon Ratchasima	591	–	3,167	3,392	18.0
Buri Ram	–	–	2,073	5,673	12.2
Surin	–	–	1,690	2,271	6.9
Si Sa Ket	–	202	5,056	617	13.6
Ubon Ratchathani	217	1,893	8,082	3,251	35.8
Mukdahan	–	–	449	1,246	2.7
Yasothon	–	147	2,460	–	6.5
Total	2,176	2,242	26,307	16,450	118.0

industrial, irrigation, public and other uses. It is usually projected separately for urban and rural areas in regional development planning. For urban areas, domestic and various public water uses are often combined to define urban or municipal water use, which is estimated generally by assuming unit water use and multiplying it with projected population. The unit water use can vary widely depending on the economic structure, income levels and other factors expected over the planning period, but it is usually set at 100–250 l/capita/day in urban areas, unless water-intensive industries are expected in the region. In rural areas; the use of 25–50 l/capita/day for domestic use is generally assumed to be the norm in developing countries. The variance depends, among other factors, on the proportion of individual house connections and public standpipes. In most developing countries, the service coverage by public water supply needs to be assumed to arrive at the total demand in rural areas.

Sometimes water demand for industrial use is separately projected rather than including it into the urban water. Common methods are (i) to determine a single unit water use reflecting economic development phase or economic structure; (ii) to determine a unit water use for each of the major sub-sectors; and (iii) to analyse production processes to set water use by sub-sector industry. In the case of developing countries and regions, actual water use data in developed countries and regions are often referred to in determining unit value. Unit water use by sub-sector may vary depending on technological innovations and water recycling ratios in the future. This is usually beyond the practice of projection, but still trends in developed countries may be analysed and incorporated in the projection.

Demand for irrigation water is estimated by determing unit water use (m^3 /ha) for crops or cropping cycles, for which established methods are available. Unit water use by crop/cropping cycle is combined with future land use for agriculture to estimate irrigation water demand. The water demand in rural areas is estimated by including water demand for livestock, supplemental water for aquaculture, and sometimes water for wildlife as well as water for human consumption.

Comparison between water resources potential and water demand is conducted by sub-area defined by combining river basin subdivision and administrative division. This can be facilitated by the use of a geographic information system (GIS). The following balances should be examined:

a) the average water resources potential versus projected total water demand,

b) water resources potential in a dry year versus projected total water demand, and

c) water balance during a critical dry period.

Depending on regional characteristics and development objectives, the following comparison may also be meaningful:

d) surface water potential versus irrigation water demand, and

e) groundwater potential versus urban or municipal water demand.

Results of water balance analysis are expressed by water utilization ratio, deficit ratio, and other meaningful percentages and illustrated by sub-area. Graphical representation may be facilitated by a GIS.

Water Balance Analysis and Project Formulation for Central Luzon

Water demand in Central Luzon for the year 2010 has been estimated in the following way. First, the population was projected by the municipality for the urban and rural population, consistently with the socio-economic framework for the region as a whole. The service coverage was set separately for urban and rural areas in different provinces in line with the long-term plan of the Ministry of Public Works. Unit water use for domestic purposes was assumed to increase from the present 125 l/capita/day to 150 l/capita/day by the year 2010 in urban areas, and from the present 50 l/capita/day to 60 l/capita/day in rural areas, based on a study on water use in the recent past.

For industrial water, it was assumed that first the average water use per establishment will decrease from the present consumption of 1,300 l/day to 1,200 l/day in 2010 due to increased recycling. In addition, the expected increase in industrial value added was reflected in the water demand projection by applying the following formula:

$$\text{Industrial water demand} = \text{Unit water use per establishment} \times$$

$$\text{Number of establishments in 1990} \times \frac{\text{Industrial value-added in 2010}}{\text{Industrial value-added in 1990}}$$

Irrigation water demand, including other miscellaneous use on the farm, was assumed to increase from 1,100 mm/ha to 1,200 mm/ha, which was multiplied by expected irrigation area to calculate the total demand by sub-basin and by municipality.

The water balance analysis covered two cases. One was to compare the total irrigation demand in 2010 with the surface water potential in the dry year. The other was to compare the total municipal water for domestic and industrial purposes in 2010 with the groundwater potential (Figure 2.2). The following were clarified:

a) The municipal water demand in coastal areas in the Bulacan Province, part of the Pampanga Province, and urbanized areas in the Bataan Province cannot be met by groundwater alone. Dependence on the surface water will increase further towards 2010, while at present only a few municipalities in the region rely on surface water for the public water supply.

b) In parts of the provinces of Tarlac, Pampanga, Bataan, and Nueva Ecija, without regulation there will be shortages of dry year river flow to meet the irrigation water demand.

Shortages of irrigation water may not occur immediately in the identified municipalities and sub-basins, as additional water may be supplied from the upstreams. In Central Luzon, two major water resources development projects conceived earlier were reformulated to resolve the foreseen water shortages (Figure 2.3). The Casecunan Phased Trans-basin Project was to divert water from the Cagayan River basin to the north of Central Luzon. This project was found to be effective in solving the foreseen water shortages in municipalities along the middle reaches of the Pampanga River. The Balintingon

(1) Comparison between irrigation water demand and surface water potential

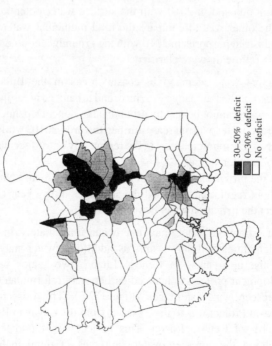

30-50% deficit
0-30% deficit
No deficit

(2) Comparison between municipal water demand and groundwater potential

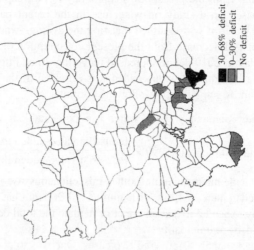

30-68% deficit
0-30% deficit
No deficit

Figure 2.2: Water supply and demand balance in Central Luzon

Source: Final Report, Volume II for the Master Plan Study for Central Luzon Development Program, September 1995, Submitted to JICA by Nippon Koei Co., Ltd. and Pacific Consultants International

Multipurpose Dam Project on the tributary of the Pampanga was proposed to solve the foreseen water shortages in downstream municipalities. For both projects, their needs were re-established from a broader regional perspective.

Water Balance Analysis and Project Formulation in Southern Sri Lanka

Reliable Surface Water Potential Evaluation: Reliable run-off estimates were available only for seven of the thirty-four river basins in the southern area of Sri Lanka. Therefore, a rainfall-run-off model was developed, using a GIS, by calibrating the model using available streamflow records and rainfall data. The digitized rainfall data, gridded into 1 km^2 cells, were correlated with measured catchment run-offs to

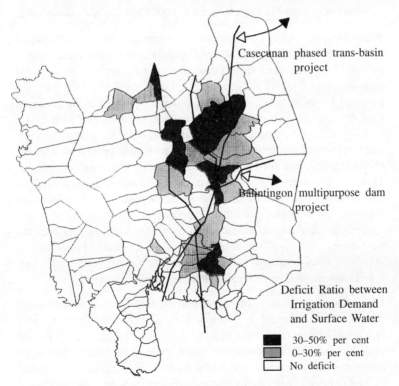

Figure 2.3: Projected water deficits and identified water resources development projects in Central Luzon in the Phillippines

produce a generalized rainfall run-off relationship. Specifically, the rainfall run-off relationship was expressed by the following linear regression:

Run-off coefficient (%) R = 0.0314 × Rainfall (mm) – 35.2,
where 15<R<85.

The 80 per cent probable monthly run-off was determined by analysing time series data from the seven river basins to derive flow duration curves for each river gauge. The 80 per cent probable run-off factors were expressed as a percentage of the mean catchment run-off. The estimated run-off factors were correlated with the mean catchment rainfall to estimate the 80 per cent probable run-off factors in ungauged catchments by using the following linear regression:

80 per cent probable run-off factor = 0.000124 × Rainfall (mm) + 0.2.

Water Demand Projection: Irrigation water demand was developed using FAO (Food and Agricultural Organization) methods and criteria published by the Irrigation Department, and checked with actual irrigation uses for existing schemes. Based on the future land use plan prepared by using a GIS, irrigation demand was derived for different cropping patterns, cropping intensities, and development areas.

Projection of municipal water demand was made by taking the urban and rural population projections together with other criteria such as per capita consumption for household connections, standpipes, and wells. These criteria were combined with predicted increases in piped supplies and house connections. Using the given population projections and assuming an increased growth in piped supplies of one per cent per annum, this could result in municipal water demand doubling over the next twenty years in the southern area as a whole. Projection of industrial water demand was based on existing water rights for industrial establishments and additional industries newly proposed.

Future Water Balance: The water balance in 2015 without any water resources development is presented in Figure 2.4. This includes provision of municipal and industrial supplies to meet the demand projected for 2015 and future irrigation water requirements. The latter consists of irrigation requirements for the existing 92,200 ha as estimated in the southern area, areas for planned irrigation schemes, and additional area of prime agricultural lands identified in 24,900 ha for further development. As seen from Figure 2.4, water shortages would further aggravate the situation in the Urubokka Oya, Kirama Oya, Malala Oya, and Kirindi Oya, which already suffer from droughts, and additional shortages would occur in a few other basins including the Menik Ganga. A gross deficit 595 million m^3 would take place annually.

Project Formulation: The water balance analysis with the identification of river basins where more serious water shortages are anticipated has led to the reconfirmation of several water resources development projects conceived earlier, including trans-basin diversion schemes. Through a cost comparison between these projects, some proposed projects were reconsidered or deferred for implementation, while others were reformulated. The Uma Oya Multipurpose Project was reformulated to reflect environmental and social concerns (Figure 2.5). It would involve diverting the Uma Oya water outside the southern area via a 24-km tunnel to the upper reaches of Kirindi Oya for hydropower generation of 150 MW by utilizing the substantial head of 750 m as well as for water uses in the downstream. The scale of diversion was reduced to fit the anticipated water shortages in the southern area, allowing the minimum monthly flow in the Uma Oya acceptable from an environmental point of view. It was proposed also that the diverted water would be utilized more effectively for soil and water conservation through contour canals linking small reservoirs by revitalizing the

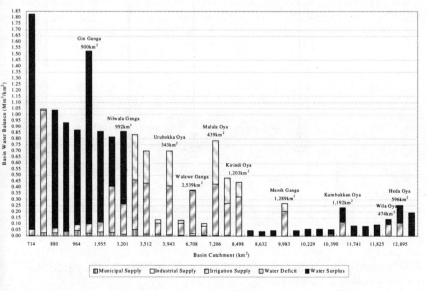

Figure 2.4: Water balance in 2015 (without project) with maximum potential lowland development and proposed cropping patterns

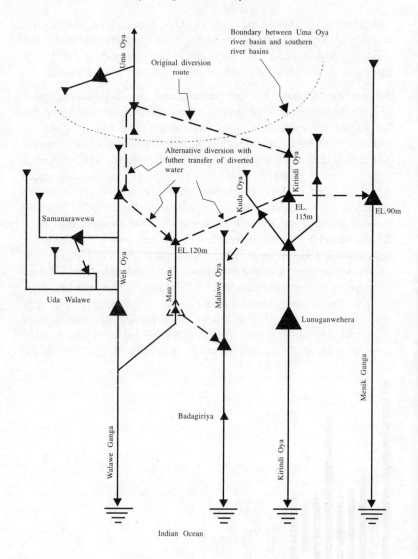

Figure 2.5: Uma Oya multipurpose project with trans-basin water diversion and contour canals for effective soil and water conservation

ancient ecosystem approach that local societies had adopted for centuries.

Opportunities and Challenges for Water and Regional Development in Asia

Regional Development in a Borderless World

Historical Perspective

The regional development approach has received renewed attention since the early 1970s as against the economic growth-oriented approach pursued during the 1950s and 1960s in many developing countries. The latter approach is reflected typically in the United Nations Development Decade (1961–70) which aimed at an annual average growth rate of 5 per cent for all developing countries. It resulted largely in excessive concentration of investment in major urban areas, typically capital regions, and resultant impoverishment of rural areas and the generation of various urban problems. These problems, in turn, caused social and political instability in some developing countries in Asia. The regional development approach was conceived to direct more attention to rural areas, to help the socially weak/deprived, to fulfil basic human needs, and to bring about inter-regional balance in economic development. This corresponds to the time when the World Bank adopted the basic human needs approach based on the Pearson Report of 1969, and also the time when JICA started the IRDP.

Development concepts have diversified in these few decades. A classic concept of development in the context of developing worlds is the liberalization from poverty and the struggle for survival. This passive concept has been generalized for a majority of the people in both the developing and the developed countries into the creation of a better living environment. A new concept of environmental development may be perceived in this broad sense. As is being increasingly recognized, the ultimate concept of development is human development or development of capacity and the ability of individual human beings. Most ordinary people can develop themselves in a society they belong to through communication for value-sharing and mutual enlightenment, or sometimes even through conflicts. Another new concept of social development is understood to be the development of the social fabric to facilitate such communication.

A regional community provides a substantive form of such a society with the totality of human relationships through which individuals develop themselves. Within the regional planning context, a regional community is the basic unit of participatory development, where needs of the local people and the community at the grass roots are identified. Specific projects or other measures to meet the identified needs may be formulated by participatory planning. Since planning is a continual process with feedbacks of monitored results of development, participatory planning represents the basic form of participatory development. The regional development approach encompasses all these concepts of development.

Issues for Regional Development

The basic issue for regional development does not change as long as the local people and the community are the focal point. It is how to meet the basic needs of the people in a regional community through the utilization and management of indigenous resources by the local people. While this issue represents an authentic regional development, pursuing it alone would not bring about desirable outcomes in the increasingly borderless world of open market economy. Sound regional development should effectively utilize indigenous resources, but should not rely only on the local market.

A new issue for regional development is how to realize self-reliant and autonomous development of a region by establishing comparative advantages in larger markets. Effective utilization of relationships with neighbouring regions and countries would be increasingly more important under the worldwide trend of open economy with free trade. The challenge for regional development now is to deal with both the basic and the new issues in a coherent way for integration, self-reliance, and autonomy in any region.

New Regional Development Approach

Economic globalization supported by information technology tends to widen disparities between developing and developed countries, and between the rich and the poor in any country. The recent financial crisis in Asia raised concerns about the instability of globalizing financial market and the vulnerability of the poor and the less developed to such a crisis. Reflecting these concerns, poverty alleviation has been receiving renewed emphasis. The poverty issue, however, is more deeply rooted, and the renewed emphasis on it proves, in fact, that past efforts to alleviate widespread poverty have not been very effective.

A major lesson learned through development efforts during the 1950s and 1960s, as mentioned earlier, was that the economic growth-oriented approach proved ineffective in alleviating absolute poverty through the trickle-down effect. An alternative conceived was the social safety net approach, which provides only necessary conditions for development and needs to be combined with support for livelihood activities. There is a danger, however, that a sustainable mechanism to support livelihood activities may end up encouraging reliance on it resulting in sustained poverty. The only viable approach seems to be to develop livelihood activities into viable economic activities by effectively utilizing a regional community for the management of indigenous resources and marketing them in larger markets. This is precisely what the regional development approach to poverty alleviation is concerned with.

The effects of globalization on individual human beings would be better received through a region and utilized for regional and human development. The borderless world, supported by information technology, should allow direct communication between human beings, irrespective of their nationalities and locations, which are the basis of all activities from cultural to diplomatic. The global flow of information, money, and commodities should be captured at the regional level and utilized for human development. This is the essence of regional development in a borderless world.

To implement the new regional development approach, the top-down and the bottom-up approaches should be reconciled in regional development planning. Macro-infrastructure projects need to be planned to maximize their positive effects on the local people, combined with complementary projects to meet the people's needs at the grass roots. Participatory planning should allow people to participate in increasingly larger regional communities, from households and villages to global communities ultimately, to realize human development at higher levels.

Issues for Water and Regional Development

The provision of safe water and assurance of proper sanitation constitute an important part of basic human needs. Satisfying these needs by itself poses a challenge in many regions, but a real challenge is how to ensure the survival of a region in an increasingly borderless and competitive world, while satisfying their basic needs. To realize self-reliant and autonomous development of a region, wider opportunities need to be sought for socio-economic interactions through marketing and resource use.

Seeking wider opportunities for regional development would call for development and management of various resources in a larger geographic context. In the case of water resources, using local springs or dug wells for rural water supply will not lead to opportunities for viable economic activities, although it may be easily manageable. Large-scale water resources development may open up opportunities for high economic growth, but may not warrant that the growth will benefit the local people unless it is properly planned and managed, probably with local participation. The challenge is how to satisfy both the need for safe water and sanitation as part of basic human needs and steady economic growth as an essential condition for survival in an open economy with free trade. This corresponds also to the challenge of reconciling regional development planning with the top-down and the bottom-up approaches.

Through the bottom-up approach, the immediate needs for water and sanitation may be more easily satisfied. The use of local sources of water such as springs and dug wells may be a solution. Wastewater may be discharged into a nearby stream, and human wastes may be disposed of by septic tanks. It is well recognized that a collection of such local solutions often leads to undesirable outcomes such as the pollution of receiving waters and health problems as well as depletion of local water sources. On the other hand, large-scale water resources development, regional wastewater treatment plants, and region-wide disposal of wastes planned by the top-down approach may result in an adverse impact on the local people and communities.

As wider opportunities are sought, it becomes more important for any region to utilize its relationships with its neighbouring regions and countries. Resources in neighbouring regions and countries may be combined in a complementary manner for mutual benefit. Cooperative regional development of neighbouring countries often involves international rivers constituting their borders, or cross-border catchment areas. The joint development and management of international waters poses another challenge for regional development in the new era. These two major issues are further discussed in the subsequent sections using Asian case studies.

Regional Approach to Water Management and Sanitation

It is well recognized that water supply and sanitation are highly interrelated within the land and water regime, and thus should be treated in an integrated way within some geographic context. This recognition is reflected, for instance, in the formulation of the integrated urban infrastructure development project (IUIDP), which has been implemented quite successfully in Indonesia since the late 1980s. Components of the IUIDP include water treatment plants and piped water supply systems, communal toilet facilities, family latrines, desludging trucks and sludge treatment facilities, and some community facilities. The IUIDP, however, has not treated the fundamental problem of polluted shallow groundwater. In addition to this problem, the following problems are typically observed not only in Indonesia but widely in Asian developing countries.

Water Supply

a) low level of service coverage;
b) low tariffs undermining financial viability of water supply operations;
c) high levels of unaccounted for water;
d) inadequate quality of water at points of delivery, unfit for drinking; and
e) low-income people relying on alternative sources of water often causing problems of hygiene.

Sanitation

a) inadequate sewerage facilities;
b) a large percentage of people using unsealed septic tanks for human waste disposal, contaminating shallow wells; and
c) desludging and sludge treatment largely lacking or not sustainable due to low awareness among people about sanitation.

Shallow groundwater is the most accessible source of water for the majority of the people in tropical and subtropical Asia. The shallow groundwater regime, in fact, represents a natural reticulation system. The fundamental problem is that the shallow groundwater is often polluted by inadequate sanitation practices. This problem cannot be overcome simply by extending piped water supply, which requires large investments and still does not assure the delivery of safe drinking water at affordable prices for various technical and administrative reasons. This is the problem structure of water supply and sanitation typical for tropical and subtropical Asia. Solving this problem would call for an integrated approach taking urban, peri-urban, and some rural areas together.

Integrated Approach

An integrated approach to infrastructure development involves both geographic or horizontal integration and functional or vertical integration. Geographic integration for water-related infrastructure refers, for instance, to coverage of service areas of both water supply and wastewater treatment facilities including water sources and waste-water discharge points. Functional integration represents inclusion of interrelated processes and activities within a project package such as proper water pricing and management arrangements for local water supply consistent with bulk water development to ensure viable operation of the entire water supply system. For both geographic and functional integration, the river basin or sub-basin generally constitutes a basic planning unit. The basin approach is recommended even for urban infrastructure, although trans-basin solutions are more common in urban areas such as trans-basin water diversion to supply a large urban demand centre or regional wastewater treatment plant serving neighbouring basins.

Basin Approach to Water Infrastructure

The integrated basin approach is particularly applicable for water and sanitation-related facilities to avoid typical problems such as the following:

a) poorly sited, designed, and constructed facilities such as unsealed septic tanks polluting groundwater;
b) discharge of wastewater, treated or untreated, threatening intakes for water supply downstream;
c) over-extraction of water upstream and lack of sufficient return flow, reducing water levels and increasing salt water intrusion downstream;
d) improper siting and sizing of drains, inviting dumping of solid wastes; and
e) leachate from solid wastes disposal sites polluting groundwater.

While the basin approach is applicable to water-related infrastructure development, care should be taken of its implications in social, environmental, and management aspects. Drainage presents a case in point. In many cases in developing countries, drainage channels are used effectively as combined sewers. A basin-wide combined sewerage network tends to concentrate large amount of discharge at a few locations, causing more serious pollution of receiving water. Drainage improvement serves a dual purpose: to reduce inundation by containing and swiftly discharging floodwater, and to protect roads from erosion by stormwater flow. The relative importance of these two purposes varies between urban, suburban, and semi-rural areas. In semi-rural areas, occasional inundation by large storms is more tolerable, while constant maintenance and repair of an extensive road system covering a large area constitute a major challenge. Under such circumstances, provision of undersized drains for longer stretches of roads may prove to be more cost-effective. Also large-sized drains may invite more abusive uses such as dumping of wastes, while its maintenance would call for greater attention. This involves important management consideration as well, since the main beneficiary areas of drainage and areas of functional drainage provision do not necessarily coincide.

Institutional Aspects

While the integrated basin-wide approach is valid for water management and sanitation, its application in practice is subject to proper institutional arrangements. To facilitate the application of the basin approach, major river basins in some countries are taken as jurisdictions of regional development authorities. Examples include the Brantas river basin in Indonesia and the Mahaweli river basin in Sri Lanka. The relatively new Southern Development Authority in Sri Lanka also has a jurisdiction largely coinciding with river basins. Such river basin authorities are relatively few in developing countries in Asia.

With or without any regional institutions, water resources development is usually undertaken primarily by state initiative in most cases. Under state initiative, privatization is progressing in the water and sanitation sector in some developing countries in Asia. The establishment of water supply enterprises is still limited; service contracts for the maintenance and repair of water facilities or water fee collection are more common. Another important direction is increasing beneficiary participation in the sector through project preparation, implementation, and operation and maintenance. Optimum development of the water and sanitation sector could take place with these multiple actors. Coordinating and integrating the activities of these actors would call for (i) rationalizing pricing to ensure financial viability of private operation, while offering basic services at affordable prices to the majority of the people, and (ii) enhancing public awareness for human and environmental sanitation.

International Waters for Complementary Regional Development

In many developing countries, regions remote from capital regions are often less developed due to poor access to major markets and neglect by Central governments. These regions, including frontier regions, are conversely closer to neighbouring countries. As the world economy becomes more open and market-oriented, closeness to neighbouring countries can be utilized for regional development of these remote and less developed regions. International rivers often constitute boundaries of frontier regions between neighbouring countries. Neighbouring regions also often share a common river basin.

Development and management of international river basins are considered indispensable tools for coping with the water crisis that might occur in Asia in the twenty-first century. While Asia is seen as a continent relatively rich in water resources, it embraces a large population and some of the fastest-growing economies. It contains several prominent rivers in the world, but most of them, except the Yang-tze and the Hwang–ho Rivers in China, are international rivers. Proper utilization of the international rivers and their basins calls not only for cooperation between riparian countries but also for consideration on social, environmental, and other domestic issues in each country and concerns of international societies.

Two international rivers are discussed here with regard to regional development of neighbouring countries. Cooperative development of the Salween River by Myanmar and Thailand was proposed recently from the regional development point of view. This involves the construction of a dam in the midstream of the Salween River within Myanmar, diversion of some water to the Chao Phraya river basin in Thailand, and related hydropower and irrigation development. The joint water resources development and management would expand opportunities for cooperative regional development in Myanmar and Thailand by creating a multiplicity of trade-off relationships (Figure 2.6).

It was shown that even if the trans-basin diversion reduces the amount of water in Myanmar, new opportunities and benefits will be generated on the Myanmar side including the following:

a) industrialization of the Yangon–Moulmein region in the downstream of the Salween and the Irrawaddy rivers by technical cooperation of Thailand;

b) development of a more robust industrial structure complementary to economic development in Thailand; and

c) reduced costs to be shared by Myanmar and additional benefits in Myanmar due to low flow augmentation and reduced floods.

For Thailand, benefits are clear, including water supply for Ayudhaya, irrigation and additional generation, and import of hydropower. Water supply for the Bangkok metropolitan area would also benefit from flow augmentation through alleviation of water pollution and reduced saltwater intrusion. Therefore, this joint development presents a win-win situation or plus-sum game for Myanmar and Thailand, which can be seen only when a broad regional development perspective is taken.

A similar approach has been proposed for the lower Mekong river basin shared by Laos and Thailand (Figure 2.7). This area, called the GMS (Great Mekong Subregion) central corridor region, includes part of the backward areas of north-east Thailand and even less developed

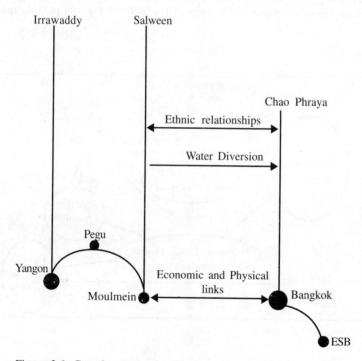

Figure 2.6: Complementary development of economic spheres with water diversion from the Salween to Chao Phraya

provinces in Laos. The proposed project concept is to construct dams on tributaries of the Mekong River within Laos and divert part of water by tunnels and siphons across the Mekong mainstream to north-east Thailand. The diverted water is to be used to irrigate agricultural land in the provinces of Nakhon Phanom, Mukdahan, Yasothon, Amnat Charoen, and Ubon Ratchathani. These provinces have a total of 1.6 million ha of agricultural land but only less than 200,000 ha are irrigated due to the lack of reliable water supply. The average yield of paddy is 1.7 ton/ha or only 60 per cent attained in Laos.

For Laos, benefits would not be confined to hydropower to be generated by the dam projects, which may be partly sold to Thailand. Agricultural development opportunities in the provinces of Khammuane, Savannakhet, Saravane, Se Kong, and Champasak, complementary to the irrigated agriculture in north-east Thailand, include cultivation of tree crops such as fruits and Para rubber and other upland crops, including feed crops, oil crops, and vegetables, and livestock, which graze on the undulating hilly areas. These activities would provide raw

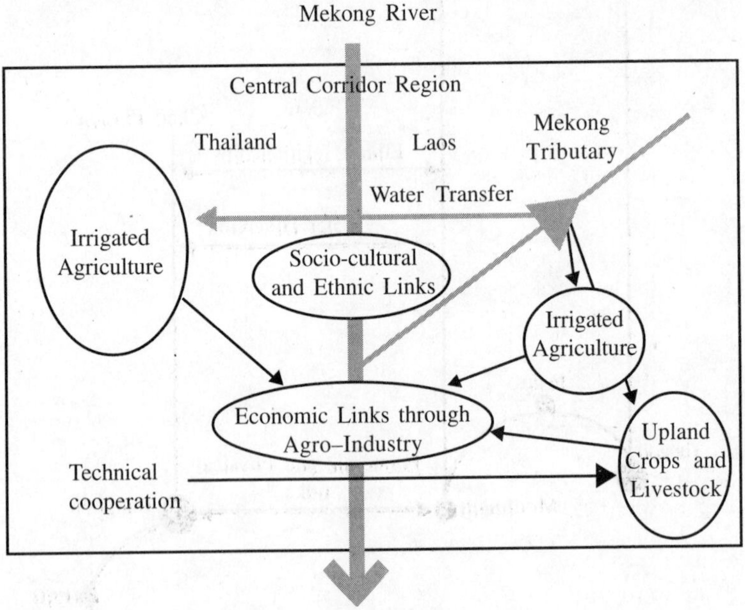

Figure 2.7: Complementary development of border regions through water diversion and technical cooperation between Laos and Thailand

materials for various agro-processing industries such as sugar, edible oils, fruit juice and puree, canned/bottled fruits, meat and dairy products, and animal feed. Thailand is expected to provide technical cooperation as well as capital for cultivation of upland crops and stock raising in return for irrigation water. Various agro-industries already exist in Thailand. Increased production of diversified crops and livestock in the GMS central corridor region would provide a stable supply of raw materials for the existing industries, and some new agro-industries are expected to establish themselves in the corridor region. The agro-industrial development may be planned jointly by both countries. The two frontier regions are expected to develop in a complementary manner through the sharing of water, land, and human resources.

Conclusion

The large population in developing countries in Asia with high population growth and density makes the development and management of water resources for sustainable uses more challenging. Large cities including old and present capitals of most countries have developed along major rivers, and economic growth-oriented development by the state in recent years has caused excessive concentration of population and economic wealth in these cities. This has caused various urban problems including water-related pollution, and impoverishment of rural areas. Inter-regional disparities have widened, especially under high economic growth in the east and south-east Asian countries. To rectify these disparities, regional development policies have been deliberately taken up in recent years by the state in many developing countries in Asia.

Under the prevailing conditions in most of these countries, chances are higher for more effective regional development to be realized by taking a carefully planned development approach, and also for the regional development to contribute to national development objectives such as socio-economic integration, resolution of ethnic conflicts, and enhancement of socio-political stability and public peace. In taking the planned development approach, water and related land resources could play critical roles.

The basic issue for water and regional development is to utilize and manage water resources as a most essential indigenous resource, by and for the benefit of the local people in order to satisfy basic needs for water and sanitation. The new issue for water and regional development in the increasingly borderless world of open economy and free trade is to develop and manage water resources from a broader regional point of view to support a wide range of viable economic activities. The challenge for water and regional development is to deal with both the basic and the new issue in a coherent way.

This challenge corresponds to reconciling the top-down and bottom-up approaches to water and regional development and planning. The bottom-up approach would ensure better satisfaction of basic needs in the immediate vicinity of the local people, such as use of local springs or dug wells for water supply and disposal of human wastes by septic tanks. This approach, however, tends to neglect possible widespread effects of such immediate solutions such as water pollution and health problems. Moreover, mere satisfaction of basic needs would not lead to self-reliant regional development with viable economic activities. The top-down approach would allow for conceiving and supporting viable economic activities by region-wide water resources development and management. The water supply and demand analysis, discussed earlier, illustrates just one possible way to reflect the regional perspective in the water sector as applied to Asian case studies.

The regional approach to water and sanitation is even more important in many developing countries in Asia under tropical or subtropical climate and which have a high population density. As reconfirmed by recent studies, a region preferably coinciding with river basins should be taken to plan for infrastructure development for water and sanitation to ensure sustainable uses of water resources in a cost-effective way. The regional approach to water and sanitation should be supported by the recent trends towards privatization, localization, and participatory development as seen in developing countries in Asia.

In seeking wider opportunities for regional development, effective utilization of socio-economic interactions with neighbouring countries becomes more important for complementary effects. Cooperative development of frontier regions among neighbouring countries may involve international rivers constituting their borders or otherwise their shared basins. It goes beyond joint development and management of international waters alone. The regional approach would expand opportunities for socio-economic interactions between the neighbouring countries not confined to regions in the immediate neighbourhood as illustrated by case studies of the Salween and the Mekong rivers. This would, in turn, enhance chances of joint development and management of international waters, which would be indispensable to cope with the possible water crisis in the current century.

References

Asian Development Bank (ADB), 'Project Performance Audit Report on Three Integrated Urban Infrastructure Development Projects: Secondary Cities Urban Development (Sector) Project, Botabek Urban Development Project, Bandar Lampung Urban Development Project in Indonesia', Reports PPA: INO 22264, 23212 and 20104 ADB, Manila, 2000.

——, 'Making Steady Progress in Water Supply and Sanitation', *Assessing Development Impact, OED's ADI Series No. 4*, ADB, Manila, 2001.

—— 'Project Performance Audit Report on the Second IKK Water Supply Sector Project in Indonesia', Report PPA: INO 22048, ADB, Manila, 2001.

BAPENAS (National Planning Agency), Repetita I (Five-Year National Development Plan), Government of Indonesia, Jakarta, 1969.

Japan International Cooperation Agency (JICA), 'The Master Plan Study on the Project Calabarzon, Final Report', Tokyo, 1991.

—— 'The Study on the Regional Development Plan for the Lower Northeast and the Upper East Regions in the Kingdom of Thailand, Final Report', 1993.

—— 'The Master Plan Study for Central Luzon Development Program, Final Report', Tokyo, 1995.

—— 'The Master Plan Study for Southern Area Development in the Democratic Socialist Republic of Sri Lanka, Final Report', Tokyo, 1997a.

—— 'Western Seaboard Regional Development Master Plan, Final Report', Tokyo, 1997b.

——, 'The Study on the Davao Integrated Development Program Master Planning, Final Report', Tokyo, 1999.

Japan Society for International Development, 'Report on Inter-regional Disparities between the Capital Region and Other Regions in Thailand' (in Japanese), entrusted by JICA, Tokyo, 2001.

Hashimoto, T., 'Regional Cooperative Development for the Salween River', in *Asian International Waters*, Delhi: Oxford University Press, 1996.

—— '*Chiiki Kaihatsu Puranningu* (Regional Development Planning)' (in Japanese), Tokyo: Kokon-shoin publishing company, 2000.

Nagamine, H., *Regional Development in Third World Countries*, Tokyo: International Development Journal Ltd., 2000.

Nippon Koei Co., Ltd., *Development of the Brantas River Basin in Indonesia* (in Japanese), Tokyo: Sankai-do Publishing Company, 1997.

Sanyu Consultants Inc. 'Land and Water Resources Management and Agricultural Modernization in the Central Corridor Region of the Mekong River' (in Japanese), project proposal submitted to Japan Bank for International Cooperation, Tokyo, 2001.

3

Water Resources, Impoverishment, and Regional Imbalances: Some Reflections from the State of Gujarat, India

Rajiv K. Gupta

Introduction

There is a powerful link between water resources development, poverty alleviation, and regional development. Experience has shown that in water deficient places the conditions of water scarcity and the resultant social and economic hardships lead to regional backwardness and stunting of regional growth, as compared to areas which have sufficient supplies of water both for irrigation and drinking purposes. This chapter broadly delineates the present water situation in the state of Gujarat in India. An attempt has been made to analyse the link between water scarcity, impoverishment, and regional imbalances. The water resources development supply-side options for dealing with the water crisis and regional growth problems, like large dams (especially the Sardar Sarovar Project) and rainwater harvesting through check dams and the like have been discussed in the light of available statistics and past experience. Additionally, demand management options have also been analysed.

Availability of Water and Constraints in Gujarat

The state of Gujarat is located in the western part of India. Covering 6.39 per cent of the geographical area and 4.88 per cent of the population of India, it is blessed with just 2.28 per cent of the country's surface water resources (Figure 3.1).

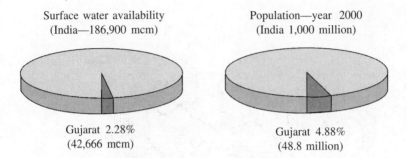

Surface water availability
(India—186,900 mcm)

Population—year 2000
(India 1,000 million)

Gujarat 2.28%
(42,666 mcm)

Gujarat 4.88%
(48.8 million)

Figure 3.1: A comparison of surface water availability and the population of Gujarat with India as a whole

Source: Narmada, Water Resources and Water Supply Department, Government of Gujarat, 1998.

Out of 185 rivers that flow through the country, the state has only eight perennial rivers and all of them are located in the southern part of Gujarat. Around 80 per cent of the state's surface water resources are concentrated in central and southern Gujarat, whereas the remaining three quarters of the state has only 20 per cent of the water resources. The average per capita availability of 980 m³/year puts the state in the water scarce category (as per the criteria set by United Nations). The intra-state variation in per capita water availability (1570 m³ in south and central Gujarat and 414 m³ in north Gujarat) is also striking (Figure 3.2).

The irony of nature is glaringly stark as far as the erratic behaviour of rainfall is concerned. On 11 September 2000, as against a total storage capacity of 4,512 million cubic metres (mcm) in all the dams of Saurashtra, Kutch, and north Gujarat, the storage available was hardly 403 mcm (8.93 per cent) (Table 3.1). On the other hand, during September 1999, as much as 24,700 mcm of water from the river Narmada flowed down to the sea unutilized.

District-wise data given in Table 3.2 shows that as many as 6,188 villages (34.3 per cent of total villages of Gujarat) do not have any definite source of water and they depend on local authorities for water which is supplied to them through tankers. This is an example of the water situation in the state. On an average, in a cycle of ten years, three years are drought years. Since 1947, the year of India's independence, Gujarat has witnessed droughts in 1951, 1952, 1955, 1956, 1957, 1962,

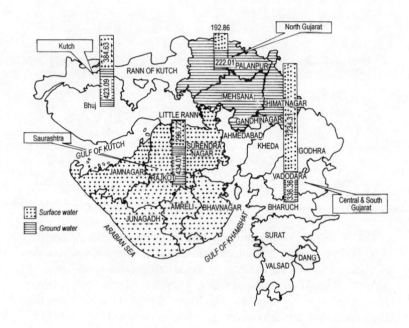

Figure 3.2: Per capita availability of water in Gujarat

Note: Based on the data available with NWR&WSD, Government of Gujarat.

Table 3.1: Water availability in dams in Gujarat

Region	Number of dams	Storage capacity		Current storage available	
		(MCM)	(%)	(MCM)	(%)
North Gujarat	13	2,018	(13.48)	96	(4.75)
Saurashtra	113	2,229	(14.90)	244	(10.95)
Kutch	20	265	(1.77)	63	(23.77)
South and Central Gujarat	28	10,452	(69.85)	4,660	(44.58)
Total	174	14,964	(100.00)	5,063	(33.83)

Source: Narmada Water Resources and Water Supply Department, Government of Gujarat, May 2000.

Table 3.2: Villages with no water sources and facilities
throughout the year (as on 1 April 1998), Gujarat

District name	Total villages in district	Villages not having water facilities for the whole year	Number of villages with no water sources
Jamnagar	694	195	690
Rajkot	844	329	682
Surendranagar	647	171	636
Bhavnagar	787	380	779
Amreli	611	172	611
Junagadh	917	121	0
Porbandar	184	27	0
Kutch	885	74	315
Banaskantha	1,243	57	315
Patan	553	9	531
Mehsana	629	18	513
Sabarkantha	1,363	149	607
Gandhinagar	212	0	131
Ahmedabad	553	0	107
Kheda	610	0	0
Anand	359	0	0
Panchmahal	896	2	92
Dahod	991	89	129
Vadodara	1,538	0	0
Bharuch	654	3	1
Narmada	548	1	0
Surat	1,184	0	0
Navsari	375	0	0
Valsad	450	0	7
Dang	309	42	42
Total	18,036	1,839	6,188

Source: Rural Development Commissionerate, Government of Gujarat, 2000.

1963, 1965, 1968, 1969, 1972, 1974, 1980, 1985, 1986, 1987, 1991, 1999,
and 2000.

During the last few years, the state has had a very bad spell of
droughts and about Rs 5,645 million (US$ 140 million) were spent on
temporary measures to provide drinking water (through water tankers)
which do not yield any permanent relief (Gujarat Water Supply and
Sewerage Boards, 2000).

The pattern of surface water availability within three different regions of the state is quite skewed from water abundant to totally water-scarce regions. Surface water available through the Narmada basin is of substantial quantity—which underscores the state's dependence on Sardar Sarovar Project built on the river Narmada, for its water requirements (Figure 3.3). As per a report prepared by Tahal Consulting Engineers Limited, (1997), the effective storage capacity of existing and ongoing major, medium, and minor schemes including lifting water from village lakes and ponds, and from check dams and percolation tanks, is roughly equal to surface water potential. Therefore, no surface water is available for further exploitation except waters of the Narmada.

An average rainfall of 25–200 cms, with a high coefficient of variance, underlines the state's reliance on dependable sources of irrigation for agriculture. As a result of over-extraction, groundwater tables are falling steadily.

Kutch, located in north-western Gujarat, has always faced the cruel vagaries of nature. At 35 cms, Kutch has the lowest annual rainfall in the state, and this too is highly unreliable. Out of the last eleven years, seven have been drought years for this district and this has had an

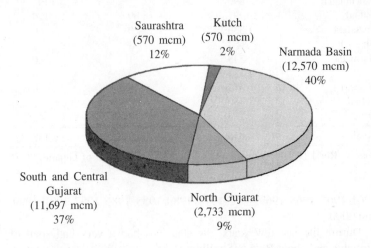

Figure 3.3: Utilizable surface water (31,500 mcm) in Gujarat

Source: Narmada Water Resources and Water Supply Department, Government of Gujarat, 1997.

adverse impact on the socio-economic development of the area and the living conditions of its people. The hardships faced by the people of Kutch is evident from the fact that while the total storage capacity of the twenty dams in this region is 265 mcm, only 23 per cent of this was available in September 2000. Given the acute drought situation and unprecedented loss of human lives and almost total destruction of the economy caused by an earthquake which struck Kutch recently, providing water for irrigation and drinking purposes would be an essential component of the reconstruction process in this district.

Relationship Between Water and Poverty

The human development index (HDI) (UNDP, 2000) measures the overall achievements in any country in terms of three basic dimensions of human development: longevity, knowledge, and a decent standard of living. It is measured by life expectancy, educational attainment (adult literacy and combined primary, secondary and tertiary enrolment), and adjusted income. While the HDI measures overall progress in a country in achieving human development, the human poverty index (HPI) reflects the distribution of progress and measures the backlog of deprivation that still exists. It measures poverty in developing countries. The variables used are: the percentage of people expected to die before the age of forty years, the percentage of adults who are illiterate and deprived in the overall economic provisioning—public and private—reflected by the percentage of people without access to health services and safe water, and the percentage of underweight children under the age of five. The World Bank[1] also emphasizes the availability of fresh water as an important constituent of world development indicators (WDI) (The World Bank, 1999). Therefore, there is an immensely powerful link between sustainable human development and water, sanitation and hygiene (Water Supply and Sanitation Collaborative

[1] The main sections of World Development Indicators (WDI) laid down by the World Bank are people, environment, economy, states, and markets and local linkages (World Bank,1999). The section on environment mentions land use and deforestation, protected areas and biodiversity, the use of energy and fresh water. The last item mentioned incorporates per capita fresh water resources, annual fresh water withdrawals, fresh water withdrawal by agriculture, industry, and domestic use, percentage of rural and urban population with access to safe water. Therefore, water becomes a cardinal constituent for development.

Council 2000). They constitute an entry point to poverty alleviation and all-round human development.

On a national scale, in India, the relationship between sources of drinking water and selected HDIs is reflected in Table 3.3.

Table 3.3: Relationship between sources of drinking
water and selected HDIs

Village variable	Per capita income (Rs)	(%) Population BPL	(%) Female literacy	(%) Enrolment (female)
Piped water	5,442	30.8	52.2	81.0
Other protected sources	4,336	39.2	41.0	64.3
Unprotected water	4,230	39.2	34.8	59.5

Source: Government of India, 1999.

Therefore, villages with better availability of drinking water report:

• improved lifestyles;
• better health;
• higher productivity and income; and
• improved female enrolments for education.

A significant correlation exists between poverty and water scarcity in Gujarat as well (Gariwala, 2001). Of the present population of about 50 million, around 65 per cent live in rural areas and are dependent on irrigation water, as their main economic activity is farming. There are around 5 million farmers, 3.5 million agricultural labourers and another 0.5 million are engaged in animal husbandary/forestry. Therefore, around 9 million people are directly dependent on water-based economic activity in 18,028 inhabited villages of Gujarat.

The Government of Gujarat undertakes 'Census of Families Below the Poverty Line' (BPL families) in rural areas. The data available from this census are analysed and given in Table 3.4 and Table 3.5. Table 3.4 provides district-wise data of BPL families in rural areas and the number of BPL families dependent on water for their livelihood. Table 3.5 provides a comparison of BPL families in rural areas during the Eighth Five-Year Plan and the Ninth Five-Year Plan.[2]

---

[2] In India, the federal government and the state governments follow the *Five-year Development Plan* system.

Table 3.4: Gujarat—Census of families below the poverty line (BPL) in rural areas (as of 1 April 2000)

District name	Total no. of rural families (in '000)	Total no. BPL rural families (in '000)	Percentage of BPL families	BPL families (in '000)			Percentage of BPL families facing water scarcity
				Small and marginal farmers	Agriculture labourers	Total	
Ahmedabad.	298	79	26	9	61	70	87
Amreli	188	51	27	9	33	42	82
Kutch	221	73	33	10	61	71	97
Kheda	298	109	36	38	67	105	96
Anand	249	74	30	13	56	69	93
Gandhinagar	161	35	22	11	18	29	83
Jamnagar	176	80	45	23	36	59	74
Junagadh	302	76	25	17	52	69	91
Porbandar	59	18	30	4	12	16	89
Dang	39	34	87	25	8	33	97
Panchmahal	318	222	70	172	36	208	94
Dahod	239	193	81	175	11	186	96
Banaskantha	399	135	34	63	65	128	95
Bharuch	214	110	51	25	81	106	96
Narmada	89	73	83	23	47	70	96
Bhavnagar	244	72	30	7	56	63	88
Mehsana	238	47	20	10	35	45	96

(Contd.)

61

Table 3.4 Contd.

District name	Total no. of rural families (in '000)	Total no. BPL rural families (in '000)	Percentage of BPL families	BPL families (in '000)			Percentage of BPL families facing water scarcity
				Small and marginal farmers	Agriculture labourers	Total	
Patan	164	56	34	14	40	54	96
Rajkot	260	78	30	20	49	69	88
Vadodara	347	132	38	54	74	128	97
Valsad	159	85	54	37	43	80	94
Navsari	149	76	51	26	46	72	95
Sabarkantha	335	146	44	101	39	140	96
Surat	406	196	48	53	127	180	92
Surendranagar	215	79	37	10	56	66	84
Total	5,767	2,329	40	949	1,209	2,158	93

Source: Rural Development Commissionerate, Government of Gujarat, 2000.

62

Table 3.5: Gujarat—Comparison of families below the poverty line in rural areas in the Eighth and Ninth Five-Year Plans

Gujarat state particulars	Eighth Plan 1992–7 BPL survey based on income up to Rs 11,000 per year (per capita Rs 184 per month)	Ninth Plan 1997–2002 BPL survey based on income up to Rs 15,250 per year (per capita Rs 254 per month) 1998	Ninth Plan 1997–2002 BPL survey based on income up to Rs 15,250 per year (per capita Rs 254 per month) 2000	Increase (+) Decrease (–)
Total rural families	4,804	5,588	5,767	+179
Total rural BPL families (No. in '000)	2,619	1,981	2,329	+348
% of BPL families (2&1)	54	35	40	+5
Total families of small and marginal farmers and agriculture-dependent on water (No. in '000)	2,278	1,818	2,158	+340
% of BPL families related to water (4&2)	87	92	93	+1

Source: Rural Development Commissionerate, Government of Gujarat, 2000.

63

As of 1 April 2000, there were 5.7 million families in rural Gujarat, out of which 2.3 million families were below the poverty line. This conveys that 40 per cent of the rural population is below the poverty line in the state at present. As per the census conducted by the government, 0.94 million families of small and marginal farmers and 2.1 million families of agricultural labourers living below the poverty line are dependent on water-related farming activities for their livelihood.

From this analysis it becomes evident that 2.1 million BPL families of farmers and agricultural labourers form 93 per cent of total BPL families in rural Gujarat. Out of the twenty-five districts of Gujarat, in seventeen districts more than 90 per cent of BPL families are involved in water-related farming activities. Corresponding figures for another 7 districts are between 70 and 80 per cent and for Jamnagar, 80 per cent. In other words, almost the entire rural economy in Gujarat is based on water-related agricultural activities.

Table 3.5 provides a comparison of BPL rural families under the Eighth Five-Year Plan (1992–7) and the current Ninth Five-Year Plan. The economic criteria for defining a BPL family are different in the two Plans. As per the current criteria, a family which has a total annual family income of less than Rs 15,250[3] is considered to be below the poverty line. As per the Planning Commission, Government of India guidelines, an average family consists of five members. Hence, for a BPL family, the average per capita income per month comes to Rs 254 (US$ 5). As per the earlier criteria of the Eighth Five-Year Plan, a family of five with an annual income less than Rs 11,000[4] (US$ 208) and hence an average per capita income per month of less than Rs 184 (US$ 4.5) was considered to be below the poverty line. As per the criteria set by the World Bank, a family having an income of US$ 1.08 per day or less, that is US$ 395 per year or less is considered to be living below the poverty line. Hence, the economic criteria set by the Government of India for defining the poverty line, that is Rs 15,250 (US$ 325 at current exchange rates) is far below that set by the World Bank (US$ 395).

Poverty Increases When Water Decreases

According to Table 3.5, BPL families accounted for 54 per cent of the rural population of Gujarat during the Eighth Plan period of 1992–97.

[3] Presently, 1 US$ = Rs 50 (rounded off).
[4] In 1992, 1 US$ = Rs 40.

This was drastically reduced to 35 per cent in 1998. The reason for this was very simple: the income criteria for defining a BPL family was raised from an annual income of Rs 11,000 or less to Rs 15,250 or less per BPL family. However, despite this accommodation, in the year 2000 this ratio increased from 35 per cent to 40 per cent. This quick increase over a two-year span is significant. This was perhaps due to the fact that the year 1999 as well as 2000 were drought years. During the drought of 1999, 8,666 villages (48 per cent of the total number of village) were declared as 'drought affected'. This number rose to 12,240 villages (68 per cent) in the year 2000. This proves the hypothesis that poverty increases when water decreases. Table 3.5 shows that total increase in the number of BPL families among the small and marginal farmers and agricultural labourers (340,000) is almost equivalent to the increase in the total number of rural BPL families (348,000). Therefore, it is this category of rural population which is impoverished due to water scarcity. This, undoubtedly, proves the directly proportionate relationship between the incidence of poverty and water scarcity in the state of Gujarat.

Water and Food Security in Gujarat

The sustainable development of water resources is critical to food security. This does not only mean sufficient agricultural production, but it also implies (i) that every individual has physical, economic, and environmental access to balanced diets, including the essential micro-nutrients and safe drinking water and to primary health care and education so as to lead a healthy and productive life, and (ii) that food originates from efficient, effective, and environmentally benign technologies that conserve and enhance the natural resources base of crop and animal husbandry, forestry, and inland and marine fisheries (Shah, 1998).

Both the World Food Summit in 1996 and World Water Vision have recognized that feeding the ever-growing world population and solving the looming water crisis are inextricably linked. By 2025, about 3 billion people, of which 1.1 billion will be in Africa, will be living in countries so short of water that they will not be self-sufficient in their food production. In many countries, the sustainable supplies of fresh water for irrigation have reached their limit. There has also been over-extraction of groundwater in these countries.

Agricultural growth has remained almost stagnant during the last two decades in Gujarat (Gariwala, 2001). When the state was constituted in 1960, the net area under irrigation was 682,900 hectares (ha), which increased almost five times by 1997. But still around 69 per cent of land in Gujarat lacks assured irrigation facilities. Groundwater contributes to more than 80 per cent of water used for irrigation in the state (Figure 3.4). Of the total irrigation potential of 6.85 million ha, 43 per cent is groundwater. Surface water from the state's basins constitute 31 per cent, and inter-state allocations 26 per cent (Agriculture and Cooperation Department, 2000).

However, the state has had to pay heavily in terms of severe depletion of utilizable groundwater resources all over the state. The cultivation of typical water-intensive crop varieties and the lack of awareness of the water problem in the so-called water-surplus regions of south and central Gujarat has also resulted in the depletion of utilizable groundwater in these regions (Table 3.6).

With a planned target growth rate of 6.04 per cent per annum (Agriculture and Cooperation Department, 2000) for the next ten years,

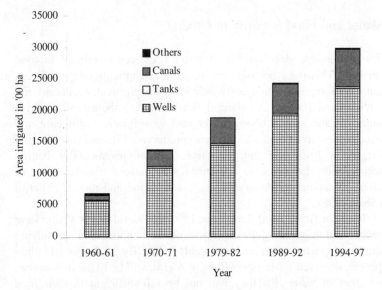

Figure 3.4: Sources of irrigation in Gujarat

Source: Based on the data available from Gujarat Water Supply and Sewerage Department, 2000.

Table 3.6: Absolute change in utilizable groundwater

	South and Central Gujarat	North Gujarat	Saurashtra	Kutch	Total
Utilizable groundwater (mcm/year)					
1984	6,783.95	4,216.87	5,682.07	682.51	17,365.40
1997	4,533.11	3,274.33	4,539.23	501.60	12,848.27
Absolute % change (1984–97)	–33.18%	–22.35%	–20.11%	–26.51%	–26.01%

Source: Based on the data available from Gujarat Water Supply and Sewerage Department, 2000.

the agricultural water demands are bound to shoot up. With an estimated rate of demand of 7,200–9,300 m^3/ha, total demand at ultimate irrigation potential would be in the range of 46,728–60,357 mcm. Thus, by the year 2010, agricultural water demands may exceed the ultimate utilizable water resources, both surface water and groundwater (Tahal Consulting Engineers Limited, 1997).

In 1998, the total food grain production in the state was 5.7 million tonnes, that is approximately 340 grams per capita per day. Thus, the state would need to raise food grain production to at least 6.95 million tonnes by 2021 in order to feed the projected population at the present rate. But the food grain production has barely increased by 0.5 million tonnes between 1980–81 (5 million tonnes) and 1998–99 (5.5 million tones). The reason for stagnancy in food production has primarily been the lack of adequate surface irrigation facilities.

The lack of irrigation water, stagnant crop production, difficulty in producing even one good crop during the monsoons, drought once every three years, increasing prices of farm implements as a result of diminishing subsidies, have resulted in the impoverishment of the small farmers and agricultural labourers. This is also reflected in share of agriculture in the state domestic product (SDP). The share of agriculture in SDP, which was 41 per cent in 1980–81, came down to 32 per cent in 1990–91 and decreased further to 21 per cent in 1998–1999. The consistent reduction in credit extended to the agricultural sector is another factor that reflects the economic conditions of the farmers. At the end of March 1995, there were 0.90 million agriculture loan accounts, which have come down to 0.74 million accounts with a reduction in outstanding credit amounts by 1999 (Dena Bank, 2000).

The agricultural credit sanctioned by 25,000 agricultural cooperative credit societies was only Rs 159,700 million at the end of March 1999, that is, a meagre Rs 4,000 per farmer which is less than US$ 100 per annum. Even in public and private sector agricultural organizations, the employment created decreased from only 25,000 jobs in 1990 to 20,000 by 1998, as compared to 1.8 million jobs in the industry and service sectors. All these indicators exhibit the plight of farmers and the agricultural sector dependent on water and further stress the direct relationship between water development and poverty alleviation. This is also ominous from the point of view of food security in Gujarat over the next few years.

Water Scarcity and Regional Imbalances

The contrast in the availability of water in the different regions of the state has had serious effects on the social development of water-deficient regions. Such intra-state regional imbalances have caused some cultural and political problems detrimental to sustainable development of the state as a whole.

It is not a coincidence that the literacy rate in the water-deficient districts is much less than the state average as compared to the water surplus districts, where it stands higher than the state average (Figure 3.5).

This may be attributed, inter alia, to the daily struggle for drinking water experienced by millions of families in these areas. The education of children, therefore, is not a priority any more. The female literacy rates in Jamnagar, Surendranagar, and Kutch are respectively 47.45 per cent, 40.65 per cent and 40.89 per cent, which is much less than the state average of around 49 per cent, again indicating that apart from other social factors, consistent struggle for water creates hurdles in female literacy in these areas. Lack of access to safe drinking water, time loss in collecting available water, the effects of carrying heavy loads of water over long distances on the health of women and girl children, and the burden of household responsibilities on women—all have a detrimental effect on the health of women and general family welfare including their income-earning abilities. There is a folklore about how poor people extract water in Vadhiar area of Kutch during

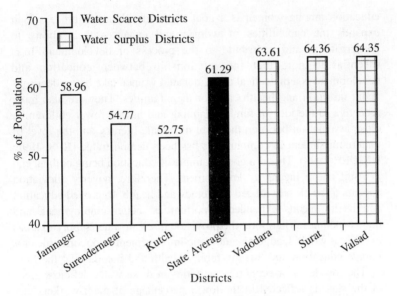

Figure 3.5: Literacy rate in Gujarat

Source: Census Handbook, 1991.

periods of water scarcity when the water is very deep in the wells. A housewife will go deep into the well, along with a drum and a pot secured with a rope. When she touches water she fills up the pot, and plays the drum to inform a person standing outside the well to pull up the water pot. This is their plight. Sometimes, their whole day is lost in search of water like this. There is a folk song on the plight faced by women due to the scarcity of water.

O Flying bird, please take my message to my grandfather,
Oh, grandfather, do not marry your daughter in Vadhiar,
As every day I am on a well from sunrise to sunset,
Even then a pot of water is not available. Oh Grandfather!.

Eventually, all developmental issues are inextricably linked. Higher literacy levels and educational attainments improve the access of people to other social services. Interlinkages between the health and education sectors have always been acknowledged at the academic level. Greater literacy and the spread of primary education increase the receptiveness of the population to improved health practices as well as increase the demand for medical care. Ending illiteracy and expanding education among women is a goal of the development process as it expands the capabilities of women and enhances their ability to participate in and contribute to the process of development. It is essential to note that the relationship between education and development is a dialectical one. Educated women take better decisions about nutrition and health care for their families. They are also more receptive to the idea of family planning and having fewer children. A close association between the level of female literacy and the level of undernourishment in communities has been demonstrated (IIPS, 1995; UNICEF, 1998). This is a good example of education being both an end in itself and a means to development. Therefore, poverty alleviation appears to be closely related to literacy and health. Improved education and health means expanded opportunities, better employment, and higher incomes. Therefore, the provision of clean water supply to areas in Gujarat would lead to an automatic improvement in opportunities for female education and thereby family health and income and the like.

The low level of economic activities in these water-deficient areas of the state is reflected in the lower percentage of main workers vis-à-vis the total population (Figure 3.6). It is very interesting to note that in the water-deficient areas of Gujarat, the percentage of females among non-workers is more than the state average and of course more than simalar percentage in water-surplus areas (Figure 3.7).

This again points to the fact that most of the productive energies of women in these areas are channelized towards the daily search for water. This is further buttressed by the fact that against the state average of 12.23 per cent female workers, the water-deficient districts have a much smaller percentage of marginal female workers (Kutch—6.99 per cent, Jamnagar—7.29 per cent, Surendranagar—11.25 per cent), implying thereby that due to their daily struggle for water, women are not left even with the time to pursue marginal working activities to augment the family income.

Gujarat has registered rapid industrial growth.With less than 5 per cent of the country's population, the state has achieved a growth rate of 6.55 per cent in net state domestic product (NSDP), at constant prices, over the past five years, and ranks third in terms of NSDP in the country after Maharashtra and Punjab (GIDB, 1999). Despite rapid industrial progress in the state, the water-scarce areas remain industrially backward in general, whereas the water-surplus districts are highly industrialized, leading to more employment generation and higher levels of economic activity in these districts (Figure 3.8). Availability of raw materials and

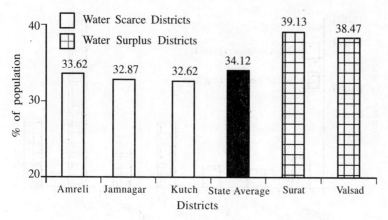

Figure 3.6: Main workers in Gujarat (as a percentage
of the total population)

Source: Report of Socio-Economic Survey, Government of Gujarat, 2000.

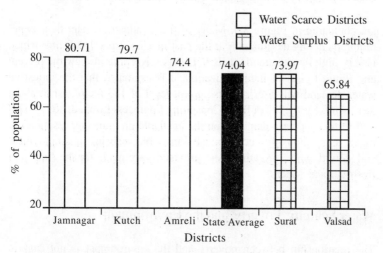

Figure 3.7: Non-workers (females) in Gujarat (as a percentage
of the state's population)

Source: Report of Socio-Economic Survey, Government of Gujarat, 2000.

physical and social infrastructure have attracted some water-intensive
industries like soda ash, fertilizers, rayon, dyes and chemicals, etc., even
in water-scarce regions. Typical water requirements of various industries

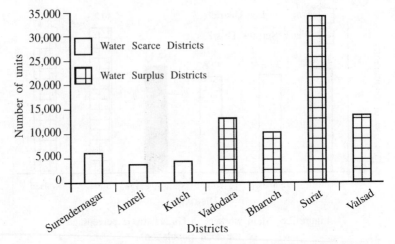

Figure 3.8: Small-scale industrial units in Gujarat, 1999

Source: *Report of Socio-Economic Survey*, Government of Gujarat.

are compared in Figure 3.9. Many of these industries meet their water requirements by desalination at the cost of US$ 2 per 1,000 litre which is very high by any standard (ICICI, 1998). Keeping in view the present and future trends of industrialization, it is estimated that the industrial water demand will rise from its current level of 153 mcm/year to about 281 mcm/year by 2021 (Tahal Consulting Engineers Limited, 1997). This will further increase demands on the available groundwater resource, or make these industries economically unviable, causing unemployment and slow down of growth, unless provisions are made for the supply of fresh surface water.

Water Scarcity, Poverty, and the Environment

The relationship between poverty and the environment is not easy to generalize. However, the fact remains that the poor are both victims and agents of environmental damage. One of the chief cause of deforestation in many parts of the world has been found to be the practice of shifting cultivation pursued by poor farmers (Reardon and Vosti, 1995). A World Bank study in Mexico suggested that poverty was directly associated with higher levels of deforestation (Deninger and Minten, 1996). During the 1980s about nineteen million hectares were deforested in Mexico, at

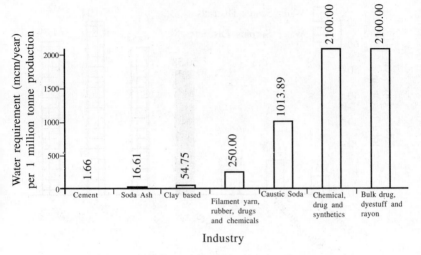

Figure 3.9: Industrial water requirements

an annual deforestation rate of about 2.9 per cent. The above study used socio-economic and physiographic data for 2,400 municipalities to examine the factors behind the higher rate of change in forest cover. And it was found that municipalities with higher levels of poverty lost a greater proportion of their forest cover during the 1980s.

In the Kutch area of Gujarat, the non-availability of water has caused advancement of the desert, environmental degradation, and national security issues along the India–Pakistan border. The emigration of people from Kutch is leading to the thinning of population in the border areas, facilitating infiltration activities from across the border. The low forest cover over water-deficient districts and the higher density of forest cover over districts with a surplus of water as compared to the state average adequately points to the detrimental effects of water scarcity on the environment (Figure 3.10). It indicates the existence of some kind of relationship between increasing poverty and environmental degradation.

In south-western Gujarat (Saurashtra), due to the over-exploitation of groundwater resources, the natural balance between sea water and groundwater level has been disturbed and salinity ingress has become a major problem. A study has indicated that the area affected by salinity will be 23 per cent by 2010 (Gujarat Ecology Commission, 1999).

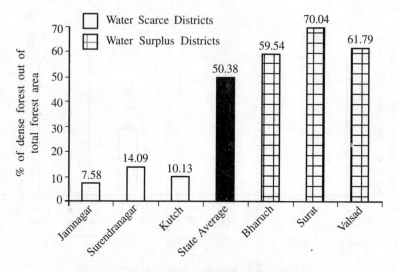

Figure 3.10: Density of forests in Gujarat

Source: *Forest Survey of India*, Government of India, 1997.

In Gujarat, due to groundwater mining, the problem of fluoride contamination has also become acute. In concentrations of 0.5–0.9 parts per million (ppm) fluoride is beneficial to human health and protects teeth against decay.[5] Dental and skeletal fluorosis is endemic in more than 2,800 habitations (out of total 18,000) in the state and cause mottled enamel of teeth in children, resulting in early ageing, permanent disability, and restrained economic activity of a large number of people. Information regarding water fluoride levels in different parts of the state have been compiled by many studies and it has been found to be between 0.2–18.0 ppm. An epidemological study of endemic fluorosis was carried out in tribal areas of the Vadodara district. The observations were as follows:

Dental fluorosis, urinary fluoride levels, and radiological changes showed significant association with water fluoride levels. Dental fluorosis was maximum among adolescents. Therefore, the young generation is worst affected by fluorosis. A health check-up programme conducted

---

[5] The standards set by the World Health Organization (WHO) permit only 1.0 ppm in drinking water as a safe limit for human consumption, whereas US Public Health Service Drinking Water Standards allow upto 1.7 ppm.

Table 3.7: Fluoride situation in selected areas of Gujarat

Water fluoride level was 0.5–4.0 ppm	
Percentage of population	Symptoms
18.2	Backache, joint pain and stiffness
35.3	Dental fluorosis
38.1	4.0 ppm urinary fluoride level
71.0	Radiological changes

on school-going children during 2000–1 revealed that sixteen districts out of nineteen are affected by fluorosis (Table 3.8).

Banaskantha, Surendranagar, Kutch, and Mehsana districts are the worst affected. These are water-scarce districts of Gujarat where the quality of drinking water poses serious threats to the future of the younger generation. This brings the question of whether the right of survival, right of protection, right of development, and right of participation, as enshrined by the UN Convention on the Rights of

Table 3.8: Fluorosis amongst children in Gujarat

District	Number of cases	Municipal corporation	Number of cases
Rural areas		Urban areas	
Ahmedabad	2,305	Ahmedabad	1,157
Amreli	994	Vadodara	185
Banaskantha	6,853	Surat	469
Bharuch	259	Rajkot	259
Dahod	32	Jamnagar	12
Gandhinagar	24	Bhavnagar	2,028
Jamnagar	220		
Junagadh	51		
Kheda	1,470		
Kutch	2,143		
Mehsana	1,922		
Rajkot	286		
Sabarkantha	857		
Surat	936		
Surendranagar	3,149		
Vadodara	898		

Source: Epidemic Cell, Commissioner of Public Health, Government of Gujarat, 2000.

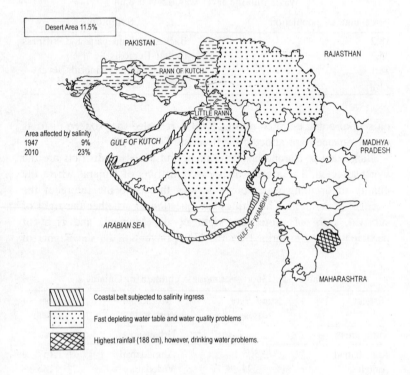

Figure 3.11: Area affected by salinity

Source: *Study Report on Salinity*, Gujarat Ecology Commission, Vadodara, 1999.

Child, the most widely ratified human rights instrument, of these children is being protected.

Options for Crisis-Solving Efforts

The emergent situation of water scarcity is seriously affecting regional development in Gujarat and is resulting in the impoverishment of a large number of people. Multi-pronged efforts, required to solve this water–energy crisis, vary significantly in concept, scale, cost effectiveness, etc. A few amongst these are as follows:

Regional Transfer of Narmada Water and Hydropower Generation from Sardar Sarovar Project

Termed as one of the 'modern eight wonders of the world abuilding' (*Time*, 1994), the Sardar Sarovar Project—a multi-state, multipurpose river valley project on the River Narmada, will provide 11.1 billion cubic metres (bcm) water annually for irrigating 1.9 million ha of land and supply drinking water to 8215 villages and 135 urban centres (having a population of around 30 million), increase the agricultural production by 8.7 million tonnes per annum (worth US$ 430 million), generate one million jobs—mostly in rural areas, and prevent the rapid processes of desertification, salinity ingress, and rural to urban migration being experienced in many parts of Gujarat. Sixty-eight per cent of the irrigation benefits and 95 per cent of the drinking water supply will go to the water-scarce regions of Saurashtra, Kutch, and north Gujarat.

On the one hand, this assured surface water supply will reduce the pressure on groundwater and, on the other hand, the return flow from irrigation will help to recharge groundwater. Around 1,350 MW of energy, which would otherwise have been used for the extraction of groundwater, would be saved. In addition, one billion units of hydropower per year is expected to be generated with the installed capacity of 1,450 MW. In order to have this 2,800 MW at the delivery point, a thermal power project of 6,000 MW is required (Vyas, 2001), which would need around 228 mcm of water per year. The water saved this way will save energy that would be required to pump it, and thus save more water. Therefore, the cumulative effect of assured water supply and energy availability will have a definite impact on the socio-economic and industrial development of backward regions in the state.

Gujarat has already begun receiving water from the Narmada at the rate of around 1,000 cusecs by pumping it from the reservoir (only as a temporary measure) in order to solve the present acute water crisis. The main dam is still under construction. Although this discharge constitutes only one-fortieth of the planned discharge from the project, it is quenching the thirst of millions. Consequently, live storage available in the reservoirs of water-scarce regions, which is presently reserved for domestic use only, is being released for irrigation, further enhancing the agricultural production in the water-scarce region.

Conceptualization and Development of the Kalpasar[6] Project

Gujarat, constrained so far by its dependence on the monsoons, has also undertaken detailed feasibility studies for another ambitious project—the Kalpasar Project. This project envisages the construction of a 64.16-km long main reservoir wall across the Gulf of Khambhat, impounding the waters of the five major rivers that flow into the gulf. Annually 1,400 mcm of fresh water is expected to be available from this gigantic project for domestic and industrial use and 1.054 million ha of land will be irrigated by it, in addition to the generation of up to 5880 MW of electricity. Apart from other inherent advantages that the initial studies have shown, this project has a significant potential to solve the water and energy crisis prevailing in the state and thereby lessen the extent of poverty in Gujarat.

Rainwater Harvesting and Aquifer Recharging

As a complementary measure to mitigate the problem of water scarcity and related regional imbalances, rainwater harvesting through micro structures such as check dams, percolation tanks, retention basins, etc., has a definite role to play. In Gujarat, such water harvesting structures have been constructed under microwatershed development programmes sponsored by various agricultural, rural, and water resources development schemes of the state government with popular participation. So far, a total of 22,697 such structures have been constructed to harvest 1,047.62 mcm of water and 2,250 works are in progress, which will store approximately 100.32 mcm of water. In spite of their limitations in terms of dependence on the vagaries of nature, carryover storage, hydrogeological factors, high evaporation due to shallow storage depths, and the like, these water harvesting measures help build up the groundwater reserves through recharge and save energy which would otherwise have been used for pumping groundwater. However, the regular failure of the monsoons restricts the effectiveness of these schemes. Thus, in Gujarat, both the alternatives of water management, micro as well as macro, have to be considered as complementary to each other for poverty alleviation and regional development. Experience has shown that these are in no way mutually

---

[6] Kalpasar, in the local language, means a lake that fulfills all wishes.

exclusive solutions. However, in the present situation and looking at the scale of the problem, regional water transfer from the Sardar Sarovar Project is essential for quenching the thirst of millions of 'naturally underprivileged' people, and to stop the vicious cycle of water scarcity and regional imbalances.

Planning for Efficiently Irrigated Agriculture

Given the water scarcity faced by many regions of the state and the fact that agriculture is going to remain the major consumer of water in the state, the efficient use of water acquires vital importance. Planning for less water-demanding crops, rotation of crops to optimally utilize the soil moisture, and improved land levelling and irrigation methods to achieve better uniformity in application of water, need immediate attention. Sustainability of irrigated agriculture needs to be ensured by proper long-term planning for the efficient use of groundwater for irrigation to avoid waterlogging as well as water mining conditions (Joshi, 1998). Innovative irrigation methods like drip irrigation, the use of sprinklers, etc., are slowly gaining popularity in the state despite the high capital investment and external energy requirements that these methods require. In order to reduce the conveyance losses, the existing canal networks of the state are being modernized (these include lining of canals, dredging of weed growth, better geometrical control, improved operation techniques, etc.). These management measures will improve water availability and agricultural production.

Public Policy Options for Preventive Measures

In the present situation, preventive measures are required to be taken for public policy provisions like:

a) legislative prohibition/control on groundwater exploitation, at least in over-exploited and grey areas;
b) enforcement of mandatory recycling of water for industrial use;
c) public–private participation in urban water supply;
d) participatory irrigation management (involving farmers in canal operation and maintenance);
e) provision of soft finance for water-efficient equipments/machines; and
f) promoting research for less water-intensive crops as well as industrial processes.

Water Pricing

Ensuring efficient consumer use of any commodity requires fixing an appropriate price for it. The rates of irrigation water supply in the state have been amongst the lowest in the country. The farmers are charged on a crop-acreage basis and the recently revised rates per hectare for the entire season varies from US$ 1.5–40 for different crops. Irrigation water supplied by pumping at the cost of the government is also charged only at the rate of Rs 1.20 (around 3 US cents) per 10,000 litres. Although the revenue generated from irrigation water is negligible as compared to the actual costs incurred to improve the water use efficiency, the government has recently decided to increase these rates every year (by 25 per cent for paddy and sugar cane and by 15 per cent for all other crops). However, this is far less than the amount paid by a farmer for securing water from a private tube well source.

With the World Trade Organisation (WTO) obligations to reduce the aggregate measurement of support (AMS) and direct production subsidies, the state has to reduce subsidies and shift to metered billing in water supply.

Consumer Education and Training

Ultimately, the message of efficient use has to percolate down to the end-users. Farmers have a tendency to irrigate their fields with excess water, unmindful of its adverse impact on the crop yield and soil conditions. Large volumes of water are wasted in the domestic sector also (for example, malfunctioning water flushes, leaking pipes, etc.). Similarly, in the energy sector, energy audit can help solve the crisis to a great extent. Farmers can be educated to use solar energy to run their drip and sprinklers.

Conclusion

Considering the cause-and-effect relationship between water scarcity and regional imbalances in the state of Gujarat, immediate measures for reducing the dependence on groundwater and harnessing available surface water resources (which otherwise flow into the sea) need to be implemented. This will require political will. The state has already entered a vicious cycle of water scarcity-poverty-regional imbalances. As the detailed analysis presented in this chapter has shown, the early completion of the Sardar Sarovar Project, the planned development of the Kalpasar Project, regulating groundwater use through public policy provisions, and tapping the untapped hydropower potential of the state can be considered as the main measures to solve the crisis. This is to be complemented by rainwater harvesting, artificial recharge, recycling of water for industrial use, and the conservation and efficient use of water and energy. All this will definitely lead to more water availability and thereby will lead to poverty alleviation and regional growth.

References

Agriculture and Cooperation Department, *Gujarat Agrovision, 2010—A Working Document*, Government of Gujarat, Gandhinagar, India, 2000.

Commissionerate of Rural Development, *Census Data*, Government of Gujarat, Gandhinagar, India, 2000.

Deininger, K. and B. Minten, *Poverty, Policies, and Deforestation: The Case of Mexico, Research Project on Social and Environmental Consequences of Growth-Oriented Polices, Policy Research*, Working Paper 5, World Bank, Policy Research Department, Washington, D.C., 1996.

Dena Bank, *State Level Bankers' Committee Data of Various Years 1995– 1999*, Dena Bank Ahmedabad, India, 2000.

Directorate of Economics and Statistics, *Statistical Abstract of Gujarat State*, Government of Gujarat, Gandhinagar, India, 1998.

Gariwala, B., *Poverty and Water: A Case Study of Gujarat*, Gandhinagar, India (unpublished), 2000.

Government of India, *National Sample Survey Data*, Government of India New Delhi, India, 1999.

Gujarat Drinking Water Infrastructure Board, *Drinking Water Scarcity 2000– 2001*, Government of Gujarat, Gandhinagar, India (unpublished), 2000.

Gujarat Water Supply and Sewerage Board, *Gujarat Drinking Water Master Plan*, Government of Gujarat, Gandhinagar, India, 2000.

Gujarat Ecology Commission, *Salinity and Ecological Degradation around the Ranns, Gujarat*, Gujarat Ecological Society, Gandhinagar, India, 1999.

Gujarat Infrastructure Development Board (GIDB), *Gujarat Infrastructure Agenda, Vision 2010*, GIDB, Gandhinagar, India, 1999.

Industrial Credit and Investment Corporation of India (ICICI), *A Study of Water Marketing*, Government of Gujarat, Gandhinagar, 1998.

International Institute for Population Sciences (IIPS), *National Family Health Survey, India: Summary Report*, IIPS Bombay, India, 1995.

Joshi, M.B., 'Software for Estimation of Irrigation: Demands and Scheduling of Canal Flows', Proceedings of the Second International R&D Conference Central Board of Irrigation and Power, Vadodara, India, 1997.

Reardon, T. and S. Vosti, 'Links between Rural Poverty and the Environment in Development Countries: Asset Categories and Investment Poverty', *World Development*, 23, 1995 pp. 1713–9.

Rural Development Commissionerate (2000), Census Data, Government of Gujarat, Gandhinagar, India.

Shah, T., Water against poverty: Livelihood-Oriented Water Resources Management, *Water Nepal*, 6(1), 1998 pp. 117–143.

Socio-Economic Review, Gujarat State 1999–2000, *Budget Publication No. 30*, Directorate of Economics and Statistics, Government of Gujarat, Gandhinagar, India, 2000.

Tahal Consulting Engineers Limited, *Water Resources Planning for the State of Gujarat—Final Report*, Vol. I, Gandhinagar, India, 1997.

Time Magazine, 24 January 1994.

UNDP, *Human Development Report*, New York: Oxford University Press, 2000.

UNICEF, *The State of the World's Children*, UNICEF, Paris, France, 1998.

Vyas, J.N., 'Water and Energy for Development in Gujarat with Special Focus on Sardar Sarovar Project', *Water Resources Development*, 17, 2001, pp. 37–54.

Water Supply and Sanitation Collaborative Council, *Vision 21: A Shared Vision for Hygiene, Sanitation and Water Supply and a Framework for Action*, Geneva, 2000.

World Bank, World Development Report 1999/2000, *The International Bank for Reconstruction and Development*, New York: Oxford University Press, 1999.

4

Water Development Potential Within a Basin-wide Approach: Ganges–Brahmaputra–Meghna (GBM) Issues

Zahir Uddin Ahmad

Introduction

There are 261 watersheds in the world which cross the political boundaries of two or more countries. These international basins cover 45.3 per cent of the land surface of the earth, affect about 40 per cent of the world's population, and account for approximately 80 per cent of global river flow (Wolf et al, 1999). One of the major stumbling blocks in water resources planning for the co-basin and co-riparian countries is the adherence to the political boundary, may be to derive political benefits. Water simply flows from a higher elevation to a lower elevation. For all the deltaic regions, the upper riparian countries are at a higher elevation and they always are in an advantageous position in terms of large-scale interventions. In the absence of a proper sharing arrangement to meet various demands for water for drinking, agriculture, fisheries, navigation, and environment, the lower riparian countries many times suffer from water shortages once a dam is built by the upper riparian country to divert the flow to their benefit. This kind of unilateral diversion without proper consultation, negotiation, and sharing arrangement may lead to long-term disputes and environmental hazards like desertification. Sometimes, the politicians gain mileage by politicizing issues related to water and by linking them to national interests. In the long run, these short-sighted, national-interest oriented plans fall apart and may even act against national interest. A unique example of this is the construction of barrages which have raised the bed levels of the upstream rivers to the danger mark. Had there been an integrated approach towards water resources planning among the co-basin and co-riparian countries, these kind of hazards could have been avoided to a large extent.

Following the Agenda 21 of Rio, it may be suggested generally that there should be an integrated approach in water resources planning among the co-basin and co-riparian countries. As we all know, water knows no boundary. Water resources planners must forget about political boundaries in order to harness and explore the water resources in a particular region in an integrated manner, making sure that it strikes a balance between water required for drinking, agricultural, fisheries, navigational, and environmental needs, not only for the nation, but most optimally for the region. It requires strong political will and wisdom to really forget political boundaries. This is possible only if the process of integrated planning is patronized with proper techno-political backing and spirit. The whole process should finally be linked to the sustainable development framework of a region. Development of water resources which cross-political boundaries have additional complexities brought on by strains in riparian relations and institutional limitations. Recent studies, particularly in the field of environmental security, have focused on the conflict potential of these international waters. Some stress the dangers of violence over international waters (see, for example, Gleick, 1993; Homer-Dixon, 1994; Remans, 1995; and Charrier, 1997) while others argue more strongly for the possibilities and historic evidence of cooperation between co-riparians (see Libiszewski, 1995; Salman and de Chazournes, 1998; and Wolf, 1998). The fortunate corollary of water as an inducement to conflict is that water, by its very nature, tends to induce even hostile co-riparians to cooperate, even as disputes rage over other issues.

There are nineteen water basins in the world which are shared by five or more riparian countries (Wolf et al., 1999). The Ganges–Brahmaputra–Meghna (GBM) is such a basin. The GBM region is unique in terms of identifying water development potentials in an integrated manner beyond the limit of political boundaries to derive a win-win scenario. The GBM region river systems constitute the second largest hydrological region in the world. The total drainage area is about 1.75 million sq km stretching across five countries—Bangladesh, Bhutan, China, India, and Nepal. While Bangladesh and India share all the three rivers, China shares the Brahmaputra and the Ganges, Nepal only the Ganges, and Bhutan the Brahmaputra.

The region is rich in natural resources including water, but the irony is that over 600 million people who live in this region are still struck by endemic poverty (one-tenth the population of the world). The development and utilization of the region's water resources had never been sought in an integrated manner by the neighbouring countries due to past differences, legacies of mistrust, and lack of goodwill. An integrated and holistic approach to the development of the region, considering water resources as a point of departure is the crying need for the region. The abundance of water in the GBM region as a shared resources can serve as a principal agent of development for millions of people living in the region to achieve a win-win scenario.

A number of options and opportunities exist for collaborative efforts in such sectors as hydropower development, flood management, dry season flow augmentation and water sharing, water quality improvement, navigation, and catchment/watershed management.

The policy environment in the region is now favourable for such cooperation. As noted in Chapter 1, a good example is the current cooperation between India and Bhutan, which benefits both the countries. Furthermore, recently, India entered into two water sharing treaties: one with Nepal which is known as the 'Mahakhali Treaty' and the other with Bangladesh in 1996, which is known as the 'Ganges Treaty'. Although these treaties were bilateral, they may be considered major ice-breaking events within the context of water sharing in this region. As mentioned earlier, water and energy can be the starting points from which regional cooperation on other issues can develop. The large multipurpose dams built in Nepal can generate huge amount of hydroelectric power and at the same time can provide water in the dry season. The present issue is how to move from bilateralism to multilateralism, and how neighbouring countries should share the costs and benefits of projects in an equitable manner. Identification of the water development potentials across political boundaries will help and facilitate policy makers and politicians to arrive at a common ground for discussions and negotiations. There are indications that the techno-political atmosphere in the region is congenial towards harnessing and exploring the water development potentials in an integrated manner. Flood damage mitigation through an improved flood forecasting and warning system in the region could be a non-conflicting starting point. The collaborative approach to water resources development should ultimately be linked to the sustainable development framework of the

GBM region which is characterized by endemic poverty—being home to about 40 per cent of the total number of poverty-ridden people residing in the developing world.

The indicators of social performance in the GBM region such as economic growth, education, and health is disappointing in comparison to other regions of the world. Though there has been a decline in poverty, the absolute number of poor has increased due to population growth. The per capita Gross National Product (GNP) of the nations of the GBM region varied between US$ 200–450 in 1998, whereas the corresponding global average was about US$ 5,000. Adult illiteracy is significantly higher compared to the average of 29 per cent for all developing countries. Public expenditure on education is about 3 per cent as against a global average of 5 per cent. Nearly 45 per cent of the land of the GBM region is arable, but per capita availability is very small (0.1 ha which is half of the global average). The urbanization growth rate is about 5 per cent in the region as against 0.5–2 per cent in the developed countries. The dependence on commercial energy is only 200–500 Kg of oil equivalent (KgOE) as against a global average of 1,680 KgOE and 5,340 KgOE for high-income countries. Despite all these poor economic indicators, the GBM region is rich in water, land, and energy.

Regional Water-Based Development Context

The GBM region has a high water development potential, if planned in a basin-wide approach. Although hydrological data for the region is not available in details in a published form, the enormity of the water resource potential of this region can be gauged in general terms. It is the singlemost important natural resource of the countries in the GBM region, and it is widely recognized that developing water resources would be the most important sector of development and could play a major role in shaping the future of millions of people living in this region (Figure 4.1).

The average annual water flow in the GBM region is estimated to be around 1,350 billion cubic metres (bcm), of which nearly half is discharged by the Brahmaputra. The three rivers constitute an interconnected system discharging into the Bay of Bengal. Compared to an annual average water availability of 269,000 m^3/sq km for the world, the availability in the GBM region is 771,400 m^3/sq km, which

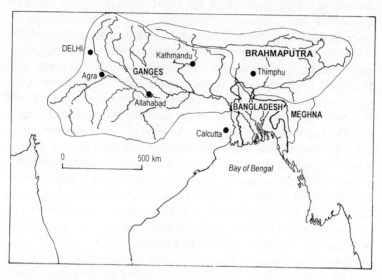

	Jamuna	Ganges	Meghna
Catchment area (km²)	573	1000	77
Average annual rainfall (mm)	1900	1200	4900
Average annual discharge (m³/s)	20,000	11,000	4600
Maximum discharge (m³/s)	100,000	78,000	20,000
Sediment transport (m ton/yr)	590	550	13

Figure 4.1: Map of the Ganges–Brahmaputra–Meghna region

is nearly three times the world average. In addition to surface water, the GBM region has an annually replenishable groundwater resource of about 230 bcm.

Water is abundant during the monsoon, but scarce during the dry season. Harnessing the bounty of the GBM rivers requires that the monsoon flows be stored and redistributed over space and time when and where required within a framework of sustainable development. The real challenge is to utilize this resource in an integrated manner. It offers the most promising entry point for achieving a social and economic transformation in Nepal, northern, eastern and north-eastern

India, Bangladesh, and Bhutan. In the new millennium, the vision concerning the development of the GBM countries should focus on options for the collaborative harnessing of water resources in the region, and in formulating a framework for multidimensional cooperation in related sectors such as energy, ecological health, flood management, water quality, navigation, and trade and commerce. Cooperation in developing the huge water resources of the GBM river systems is not a zero-sum game. Instead, all the regional partners can and should overcome the current mindset to construct the future scenarios on a win-win dispensation for all by working on all relevant perspectives, concerns, options, and trade-offs with the aim of achieving optimal benefits for all. In the absence of such a long-term vision, the GBM region would continue to stagnate, and millions of people would remain in a state of deprivation while other parts of the developing world march ahead towards prosperity. Water resources development can play a catalytic role in bringing about wider changes and in promoting sustainable development in the GBM region.

Fortunately, a climate of goodwill and confidence has been created over the past few years through the signing of the Mahakali Treaty between India and Nepal in January 1996 and the Ganges Water Sharing Treaty between Bangladesh and India in December 1996. These treaties are acclaimed as landmark events, offering a window of opportunity for water-based collaborative development endeavours in the region. The time is, therefore, conducive for formulating a broad framework of regional cooperation among the countries of the GBM region for the optimal integrated development of the region. At the same time, one should not remain complacent and euphoric in the wake of these landmark treaties; it is necessary to strengthen the process through renewed commitments to a wider vision of sustainable regional development and commensurate collaborative efforts.

The declarations made at the South Asian Association for Regional Cooperation (SAARC) Summits at Male in 1997 and at Colombo in 1998 endorsed sub-regional cooperation among SAARC countries by accepting the idea of two or more countries collaborating in project-based cooperative activities within the SAARC framework, which consists of India, Bangladesh, Nepal, Pakistan, Sri Lanka, Bhutan, and the Maldives. Therefore, the efforts of the countries of the GBM region to forge a common water-based development vision is consistent with the principles upheld by the SAARC countries. It may be noted that in the report of the SAARC Group of Eminent Persons, entitled *SAARC*

Vision Beyond the Year 2000, it has been envisioned that the enhancement of the quality of life and the welfare of the people of south Asia should be pursued through 'the creation of dynamic complementarities transcending national boundaries, development of managerial and productive capacity so as to exploit the internal and external economies of scale, resource complementarity and cost reduction'.

The development of water resources should trigger wider development in the region. As opportunities unfold, it is expected that the emphasis will shift from irrigation to sustainable agricultural productivity, from electricity production to energy grids and industrialization, from flood control to flood management, and from inland navigation to inter-modal transport. The ultimate goal is to attain a mutually beneficial synergy between national interests, people's well-being, and regional prosperity, initiated through the best possible utilization of the huge potential of the region's water resources.

Issues Related to the Utilization of Regional Water Resources

Floods, Riverbank Erosion, and Sedimentation

The countries of the GBM region are severely handicapped by recurrent floods, which cause damage to life, property, and infrastructure. It is the poor who occupy the more flood-prone areas and constitute the bulk of the victims. The general flooding pattern is similar in all the three countries, characterized by some 80 per cent of the annual rainfall occurring during the monsoons which last for four to five months. The rainfall is often concentrated in heavy spells of several days.

In Nepal, the run-off generated by heavy precipitation cannot quickly drain out, often because of the high stage of the outfall river. Flooding in hill valleys occurs due to sudden cloudbursts which are localized in nature, but may be heavy for several days. In the higher mountains, floods induced by glaciers, that is glacier lake outburst floods (GLOF), are also experienced. The Nepalese *terai* region is prone to flash floods, which also produce spillover effects in northern India.

Floods have become an annual feature in the GBM plains of India. Of the total estimated flood-prone area in India, about 68 per cent lies in the GBM states, mostly in Assam, West Bengal, Bihar, and Uttar Pradesh. The Ganges in northern India, which receives water from its northern tributaries originating in the Himalayas, has a high flood damage potential, especially in Uttar Pradesh and Bihar. Likewise, the Brahmaputra and the Barak (headwaters of the Meghna) drain regions of very heavy rainfall, and produce floods from overbank spilling and drainage congestion in north-eastern India.

Bangladesh, being the lowest riparian, bears the brunt of flooding in the GBM region. Even in a normal year, up to 30 per cent of the country is flooded, and up to about 80 per cent of the land area is considered flood-prone. Flooding in Bangladesh is caused by a combination of factors like flash floods from neighbouring hills, inflow of water from upstream catchments, overbank spilling of rivers from in-country rainfall, and drainage congestion. The conditions could be disastrous if flood-peaks in all the three rivers synchronize.

A natural corollary of flooding is riverbank erosion, especially in the Brahmaputra system. Large seasonal variations in the flow of rivers and the gradual loss of channel depth cause banks to erode and river courses to change. Wave action during the high stage further accelerates the process. Riverbank erosion is manifested in channel shifting, the creation of new channels during floods, bank slumping due to undercutting, and local scour from turbulence caused by obstruction. Riverbank erosion is responsible for the destruction of fertile agricultural lands, homesteads, and sometimes, entire clusters of villages.

The GBM rivers convey an enormous amount of sediment load from the mountains to the plains, which compound the adverse effects of floods. The Kosi and some of the tributaries of the Brahmaputra are particularly notable in this regard. Bangladesh is the outlet for all the major rivers and receives, on average, an annual sediment load varying between 0.5 billion and 1.8 billion tons. Most of this sediment load passes through the country to the Bay of Bengal, but a part of it is deposited on the floodplain during overbank spilling. This process gradually changes the valley geometry and floodplain topography, often reducing the water conveyance capacity and navigability of the drainage channels.

Water Quality

In the past, the emphasis in water resources planning was largely related to water supply or quantity rather than to quality. Meanwhile, water quality has progressively deteriorated due to increasing withdrawals for various uses, progressive industrialization, and insufficient stream flows to dilute the pollutants during lean flow periods. The increased use of agrochemicals and the discharge of untreated domestic sewage and industrial effluents into rivers have aggravated the problem. Given that water flows down the river systems and across countries and that there are interactions between surface and groundwater, water pollution in the region has been spreading across water sources, rivers, and countries and has already reached alarming proportions.

In Nepal, the water quality has deteriorated mainly due to industrial pollution. The volume of effluents generated by most industries is not large, but the concentration of pollutants is remarkably high. India had initiated an elaborate water quality monitoring programme under the 'Ganga Action Plan' in the late 1980s. This has recently been incorporated into a larger National Rivers Conservation Programme. In Bangladesh, the magnitude of the problem of water quality deterioration from the causes mentioned earlier is further compounded by salinity intrusion in the south-western region. The reduced flow of the Ganges in the dry season, coupled with the silting of distributary mouths, has exacerbated the process of the northward movement of the salinity front, thereby threatening the environmental health of the region. An additional problem is the detection of high concentrations of arsenic in the groundwater of fifty-nine of the sixty-four districts of Bangladesh and in some of the adjoining districts of West Bengal. Since the detection of arsenic and the recognition of the potential hazard of ingesting arsenic-contaminated water, water resource planners have realized that a radical shift in strategy is necessary to ensure affordable safe domestic water supply.

Climate Change

The impact of climate change due to global warming on the water resources of tropical Asia could be very significant. General circulation models have revealed that the mean annual rainfall in the north-eastern part of the south Asian subcontinent could increase with an increase in temperatures. The 'best-estimate' scenario for the year 2030 is that the monsoon rainfall could increase by 10 to 15 per cent. It is believed that increased evaporation (resulting from higher temperatures), in combination with regional changes in precipitation characteristics (for example, total amount, spatial and seasonal variability, and frequency of extremes), has the potential to affect mean run-off, frequency, and intensity of floods and drought, soil moisture, and surface and groundwater availability in the countries of the GBM region.

Increases in temperatures related to the change in the climate could also increase the rate of snowmelt in the Himalayas and reduce the amount of snowfall if the winter is shortened. In the event of climate change altering the rainfall pattern in the Himalayas, the impact will be felt in the downstream countries, that is, northern India and Bangladesh. The impact of any change in the length of the monsoon would also be significant. If the monsoon is shortened, soil moisture deficits in some areas might get worse, while prolonged monsoons might cause frequent flooding and increase inundation depths. By and large, any change in the availability of water resources as a consequence of climate change could have a substantial effect on agriculture, fishery, navigation, industrial and domestic water supply, salinity control, and reservoir storage and operation. Besides, the anticipated rise in the sea level in the Bay of Bengal would further compound the problem in Bangladesh through coastal submergence and enhanced drainage congestion in the floodplain.

The Gender Dimension

The status of women in the GBM region has special significance with respect to water supply and sanitation because they are the principal managers of domestic water use and family health care. Women play a vital role as water collectors and water managers. It is the women who possess the knowledge of the location, dependability, and quality of the local water resources, and their indigenous knowledge of the local water conditions is passed on to the successive generations. Collecting water for the family is an arduous and tiring task, especially in the hilly and semi-arid regions, and not only adult women but also female children are involved in this life of hard labour and drudgery. Women are also the worst sufferers of floods which affect this region regularly.

Despite playing a very important role in the collection and management of water for domestic use, women enjoy little or no authority in decision making in water resources management. The knowledge and perceptions of women can be gainfully utilized in planning the water distribution network, designing and locating water pumps, and organizing the management of water supply facilities by the community. The ultimate goal concerning the gender dimension in water management is to attain and ensure equitable participation of both men and women in the allocation and use of water.

Demand Management

Sustainable water management calls for a comprehensive, cost-effective, market-oriented, and participatory approach to water demand management. Nepal has formulated liberal policies for strengthening its economy and has made corresponding changes in the role of the state and the market in its water resources policy. The National Water Policy of India, adopted in 1987, defines priorities for different water-using sectors, treats water as an economic good, and proposes the use of water pricing in a manner that would cover the costs of investment, operation, and maintenance. The National Water Policy of Bangladesh, approved in January 1999, emphasizes the principle of accessibility of water to all, and proposes to develop sustainable public and private water delivery systems, including delineation of water rights and guidelines for water pricing.

Two types of demand-side approach are feasible. The first is entirely market-based, dependent on a market-determined price mechanism for the economic use of water. This requires certain prerequisites like an efficient water distribution system, the complete dissemination of information relating to water demand and supply, and an appropriate regulatory framework—conditions which are lacking at present in the countries of the GBM region. The second approach, which is more realistic and partly in operation in the region, is through a system of administered control which determines water allocation and pricing according to given or chosen social, economic, and environmental criteria. An administered system has the advantage of laying down the priorities concerning access to water resources, especially of the poor and underprivileged, involving the users in conservation and quality monitoring, and determining the changes or prices to be borne by different categories of users.

Institutions and Governance

Institutions and the manner in which they foster good governance determine the long-term ability of a country to manage its water resources. Institutions which are responsible for implementing water policies and strategies suffer from serious deficiencies and drawbacks in the region. They lack efficiency or perform sub-optimally in respect of such components as legal and regulatory aspects, implementation of rules, accountability, and responsiveness to user needs.

Water sector planning is now changing from a top-down technocratic approach to a bottom-up grass-roots approach. The goal is to establish a genuine participatory water management environment. Along with the participatory approach comes the steps to develop a nexus between the public and the private sectors in water development and management. Public sector water institutions have a poor record of cost recovery. The involvement of the private sector under the build-own-operate (BOO) mode may help reduce public sector deficiencies, improve the level of governance, and attract new investments for infrastructure.

Regional Water-Based Development Potentials

The enormity of the development potential of the huge water resources of the GBM region stands out in stark contrast to the region's socio-economic deprivation. It is important to formulate a long-term vision in order to develop a sustainable framework for water utilization. As has already been discussed in the preceding section that, owing to the seasonal availability of water in the Himalayan rivers, harnessing the resource requires that it be stored for meeting year-round demands. Run-of-the-river projects may produce valuable energy, but do not store water. Flood control benefits also cannot be ensured without storage facilities. In principle, there is no conflict between small, medium, and large projects since each has its unique place depending on the quantity of discharge, valley traits, and other technical and socio-economic and environmental considerations. Storage reservoirs are custom-built, and each can be designed to meet specific parameters.

The terrain of the northern and middle belts of Nepal offer excellent sites for storage reservoirs. One study in Nepal, based on available information and past studies, has identified twenty-eight potential reservoir sites in the country. Nine of them are classified as large, with an aggregate gross storage capacity of 110 bcm, and each site has a gross storage capacity of over 5 bcm. The Brahmaputra Master Plan of India (1986) has identified eighteen storage sites in north-eastern India, five of which are classified as large, having a total gross storage capacity of 80 bcm. In the Meghna (Barak) system, one large storage site (Tipaimukh) with a gross storage potential of 15 bcm has been identified.

The potential sites, referred to here, provide the opportunity to construct dams for storing excess water in the Himalayas for a variety of downstream uses. Hence, by definition, they are multipurpose in nature, providing benefits (beyond national borders) in such areas as power generation, flood moderation, dry season flow augmentation, irrigation, and navigation. The hydropower potential of these reservoir sites is the most significant aspect of water development in the GBM region, especially since per capita energy consumption in the region is among the lowest in the world. However, the construction of such storage dams involves high costs and has a long gestation period.

High dams and other large water resources development programmes have encountered severe criticism and opposition in recent years on a variety of technical, social, and environmental considerations. This sensitivity ranges from concerns for seismic hazards, submergence, population displacement, loss of farmlands, forests and biodiversity, and downstream physical impacts. At its extreme, the opposition to large projects is rooted in a subjective belief that they are like demons and wicked artifacts—a belief which ignores the long-term benefits of development and prosperity as well as the potential for trans-border cooperation. Some environmental activists are so zealously opposed to large dams as to evoke a vision of a perpetually stagnating pre-industrial society. The movement against large dams is obviously motivated by sympathy for the population which may have to be displaced due to the submergence of land in the vicinity of the dam. No one denies the rights of the displaced or affected people to appropriate rehabilitation and compensation packages or their right to claim and enjoy equitable shares of the accrued benefits. However, blind opposition to each large water development project demonstrates a subjective appraisal psyche and the failure to appreciate that development and environment are complementary aspects of the agenda for poverty eradication. In the past, things have gone wrong in certain instances due to the lack of knowledge, experience, and coordination, use of wrong technology, inefficient/poor implementation and management, corruption, and insensitivity towards project affected persons (PAPs). The key does not lie in doing nothing, but in doing things differently and wisely. Lessons learned from past mistakes can serve as building blocks in the context of promotion of sustainable development.

With respect to the construction of dams in the Himalayas, which is a dynamic tectonic region, the issue of seismicity deserves serious consideration, The countries of the GBM region should monitor seismicity and understand Himalayan tectonics comprehensively. That would help in identifying the potential zones of seismic activity. Careful and rigorous geophysical and geophysical investigations are now done prior to the siting and designing of high dams. Construction of earthquake-resistant high dams, in terms of both design and construction, especially rockfill dams, with greater strength against seismic forces, are now possible.

Similarly, the socio-economic and environmental impacts of large dams and water projects must also be addressed adequately. The national guidelines of the countries of the GBM region and the norms of international funding agencies are both specific and stringent in matters of resettlement and rehabilitation of the PAPs and the mitigation of potential negative impacts on the environment. The basic rule for the resettlement and rehabilitation is that the PAPs should preferably be better off after the project. As a compensation measure, 'land for land' is not always possible in the GBM region, where the population pressure on land has been rapidly increasing. Employment creation, capability improvement to shoulder new responsibilities in the work place, and self-employment (income-generating) opportunities with an emphasis on education and skill development may therefore constitute the areas of crucial focus as the means of rehabilitation of the PAPs. Resettlement and rehabilitation could also be accomplished in situ, that is, above or near the submerged area, and the programme could be a part of a broader community and area development concept. It would relieve the PAPs from the cultural trauma of relocation in alien surroundings. Besides, the dam sites which are generally remote and inaccessible would witness the development of transport routes and other infrastructure that would open up the area and, in turn, foster mobility, market access, and all-round development. Large water resource projects could thus find their image transformed from apparent monsters to harbingers of economic growth, social change, and improvement in the quality of life of the people of the region.

A long-term regional water vision for the GBM region should be built on the premise that the supply is likely to remain more or less finite, while the demand will continue to rise in the coming decades and centuries at a rapid pace. In order to develop a framework for water utilization as a trans-border challenge, it is essential for the countries of the GBM region to identify sectors and issues in respect of which cooperative strategies and action plans can be formulated, using water as the focal take-off point.

A number of options and opportunities exist for collaborative efforts in such sectors as hydropower development, flood management, dry season flow augmentation and water sharing, water quality improvement, navigation, and catchment/watershed management. The policy environment in the region is now favourable for such cooperation, and the remaining roadblocks can be cleared through mutual confidence-building measures.

Hydropower Development

Energy consumption is often a useful index of a country's level of development and standard of living. The GBM region's consumption of energy is very low. The energy economy of the region's countries is highly dependent on non-commercial sources, mainly biomass. This is not a sustainable situation, especially in view of the growing energy demands of a rising population and expanding economic activity. Yet, the hydropower potential of the region is vast. In the past, efforts have been made by each of the regional countries to develop hydropower within its own borders to meet domestic needs. But cooperative efforts to produce and trade hydropower have not been pursued.

Nepal's theoretical hydropower potential is estimated at about 83,000 MW. However, the identified economically feasible potential is about 40,000 MW. Given its modest load curve, Nepal's energy market lies in the northern and eastern regions of India as well as in Bangladesh, and possibly even in Pakistan. Nepal's hydropower could serve as valuable peaking power to the adjacent thermal-based load in India. The country's three-pronged approach to hydropower development envisages small decentralized projects to meet local needs, medium-scale projects for national needs, and large-scale multipurpose and mega projects to meet trans-border regional demands. The installed capacity of hydropower generation in India is about 22,000 MW, which is only 25 per cent of the country's total installed power capacity. The demand for electricity in India is growing at an average annual compound growth rate of 8–9 per cent. In order to reduce the current imbalance in the hydrothermal mix and the general consensus in favour of environment-friendly water-based power, future planning would incorporate a need to exploit maximally the GBM region's hydropotential through a regional grid. Bangladesh had an installed power capacity of about 3,000 MW as of 1997–8. The country's hydropower potential is limited by its flat terrain except in the Chittagong Hill tracts. Its lone hydro plant, the Kaptai Dam, which is outside the GBM catchment area, was constructed in 1964. It has an installed capacity of only 230 MW. Thermal power based on gas and fuel oil, accounts for the balance.

Bhutan and India are an exemplary case of collaboration in sharing hydropower. Unquestionably one of the best examples of successful collaboration in the development of the Asian international rivers is the one between Bhutan and India. The example of collaboration between India and Bhutan is discussed in Chapter 1.

The energy consumption of the world provides a fair index of a nation's level of development and standard of living. The GBM region's consumption of energy is markedly low. The countries of the region are variously endowed with energy resources: Nepal, Bhutan, and India have considerably high power-generation potentials; India is rich in coal and has oil and gas reserves in the north-east; Bangladesh is poised to become a major gas producer. In each case, the import of fossil fuel is constrained by the availability of foreign exchange.

Overall, however, the region is still critically dependent on non-commercial energy sources, mainly fuelwood from depleting forests and agricultural residues. This is not a sustainable situation, especially in view of the growing energy demands of a rising population.

Hydroelectricity has many advantages. It is a renewable source of energy without any recurring fuel cost and is therefore not subject to inflation during its long operational life. On the contrary, it exhibits a declining unit cost of generation over time with amortization of the initial capital expenditure. Its highly flexible response time makes it ideal for peaking purposes. It is clean and emits no greenhouse gases or other pollutants. Its operational and maintenance costs are about half that of thermal power systems. It can be integrated with multipurpose development and often opens up remote areas that would otherwise have suffered continuing neglect. These gains should of course be weighed alongside possible adverse social and environmental impacts such as submergence and displacement. Costs and benefits must be carefully assessed.

All things considered, the eastern Himalayan region is well suited for the development of hydropower projects which can exploit the enormous power potential to this region. In addition, these projects could help store water and also help in flood moderation.

Within the span of a hundred kilometres, Nepal's rivers lose a height of 1,000 to 4,000 m. This, coupled with the flow of around 225 bcm of water, invests its many valleys with a large hydropower potential. Even though Nepal's first hydropower station was commissioned some ninety years ago, further development has been slow. The current hydropower capacity is hardly 250 MW and load shedding is a regular feature in the dry season.

The theoretical energy potential of Nepal is 83,290 MW. However, the identified power potential, based on a study of sixty-six sites by the Water and Energy Commission, is 42,133 MW. Given its modest load curve, Nepal's energy market lies primarily in India and to some extent in Bangladesh and even in Pakistan. The kingdom's three-pronged approach to water resource development envisages small decentralized projects for meeting local needs, medium-sized projects for meeting national demands, and large-scale multipurpose and mega projects to meet regional demands beyond its borders. The last category of projects will be largely sited on the four snowfed rivers, namely, the Mahakali, Karnali, Gandaki, and the Sapta Kosi.

Nepal has identified certain projects for multipurpose benefits and a few of these have been studied in some detail. It has been agreed that detailed project reports of the Pancheswar and Sapta Kosi dams be jointly prepared with India as a first step towards their joint execution for mutual benefit. Under the Power Trade Agreement of 1996, India has agreed to buy power from Nepal and also allow it transmission access for power trading by both public undertakings and private agencies.

Economic development entails intensive use of energy in modernizing societies. But whereas per capita electricity use in 1996 in the industrial countries (accounting for 15 per cent of the world's population) was 9,491 kilowatt hours (kWh), that for the least developed countries (LDCs), including Bangladesh, Bhutan, and Nepal, which account for about 10 per cent of the world's population, was only 81 kWh. Several Indian states in the GBM region and the Tibet region of China also belong to the 'least developed' category (Table 4.1).

Indo-Nepalese water cooperation has faltered on mutual mistrust. The small country–big country syndrome apart, other political and security issues have vitiated the relationship from time to time. Nepal believes that it did not get a fair deal from India with regard to earlier projects on the Sharda, Kosi, and Gandak rivers, which were initiated earlier or in respect of certain other proposals requiring its concurrence

Table 4.1: Electricity consumption, 1996

Region/country	Population (million)	Total electricity GWh	Per capita kWh
World	5,743.7	13,338.04	2,370
Developing countries	4,502.9	3,742.25	845
LDCs	568.4	43.59	81
Industrial countries	842.0	7,941.17	9,491
India	966.2	433.91	459

Source: *UNDP, Human Development Report 1999*, New York.

as a lower riparian country. It also feels it has not got a fair deal as a market for power and water. India has another view of the matter. Be this as it may, perceptions mould opinions and constitute the political reality with which negotiators must contend.

As a result, progress has been agonisingly slow and, looking back, the mutual benefits foregone have been significantly large. The sheer size and technical complexity of some mega projects discussed are also daunting, with total outlays in each case considerably exceeding Nepal's GNP. This has engendered the view among some activists that Nepal should proceed with caution, learn to walk before it runs, and move gradually from the simple to the complex, the modest to the mega. Environmental concerns have also been raised in Nepal, as in India. These are not to be brushed aside but merit the most careful examination and transparent technical and thoughtful solutions. Nevertheless, time does not wait and the environment too has suffered degradation for lack of development.

Bhutan, a smaller and, until recently, even more hermetically sealed and landlocked Himalayan kingdom, exhibits another model. Having started later with respect to the development of its water resources, it has pulled well past Nepal with Indian assistance. The 370 MW run-of-the-river Chukha project on the Wangchu, commissioned in stages from 1988 onwards, has transformed Bhutan. It had paid for itself by 1993. As much as 45 per cent of Bhutan's annual revenue now comes from Chukha, with four upward tariff revisions since its inception. Kerosene imports have been held in check and deforestation prevented. The first steps in Bhutan's industrialization have been taken and surplus power fed into the Indian grid—a major export earner that has helped fund development.

In 1980, both Bhutan and Nepal used 17 kWh of electricity per capita. By 1996, Bhutan's consumption had reached 144 kWh whereas Nepal's consumption stood at 56 kWh. What is striking is that Bhutan's per capita income also has shot past that of Nepal, Bangladesh, and India, and now stands at US$ 545, second only to Sri Lanka within the SAARC region. With the ongoing lower Chukha or Tala project (1,020 MW) due to be commissioned in 2003, Bhutan's per capita income could cross US$ 1000. Energy-intensive industry is being fostered along the southern border with the resultant value-added products and remaining surplus power being absorbed in India. Possibilities for the initiation of other hydro projects are under investigation.

The King of Bhutan places gross national happiness (GNH), including preservation of the kingdom's cultural values and relatively unsullied environment, above its GNP. He would like to see the two march in step, with GNH taking the lead role. Yet, even to pursue this path, Bhutan's Vision-2000 sees the careful exploitation of Bhutan's 20,000 MW hydro potential as the means of leveraging the country forward and upward in its endeavour to attain a better quality of life.

The installed capacity of power plants in India increased from 1,713 MW in 1950 to 89,000 MW in 1998 and generation from about 5 TWh to around 420 TWh. Per capita electricity use rose correspondingly from 15 to 459 kWh (Table 4.2).

The demand for electricity in India has been growing at an average annual compound growth rate of 8–9 per cent. The Fifteenth Electricity Power Survey (1997) forecast the electric energy requirement by 2012 at 1058 TWh and the installed capacity to meet peak demand at 176,647

Table 4.2: India—Installed power capacity, 1998

Category	Installed capacity (MW)	Percentage of total
Steam	55,970	62.9
Gas	7,660	8.6
Diesel	376	0.4
Sub-total—Thermal	64,006	71.9
Hydro	21,891	24.6
Nuclear	2,225	2.5
Wind	900	1.0
Total	89,002	100.0

Source: Ahmad et al., 2001.

MW. Another study, assuming higher efficiencies all round, projects the installed capacity required by 2050 to be in the range of 800,000 to 1,000,000 MW (ten times the present capacity) (Table 4.3).

The share of hydropower in India's power mix declined steadily from 44 to 25 per cent between 1970 and 1998, even though the ideal ratio advocated is 40:60. Because of the growing imbalance in the hydrothermal mix in the eastern and western regions of the country, many thermal stations are presently required to back down during off-peak hours. Even if the present hydro ratio of 25 per cent is maintained in future, the installed hydro capacity in the year 2050 should still be of the order of 200,000 MW. This in effect implies the need to fully exploit the GBM region's hydro potential through a regional grid.

India has a nuclear power option and large coal and lignite reserves. However, the present stage of commercial energy use and its increase since 1980 are well below the world average. Other forms of energy, including non-conventional sources, and improved systems of generation and transmission will undoubtedly come into being over the next several decades. Extra-high voltage transmission and superconductivity and better means of demand management and conservation can be expected. Nevertheless, the role of hydropower will remain undiminished, especially with growing pressures to avoid burning coal which emits large quantities of greenhouse gases, which are a cause of global warming. 'Green' Himalayan power, especially that generated in Bhutan and Nepal, could be considered for emission trading under the 'clean development mechanism' being sought to be put together under the Kyoto Global Climate Convention—with safeguards to protect the long-term interests of smaller developing nations (Table 4.4).

The GBM region extends into the Tibet region of China. This part of the basin, lying north of the Himalayas, constitutes a discrete entity in itself. But this is no reason to exclude it from a longer-term cooperative framework. While some water resource development has taken place, Tibet has limited irrigation possibilities but could develop hydropower and navigational facilities, especially on the Brahmaputra/Tsang-po.

The *Scientific American* reported in 1996 a statement purported to have been made before the Chinese Academy of Engineering Physics regarding the technical feasibility of using nuclear explosions to divert the waters of the Tsang-po north through the intervening mountains to the arid Gobi region. A less distant possibility lies in tunnelling through the giant U-bend of the Brahmaputra between Tibet and

Table 4.3: Electricity consumption in the countries of the GBM region, 1996

Country	Population (million) 1997	Electricity consumption 1996 (MkWh)	Per capita consumption 1980 (kWh)	Per capita consumption 1996 (kWh)	Ratio ((col. 5/col. 4) x 100) (1997)	Per capita GNP (US$)
India	966.2	433,914	173	459	265	370
Bangladesh	122.7	12,404	30	103	343	360
Nepal	22.3	1,243	17	56	329	220
Bhutan	1.9	261	17	144	847	430
China	1,244.0	21,078,910	307	891	290	860

Source: UNDP, Human Development Report 1999, New York.

103

Table 4.4: Commercial energy use (oil equivalent)

Country	Per capita 1980 (kg)	Per capita 1999 (kg)
India	352	476
Bangladesh	172	197
Nepal	322	320
Bhutan	na	na
China	604	902
World average	1,625	1,681
Industrialised countries average	4,889	5,388

Note: na: not available
Source: *UNDP, Human Development Report 1999*, UNDP.

Assam to utilize the 3,000-metre drop in the river to generate 50,000–70,000 MW of power. China and India would need to cooperate in such a venture, and it is not an impossible idea.

Interconnecting the various national power systems through a regional grid could effectively improve overall system reliability and efficiency. Progress in this regard is limited. Cooperation between India and Nepal began with the Gandak and Kosi projects, with arrangements for border power supply to Nepal. The more recent Tanakpur agreement carries this forward but it is the Pancheshwar project that will give a massive impetus to the creation of a grid. Transmission links with Bhutan will also be further developed with new projects coming onstream.

It is sometimes argued that Nepal, India, Bhutan, and Bangladesh are inefficient consumers of electricity owing to system loss through transmission/distribution anomalies, and pilferage; and, hence, production of more power from large hydroelectric projects are both socially and economically undesirable. Yet, the per capita electricity consumption in these countries is minuscule in comparison to countries like Canada, the United States, Norway, Sweden, or Switzerland. It is also difficult to accept the contention that Nepal, India, and Bhutan should refrain from undertaking large storage schemes to produce electricity, when all the identified future storages would together harness a little more than 10 per cent of the annual flows. A more striking comparison would relate to the proportion of the installed hydropower to total hydro potential, which is only 0.6 per cent in Nepal compared to 56 per cent, 73 per cent, and 87 per cent respectively in similar mountainous countries as Norway, Sweden, and Switzerland.

In view of the likely financial constraints on the development of large projects, there is a clear need to promote private investment in the hydropower sector through joint ventures and foreign direct investment (FDI). Interconnecting the various national power systems through a regional grid could open up the power market, and enable Nepal and Bhutan to export surplus electricity to India and Bangladesh.

Flood Management

The recurrent floods in the GBM region demand an integrated approach involving cooperation among all the co-basin countries. Both India and Bangladesh have undertaken certain in-country measures for flood mitigation during the past four decades. These include embankments, river training, and channel/drainage improvement. Upstream storage reservoirs can play a vital role in flood management. Multipurpose reservoirs on the Ganges and Brahmaputra systems, with provision for a dedicated flood cushion, and reservoir operation and regulation instructions spelt out clearly, will be beneficial in moderating floods in northern, eastern, and north-eastern India (particularly Uttar Pradesh, Bihar, West Bengal, and Assam) as well as in Bangladesh.

Among the nonstructural flood management approaches, the greatest potential for regional cooperation lies in flood forecasting and warning. Currently, bilateral cooperation exists between Nepal and India and between India and Bangladesh for the transmission of flood-related data, which needs to be strengthened further. More reliable forecasts with additional lead time would be possible in Bangladesh if real time and daily forecast data are available from additional upstream points on the three rivers. Such effective arrangements to share data on floods are also necessary with upper riparian countries, namely, Nepal and Bhutan, for providing Bangladesh with greater lead time to undertake measures to prepare for disasters. A review of the current status of methods for forecasting floods in India and Bangladesh shows that both countries are using similar technologies for data observation and transmission. This provides an excellent opportunity to exchange expertise and experiences between the two countries for mutual benefit.

As a broader vision, the flood forecasting and warning system needs to be integrated with overall disaster management activities— both nationally and regionally. The co-riparians should agree on free flow of data relevant to flood forecasting amongst themselves on a real-time basis. The importance of satellite observation, especially for early warning of heavy rainfalls, should be recognized; and, for that purpose, the installation of adequately equipped satellite groundstations throughout the region could be considered.

Flow Augmentation and Water Sharing

The dry season flows of the GBM rivers, particularly of the Ganges, are inadequate to meet the combined needs of the countries of the GBM region. As early as in 1974, the prime ministers of India and Bangladesh had recognized the need for the augmentation of the dry season flow of the Ganges. The Ganges Water Sharing Treaty of 1996 included a provision for the two governments 'to cooperate with each other in finding a solution to the long-term problem of augmenting the flows of the Ganges during the dry season'. With Uttar Pradesh, Bihar, and West Bengal seeking additional water to meet their requirements, the issue of augmentation deserves serious attention. The Calcutta Port authorities are concerned that the Ganges Treaty has diminished lean season diversions into the Bhagirathi, which would affect drafts requiring increased dredging.

One possible option for a substantial augmentation of the flow of the Ganges, which could benefit Nepal, India, and Bangladesh, would be to construct large storages on the tributaries of the Ganges originating in Nepal. A highly favourable project from this perspective is the Sapta Kosi High Dam in Nepal, the revived third phase of the original Kosi project. It is likely that the Indo-Nepalese detailed project report on the Sapta Kosi Project will soon start moving forward. The Kosi Dam will have a significant storage capacity that should provide both north Bihar (India) and Bangladesh with flood cushion and augmented dry season flows after meeting Nepal's irrigation requirements.

Another option for augmenting dry season flows could be the proposed Sunkosh Dam in Bhutan with a power potential of 4,000 MW. It is proposed that water stored behind the dam could be released into a canal, designed to provide a two-stage link to the Teesta and Mahananda barrages in West Bengal. Augmentation of about 12,000 cusec (340 cumec) is expected—a part of which could supplement the water needs of the two Teesta barrages (one in West Bengal and the other in Bangladesh) and a part could reach the Ganges at Farakka. This option still awaits full environmental assessment and Bhutan's concurrence.

The issue of augmentation has a direct relationship with concerns for trans-border water sharing among the co-riparians. The Ganges Treaty of 1996 calls on India and Bangladesh to make efforts to conclude water sharing agreements with regard to other common rivers. One river which has received priority in the water sharing negotiations is the Teesta—especially because its lean season flows are inadequate to meet the requirements of both countries and each country has constructed a barrage on the river. Although some ad hoc sharing ratios were proposed earlier, it may be useful to examine seriously the option for the augmentation of the waters of the Teesta as well as whether some arrangements could be arrived at to operate the two barrages in tandem. In such a case, parts of Bangladesh's land lying outside its barrage's command area could be irrigated by extending canals from the barrage in India.

In the same track of regional cooperation, various other arrangements for augmentation and sharing could be conceived in the backdrop of probable trade-offs between the two countries. One such possibility is the westward diversion link (through Indian territory) between the Brahmaputra and the Ganges, with provisions for diversion along a lower alignment to augment the water of the Teesta in Bangladesh, or a further alignment southward to revive derelict streams and link them with the Ganges above the proposed barrage site at Pangsha. Some of these options are futuristic in nature, yet they deserve consideration within a long-term time frame for the region.

Linked to the issues of water sharing, lean season water availability, and augmentation options is the state of ecological health of the rivers. Environment is now recognized as a stakeholder in the water demand nexus. Hence, apart from meeting the requirements of irrigation, power generation, domestic supply, and other consumptive uses, a reasonable quantity of water must be available in the rivers in order to sustain the channel equilibrium as well as to maintain acceptable standards of the quality of water. The question of setting aside a proportion of water in the rivers received attention in Indo-Bangladesh negotiations relating to the sharing of the water of the Brahmaputra and Teesta. All future planning for water resources development needs to take special note of this aspect.

Following the 1996 Ganges Treaty, Bangladesh now has the opportunity to plan for environmental regeneration of its south-western hydrological system. One option is to construct a barrage on the Ganges at Pangsha to pond the river and force its backwaters into the

Gorai river (the principal distributary of the Ganges in Bangladesh). India has offered to assist in the feasibility study for such a venture and extend whatever technical support it can towards its construction. However, several international funding agencies have expressed reservations about such an intervention and have stressed that the resuscitation of the Gorai through dredging with the aim of rejuvenating a network of moribund channels, ox-bow lakes, and other wetlands in the south-west would be sufficient. Work on the restoration of the Gorai and associated studies are now in progress. An options study for the best utilization of the water available as a result of the Ganges Treaty, including a barrage on the Ganges, has just been initiated. In spite of the dredging of the Gorai, its proneness to siltation at its intake point from the Ganges necessitates additional measures like the Ganges Barrage to supplement the flows in the Gorai and other channels for achieving long-term environmental sustainability.

Water Quality

In all the countries of the GBM region, the deterioration in quality of both surface and groundwater is now a matter of serious concern. Water is essential to sustain agricultural growth and productivity; it is also vital for life and healthy living. More than half the morbidity in the GBM region stems from the use of impure drinking water. Safe water supply and hygienic sanitation are basic minimum needs which the countries of the GBM region are yet to meet in both rural and urban areas. A holistic approach is required to monitor the quality of water in each country together with regional initiatives to prevent further deterioration and to bring about improvement in the quality of water.

The mitigation of the additional problems of salinity and arsenic contamination in Bangladesh involves special action plans. Saline intrusion in coastal areas could be addressed through dry season flushing of channels by means of such methods as storing monsoon water and resuscitating moribund channels. The Bangladesh Arsenic Mitigation Water Supply Project (BAMWSP) funded by the World Bank and Swiss Development Corporation is presently engaged in assessing the extent, dimensions, and causes of the problem of arsenic contamination with a view to developing a long-term strategy for supplying arsenic-free water.

The monitoring of the quality of water in the GBM rivers is not as extensive as it should be except in the case of the Ganges in India and the Buriganga in Bangladesh. The countries of the GBM region need to set uniform standards relating to water quality parameters along with establishing an effective network to monitor the quality of water. The countries should review their existing laws relating to quality of water and water pollution, and make efforts to enforce the 'polluter pays' principle. At the regional level, they should also coordinate their actions to deal with trans-border transmission of pollution, and evolve a mechanism for real-time water quality data exchange.

Inland Navigation

The Ganges, the Brahmaputra, the Meghna, and their principal tributaries had served as major arteries of trade and commerce for centuries. However, in recent years, their importance has diminished as traffic has moved away from waterways to road and railway nodes. Yet, even today, the lower part of the GBM system is dependent on waterways, especially in Bangladesh and north-eastern India. With a view to reviving the past significance of inland water routes, India has already designated the Ganges between Allahabad and Haldia (1,629 km) as National Waterway Number1, and the Brahmaputra between Sadiya and Dhubri (891 km) as National Waterway Number 2.

For landlocked Nepal, Bhutan, and north-eastern India, an inland water outlet to the sea is of great significance. The establishment of links with the inland water transport networks of India and Bangladesh would provide Nepal access to the ports of Kolkata in India and Mongla in Bangladesh. In Nepal, all the three major rivers—the Karnali, Gandaki, and Kosi, which are tributaries of the Ganges—can be used for developing a water transport system. Construction of high dams on these rivers would improve their navigability.

The Karnali (known as the Ghagra in India) has the maximum potential for navigation—all the way from the Indo-Nepalese border to the confluence with the Ganges. The Gandaki is an important waterway serving central Nepal and has the navigation potential to serve eastern Uttar Pradesh and eastern Bihar in India if it is linked with India's National Waterway Number 1. The upper reaches of the Kosi is too steep for navigation, but river training works could facilitate the operation of shallow draft barges. Among the multiple benefits to be derived from the proposed Sapta Kosi High Dam is the provision for a navigational channel with a dedicated storage. The principal focus for Nepal's navigational development would be to gain an exit to the sea through the Ganges, and obtain linkages with the inland ports of India en route. The strategy should be designed to ensure that structures constructed under water development projects do not impede the development of inland water routes.

The maintenance and further development of the requisite minimum navigable width and depth coupled with the provision of navigation aids and terminal facilities would enhance the navigation potential in the GBM region. India and Bangladesh have a bilateral protocol, renewed every two years, for India to use the Ganges–Brahmaputra–Meghna riverway for water transit between West Bengal and Assam. The potentials of these routes—not optimally used at present—could expand through channel improvement, better pilotage and navigation aids, and simplification and standardization of rules and regulations. A dedicated willingness to integrate the waterways network in the GBM region would benefit all the countries in the region in the long run.

The development of the railways under state patronage led to a decline in IWT traffic. The development of the roadways network and its improvement in this century accelerated the decline. Kolkata emerged as the premier port in the eastern region not least because of its extensive IWT connections. The partition politically divided this great waterway which fell into virtual disuse in India despite navigation protocols with Bangladesh.

The waterways of Bangladesh survived and can be divided into three groups—deep rivers, navigable round the year, shallower rivers, navigable only in the monsoon, and smaller channels only capable of handling country boats. The IWT system is served by two seaports—Mongla and Chittagong.

The Ganga, the Brahmaputra and the Meghna flow into Bangladesh from three directions and merge into a single outlet. This constitutes a vast water network which can be revived as an integrated IWT system for the entire region. Out of a total freight traffic of 550 million tonnes carried by all modes of transport in India in 1989, the share of IWT was 17 million tonnes or just 3 per cent. In terms of tonne-km, IWT accounted for a mere one billion tonne-km whereas road and rail carriage accounted for 817 billion tonne-km.

The Inland Waterways Authority of India (IWAI) estimates that even if 20 billion tonne-km of traffic shifts to the IWT mode, it would result in an annual saving of Rs 500 crore on the import of crude oil and Rs 900 crore in transportation economies.

The IWAI was set up in 1986 to develop, maintain, and administer 'national waterways'. The national waterways are being developed for a minimum navigable depth of 2 m and a minimum width of 45 m, with navigational aids and terminal facilities.

For landlocked Nepal, an inland water outlet to the sea would assume great significance. At present, country boats carry some 130,000 tonnes of freight in the lower reaches of the Karnali, Gandak, and Kosi. However, studies have been carried to establish the feasibility of navigating the Kosi up to the Ganga below Patna as part of the Sapta Kosi High Dam project.

Catchment Management

The geographically interlinked character of the major rivers in the GBM region warrants an integrated regional approach to the care and management of the catchments. Sound basin-wide catchment management is an essential long-term strategy to combat the threat of floods and erosion and to preserve the ecosystem. The sediment load in the rivers, which is largely the consequence of geomorphologic processes in the upper catchment areas, tends to increase with the progressive removal of vegetative cover on the hill slopes.

Soil conservation and reforestation in the upper catchments of Nepal and India and also within Bangladesh could help in substantially reducing sedimentation. In most instances of water resources development programmes at higher elevations, soil conservation practices are initiated as a follow-up step. This need not be so. Soil conservation and management programmes could be taken up independently in vulnerable sites as well as through integrating them in the environmental management plans for water-related interventions. Soil conservation strategies should be both rehabilitative and preventive, and can succeed only with people's participation in the whole process of strategy formulation and implementation.

In the context of the fragility of the Himalayan ecosystem and burgeoning population pressure on hilly slopes, an integral part of water resource planning should be to adopt rational land use and cropping patterns, including contour ploughing, in the upper catchments. Measures to conserve soil quality and improve the ecological health of the land might be highly desirable in the context of area development programmes in upland regions which tend to be neglected or are less accessible.

Conclusions

The water-based development contexts, issues, and potentials presented and briefly described in the previous three sections reflect vast scopes and opportunities for the future implementation of water-based regional projects. The true spirit of regional cooperation lies in moving from bilateralism towards multilateralism. Congeniality in the political atmosphere may lead to a visionary approach among the politicians of the countries of the GBM region to create an enabling environment of multilateralism. Civil societies of the GBM countries have a vital role to play in this context. They should voice the need for the integrated basin-wide approach in water resources planning in their respective countries and eventually across the region. Various institutions and 'think tanks' of the region should try to link up structurally with each other with a long- term commitment to cooperate and collaborate with one another.

Once a relationship of cooperation and collaboration is developed in the region, water-based projects should be implemented with a clear understanding of the issue of prioritization and institutional arrangement. For example, a non-conflicting and nonstructural issues like technically and regionally improving the flood forecasting and warning system could be a starting point both in terms of piloting the institutional linkages and institutional arrangements.

Under the purview of the institutional arrangements, the countries of the GBM region should carry out detailed feasibility studies of regional projects, which will have a bearing both on cost sharing and benefit sharing between them. Linked to this, institutional arrangements for long-term cost recovery is also very crucial. These are issues essential to attaining and ensuring sustainability of integrated water resources management in the region. The active donors in the water sector of the region have an important role to play in this regard.

Bilateral funding plays a major role in water resources management. World Bank financing helped resolve the Indus dispute, while the United Nation-led investments helped achieve the Mekong Agreement. Cooperation-inducing financing has not always come from outside the region. Thailand helped to finance a project in Laos, as did India in Pakistan, in conjunction with their respective watershed agreements. A provision of the Nile Waters Treaty has Egypt paying Sudan outright for water to which they both agreed Sudan had rights, but was not able to use.

As a general recommendation, proposition, and thesis towards improvement, it is suggested that the active water-sector donors of the region persuade and promote agenda of water-based regional cooperation. As development partners of the region, the donors may be involved in the cost-sharing arrangements along with the respective governments, maybe in a sector-wide approach. A more coordinated, harmonized, and synchronized approach may be adopted from the donor's side in order to take care of the long-term vision. As a proposition, the donors may create regional funds for the regional projects and attach a general conditionality that all relevant countries in the region should act together in a joint spirit both in terms of cost and benefit sharing in order to realize the funds.

References

Adhikari, K.D., Q.K. Ahmad, S.K. Malla, B.B. Pradhan, Khalilur Rahman, R. Rangachari, K.B.S. Rasheed, and B.G. Verghese, *Cooperation on Eastern Himalayan Rivers—Opportunities and Challenges*, Konark Publishers Pvt. Ltd, Delhi, 2000.

Ahmad, Q.K., B.G. Verghese, R.R. Iyer, B.B. Pradhan, and S.K. Malla (eds), *Converting Water into Wealth: Regional Cooperation in Harnessing the Eastern Himalayan Rivers*, Academic Publishers, Dhaka, 1994.

Ahmad, Q.K., A. Ahsan, K.B.S. Rasheed, and A.H. Khan (ICID, CIID), *Management of International River Basins and Environmental Challenges*, Academic Publishers, Dhaka, 1994.

Ahmad, Q.K., R. Rangachari, and M.M. Sainju, *Ganges-Brahmaputra-Meghna Region—A Framework for Sustainable Development*, University Press Limited (UPL), Dhaka, 2001.

Malla, S.K., *Three Country Study on Water Resources Development of the Ganges-Brahmaputra-Barak River Basins: Nepal Country Report*, IIDS, Kathmandu, 1989.

Shrestha, A.P., *Hydropower in Nepal: Issues and Concepts in Development*, Kathmandu, 1991.

Verghese, B.G., *Water of Hope: From Vision to Reality in Himalaya-Ganga Development Cooperation (Second Edition)*, Centre for Policy Research, New Delhi, 1998.

5

Water and Regional Development in the Yellow River Basin

James E. Nickum

Introduction

Rivers are often the backbone for regional development. They can also be channels for transmitting stress between regions. The terms 'upstream' and 'downstream', put together, immediately evoke the word 'conflict'. Rivers slice through, bound, and, in many cases, create regions. Rivers bring water, floods, silt, and pollutants across administrative and regional boundaries, and when they do not, the result is often disaster in the form of drought.

The conflict between regions is made more complex by asymmetries of development. Stephen McCaffrey (1993: 99) has stated a general (although, as we shall see, not necessarily universal) case of this problem:

First, it seems apparent that downstream countries, whose topography generally lends itself more readily to agriculture than that of countries at the

I am grateful to Hiroshi Kameyama for asking me to join him on his study of the Yellow River; to the Ministry of Education, Culture, Sports, Science and Technology of the Government of Japan for the funding; to the Chinese Academy of Agricultural Sciences (especially Zhang Linxiu and Huang Jikun), my old stomping grounds, for their in-country sponsorship; to Chaolunbagen and his colleagues, including but not limited to Liu Huizhong and Xiao Fei, for their warm and enlightening introduction to Inner Mongolia and its water system; to Hu Tao of the State Environmental Protection Administration for his insights on Yellow River management issues; and to Paul Bacon for his astute review of the manuscript.

headwaters of international rivers, make intensive use of their water resources before their upstream neighbors. ... Second, efforts or even plans by later-developing upstream countries to begin to utilize their water resources often give rise to strenuous objections by their downstream neighbors. The downstream countries often take the position that they have acquired rights to the quantity (and perhaps quality) of the water they have used in the past. ... Third, the later-developing upstream countries typically take the position that they have not heretofore had the need or the capability to develop their water resources....

There is much truth to McCaffrey's observation that can be applied inside nations as well, to provinces and regions within a large continental country such as China, which contains entirely within its borders the 5,463 km-long Yellow River (Huang He).[1] At the same time, actual upstream–downstream relationships, as illustrated by the Yellow River, are considerably more complex than this. An important qualification to McCaffrey is to note that 'downstream', defined as the rich, flat, river-bottom land below the sloped 'upstream' watershed areas, is often quite far inland from the estuary, especially in a closed economy. The region of the Yellow River, and therefore China, that developed first, including the construction of large-scale surface diversion irrigation works over two millennia ago, was in the 'middle reaches' in the present provinces of Shaanxi and Henan.

Indeed, the general pattern appears to be for the lowest reaches, especially delta regions, to develop late, if at all, often in response to an opening of the economy to foreign trade, requiring ocean ports. The Yellow River delta still contains no major city, but has economic significance due to the presence there of China's second-largest oil field, the Shengli. By now, the lower reaches of the Yellow River and attendant areas, encompassed by the province of Shandong, have indeed become the most economically developed area along the river course, and are commensurately one of the heaviest users. While this gives them considerable political clout, it does not give them a strong historical claim over the resource.

Despite this caveat, the experience of China is that even when the upper reaches develop early, they suffer the dual disadvantages of

---

[1] And the even longer (6,300 km) Yangtze River (Chang Jiang), with a catchment basin twice as large (1,808,500 sq km compared to 752,443 sq km) and an average annual flow nearly fifteen times as large (951,300 million cu m compared to the Yellow River's 66,100 million cu m).

geography and poor access to trade. Significant irrigation has a long history in the 'lower' part of the upper reaches of the Yellow River (present Gansu, Ningxia, and Inner Mongolia) (*Shuilibu Huanghe ... Ministry of Water Resources*, 1982: 81–3). These areas, however, have tended to remain on the geographical, military, and economic margins of China.

One consequence of the staggering of the stages of development between upstream and downstream areas is that the latter may have more water in total but more stress, because the demands on the resource are greater. Upstream areas may have relative abundance but have to contend with prior claims to the water by the richer, hence politically more powerful downstream areas, who may also have better access to the systems of rights and adjudication. With a more developed economy, they can also make compelling arguments on the grounds of economic efficiency, that their use of water on the margin will yield higher value than upstream areas. Here I focus on the interplay between a downstream province, Shandong, and Inner Mongolia upstream.

The water that could enable development is at hand upstream, but contested by the already relatively well-developed downstream area. This can lead to ironic solutions, such as a recent proposal to build a massive (20 cu km/annum) and technically complex interbasin transfer to take water from the abundant head reaches of the Yangtze River and put it into the abundant head reaches of the Yellow River in order to promote development of western China (Zhang Guangdou, 1997: 2).[2]

Case of the Yellow River

Budao Huanghe xin busi
(Do not lose heart before you reach the Yellow River—Do not give up until you reach your goal)

—Traditional Chinese saying.

The Yellow River (Huang He) bears many appellations—the cradle of the Chinese nation (*Zhonghua minzu di yaolan*) being perhaps the most common. The early dynasties located their capitals along a narrow band of the middle reaches, stretching a few hundred kilometres from

---

[2] The focus in this chapter is on the interplay between regions to date. Hence this new programme to develop western China and its waters will not be addressed here.

Xianyang in the west on the Wei River tributary to Kaifeng in the east, at the apex of the region where the river has changed its course, switching in geological time from the north to the south sides of the north China plain like a flickering filament. Together with the Great Wall, the Yellow River has come to epitomize Chinese identity (See, for example, Zhong Bin and Zhang Xuejian, 1998: 5), even though fewer than one in ten, about 100 million people, now live in the basin.[3]

The Yellow River is also known as a 'suspended river' (*xuanhe*), its highly silted bed lying above the land to either side of its wide embankments in the lower reaches. It is the most silt-laden major river in the world. Its total annual sediment discharge of about 1,080 million tonnes is second only to that of the Ganges/Brahmaputra's 1,670 million tonnes, but with 5 per cent of the latter's run-off (Gleick, 1993: 158). Ironically, then, for most of its lower course the river is itself a watershed divide between river systems to the north (*Hai*) and south (*Huai*).

Regional development on the north China Plain itself has been precarious historically, as the main surface water source cannot be relied on to stay in place, and presents a constant threat of flood due to its high silt load. In the past, agriculture relied on a highly uncertain supply of precipitation. Serious droughts affected Shanxi province 254 out of the 1,800 years between 100 and 1899. Ninety-seven severe droughts devastated Henan province during the same period. Compounding the problem, dry years tend to come in strings (*Huanghe liuyu ji xibei pian shuihan zaigai bianweihui*, 1996: 270–5, 304). Recently, especially since the 1970s, irrigation on the plain has expanded dramatically with the drilling of tube wells and opening up of diversions from the Yellow River (Nickum, 1995).

The battle to keep the river from switching its natural regime has been a central problem of many historical political regimes and has led

---

[3] The imagery has been used for cultural critiques as well. *He Shang* (*River Elegy*, with a word play on '*River Wound*' or '*River Sorrow*'), a popular and controversial 1988 television series, used the symbolism of the Yellow River to indict contemporary Chinese as 'a servile, static, and defensive people who always meekly hug to mother earth to eke out a miserable living, rather than boldly venturing forth on the dangerous deep blue sea.' (Chen and Jin, 1997: 221–2) 'China's Sorrow' is commonly used to refer to the Yellow River in English-language writings (for example, Ongley, 2000: 227), but appears to be rare in the Chinese-language materials consulted here.

to a constant cycle of diking, silting and elevation of the river bed, further diking, and eventual overtopping and victory for the river. Near Jinan, the largest city on the lower course, the current bed of the Yellow River is 3–5 m above the land outside the dikes. It is 13 m above the medieval capital of Kaifeng and 20 m above the city of Xinxiang (Li Guoying, 2002: 21). One of the hydrological consequences of this diking and elevation process is that the river receives very little natural run-off in its 768 km of lower reaches, only 2,100 million cu m out of 58,020 million cu m for the river as a whole (Zhong and Zhang, 1998: 11, 40). At the same time, and perhaps partially in consequence, groundwater on the plain on either side of the river tends to be quite abundant.

Being downstream on the fertile plain was definitely a mixed blessing historically, because silt and seasonality led to occasional course shiftings, usually with enormous loss of life, property, and, for those who depended on the river, of water source. In historical times, major changes of channel, often by hundreds of kilometers, have occurred in 602 BC, AD 11, 1048, 1128, 1855, 1938 and 1947. The shifts in AD 1128 and almost certainly in 1938 were carried out deliberately by the Chinese for military purposes. In the latter case, the river was restored to its former course in 1947. In the other four instances, the river itself decided when to move its bed as its existing ones silted up.

Even Kaifeng has not been immune. The history of the city is buried underneath it, sometimes quite deep. The relics from one to two millennia ago (from the Han dynasty to the Tang dynasty) lie under 13–15 metres of silt. City walls from the Ming dynasty (1368–1434) press down on walls from the Jin dynasty (1127–1234) that are atop walls from the Northern Song dynasty (960–1126), under which one can find walls from the Tang dynasty (649–712) (Zhong and Zhang, 1998: 55–6).

Indeed, the management problem for the lower Yellow River was not how to use its water but how to speed its muddy course to the sea penned within a given channel. About 20 cu km of flow are necessary annually to maintain an adequate discharge of silt to the ocean. Flood control is thus one of the largest 'consumptive users', but without an effective human agency to enforce this environmental 'water right', it is also one of the most dishonoured in the breach, and the most likely to lead to a breach of the banks. One irony is that the downstream area, which is most likely to be hit by flood, is also tapping the water most heavily.

Hence, the downstream was a 'late developer' as much, or perhaps even more than many parts of upstream (for example, Inner Mongolia and Ningxia). Historical development tended to be centred along the midstream, often on tributaries (not unlike other historical rivers such as the Indus, the Danube, the Nile, the Ganges, or the Mississippi). In recent decades, downstream water development has increased in significance, accelerated by the rapid changes in the economy and the growth of urban areas brought about by economic reforms and the opening up of the economy that began in the early 1980s. Energy needs have also become increasingly intertwined with the water of the Yellow River and its tributaries with the development of a major oil field—Shengli—in the river's delta and the mining of one of the world's largest deposits of coal in the arid Shanxi province.

The Yellow River basin is also known to be notoriously short on water. Its run-off is only about 2 per cent of the national total, or about 590 cu m/capita, about one-quarter the national average. Even adding a bit for groundwater, about one-quarter of run-off, the area qualifies as a seriously water-short one, about the same as Syria.[4] This water is spread thinly over crops, at 4,860 cu m/ha, which is only 17 per cent of the average national application rate (Wang Jianzhong, 1999: 10).

Regions

China as a whole is conventionally divided into east, west and middle China. The east, with twelve of China's thirty-one provinces,[5] has the smallest land area (13.5 per cent) but the largest population (over 40 per cent), and generates 60 per cent of the country's economic output. With its access to the coast and foreign trade and investment opportunities, the east has benefited far more than the other regions from the rapid expansion of the economy over the past two decades.

The strategic middle, with nine provinces, is the most balanced region, with 30 per cent of the land area, 35 per cent of the population, and about 30 per cent of the economic output. It contains nearly half

[4] International comparisons of per capita water availability are notoriously problematic for reasons of definitions and data quality, so this is just a rough approximation.

[5] Technically, the thirty-one administrative units ranked below the centre are provinces, autonomous regions, and directly administered municipalities. In the Yellow River basin, Inner Mongolia and Ningxia are autonomous regions; the remainder are provinces.

Table 5.1: Three reaches of the Yellow River

	Mainstem length (km)	Inflowing tributaries	Catchment area (sq km)	Cropland (million ha)
Upstream	3,472	43	385,966	3.173
Midstream	1,224	30	344.070	7.752
Downstream	768	3	22,407	1.018
Total	5,464	76	752,443	11.943
	Irrigated land (million ha)	Pasture (million ha)	Surface water (cu km)	Groundwater (cu km)
Upstream	1.379	21.587	36.22	2.20
Midstream	2.314	6.346	26.74	4.36
Downstream	0.685	0.010	2.92	1.03
Total	4.378	27.943	65.88	7.59

Notes: Irrigated land in lower reaches does not include land in other basins; surface and groundwater are the amounts generated within the catchment of each reach.
Source: Huanghe liuyu ji xibei pian shuihan zaigai bianweihui, *Huangshe liuyu shuihan zaigai*, 1996: 12–14.

of China's farmland and produces over two-fifths of its grain, meat, and aquatic products.

The west, with ten provinces, covers 56 per cent of China's land area, having only 23 per cent of the population, and contains a significant share of China's minority populations and energy reserves (Yang Jin, 2000: 57).

The Yellow River, like the Yangtze, flows from west to east through all three regions. Most of China's provinces lie at least partly in the basin of one or the other of these two great flows.

The Yellow River is conventionally divided into upstream, midstream and downstream reaches (Table 5.1). As shown in Table 5.2, there is a close but not exact correspondence between these reaches and the economic regions. Upstream originates in Qinghai and a small part of Sichuan provinces and flows through the western provinces of Qinghai, Gansu, Ningxia, and almost all of Inner Mongolia, a central province.[6]

---

[6] The lightly populated upper reaches can be clearly seen on a cloudless night from space, from the lights of the settlements concentrated along its banks.

Table 5.2: Areas of Yellow River provinces in the Huang–Huai–Hai basins

Province	Total area	in Hai basin	in Huai basin	in Yellow basin
West				
Qinghai	721,000			152,500
Sichuan	570,000			11,080
Gansu	454.300			144,750
Ningxia	66,400			51,820
Central				
Shaanxi	205,600			133,000
Inner Mongolia	1,183,000	12,580		153,600
Shanxi	156,300	59,130		97,090
Henan	167,000	15,300	88,240	36,040
East				
Shandong	153,000	29,710	111,570	12,020

Note: Shaanxi and Shanxi are midstream provinces. Henan has some downstream portions as well.
Areas are in sq km
Source: *China: Agenda for Water Sector Strategy for North China*, Vol. 1, 2001, World Bank.

For most of its course, the mainstem of the middle reaches forms the boundary between the central provinces of Shanxi and Shaanxi provinces and receives the flows of two major tributaries (the Fen and the Wei) from them. The central province of Henan straddles the middle and lower reaches, which begin at Taohuayu near Zhengzhou in the plains part of the province. Downstream Shandong is the sole eastern province directly in the Yellow River basin, although inter-basin diversions provide supplementary Yellow River water to Hebei and Tianjin as well.[7]

Within and adjacent to these larger regions, there are a number of distinct sub-regions, which may be delineated roughly, beginning from the source, as:

a) *Headlands*. High, cold, sparsely populated, and with little economic activity, this area is of little significance for our purposes, except that it is the source of 32 per cent of the river's flow. As with the Nile, most of the water of the Yellow River originates far upstream. Recent years have seen a reduction in source waters, perhaps due to global warming.

[7] For the division into upstream, midstream, and downstream, see Huanghe liuyu (1996: 11–3)

b) *Power centres.*   This area contains four major reservoirs, notably the Longyang Gorge in Qinghai and Liujia Gorge in Gansu, that produce hydroelectricity for regional mining and industrial areas, exemplified by Lanzhou. The multipurpose uses for these dams to control floods and release water for river maintenance or to relieve shortages far downstream creates a potential for conflicts with widely separated regions. At the same time, these reservoirs provide most of the storage along the mainstream of the river and therefore have been relied on heavily in recent years for recurrent emergency releases of water to the lower reaches and estuary.

c) *'Minority' areas.*   This dry area contains extensive grain areas in Ningxia and Inner Mongolia, both 'autonomous regions' with significant minority populations (34 per cent, mostly Moslem Hui; and 19 per cent, mostly Mongol, respectively). These have seen extensive migration over the years from the majority Han population, who constitute over 90 per cent of Chinese. Since the Han are more likely to be farmers, while the Mongols are traditionally herders, there is an ethnic undertone to irrigation farming as well.

These grain belts along the river, among the largest individual irrigation districts in China, depend almost entirely on the Yellow River for irrigation. Flat and with high rates of evaporation, they are prone to salinization. In response to these conditions, but in the long run aggravating them as well, gross water application rates are quite high. Because consumptive use of Yellow River is high in this region, for economically low-value uses, conflict with downstream users is also high.

Without a reliable source of recurrent funding, irrigation and drainage facilities have aged rapidly and often no longer function well. While deferred maintenance [or, as the Chinese put it, 'focus on construction to the neglect of management' (*zhongjian qingguan*)] is hardly unique to this locality or even to China, it does aggravate the problem of reducing water use rates in these large surface systems.

d) *Loess plateau.*   The Yellow River gains most of the yellow sediment load that gives it its name as it flows through the world's largest loess deposit between Hekouzhen and Longmen. This is a largely inhospitable region of gullies, dunes, and low-yielding farmlands. Erosion control measures here not only reduce flooding danger from sedimentation downstream, but also free up water otherwise necessary for flushing the sediment to the sea. Some progress has already been made in the past fifty years, but it has been slow. In the long run, it should be possible to reduce sediment loading by half (Ongley, 2000: 228).

e) *Historical heartland.* The Wei River Valley, incorporating and irrigating the Guanzhong Plain, is a major grain producing area as well as the site of the ancient capital of Xi'an. Groundwater extraction for urban and industrial uses has led to significant overdrafts. Accumulated silt behind the Sanmen Gorge dam at the confluence of the Wei and the Yellow River reduces the silt burden in downstream channels but aggravates the potential flood threat along the Wei.

In recent years, the banks of the Wei have been the centre of rapid economic and urban growth. These growth stresses, combined with a rapid decline in water flowing in from upstream areas, have led to maladies that are reminiscent of those further down on the mainstream of the Yellow River: flow stoppages and consequent silt accumulation in the river bed, degradation of water quality (the greatest peril at present), and overdrafts of groundwater on adjoining plains areas. Irrigation remains the major water user, with aging projects falling into disrepair because of inadequate cost recovery, like their counterparts upstream and downstream. (Qian, 2002: 6).

f) *Coal area.* The densely populated valley of the Fen River tributary in Shanxi is a major industrial and coal-mining area. The water stress in terms of water available per capita or per unit of irrigated land is among the most critical in China. Droughts are common, groundwater aquifers are overtapped, and the quality of water has deteriorated.

g) *Downstream plains, including inter-basin diversion areas.* Historically, the Yellow River has been interconnected with the Hai basin to the north and the Huai basin to the south as its channel has shifted back and forth across the north China (Huang-Huai-Hai) plain. As indicated in Table 5.2, planning is still done considering the three basins as one larger unit. Here, the focus is on the Yellow River basin, but it must be kept in mind that it is not isolated from its neighbours on the plain, as its water is increasingly transferred to them.

h) *Delta area.* The Shengli oilfields and Dongying municipality in the delta area are in one of the most economically vulnerable positions along the river. At the same time, they are politically potent and economically strong, and therefore can more readily appeal for action by the national government.

Regional Development Policies

Before the initiation of economic reform in 1979, China's regional development policy was dominated by concepts of self-reliance, planned economy, and strictly limited mobility of labour and capital. Some transfers were made from the relatively prosperous east to the lagging west, but without appreciably closing the gap. The opening up of the economy to foreign trade and investment began with the setting up of special economic zones along the coast, boosting the locational advantage already enjoyed by the east and maintained even after the entire economy was opened up. In 2000, a major programme was launched to develop the west, raising the issue of where to obtain the water to do so.

The economy of Shandong in particular has prospered from the opening up of the economy due in large part to the proximity of its dry eastern peninsular cities to South Korea and Japan and its location along major transportation routes from the economic centres in the north to those in the south. As Table 5.3 indicates, the prosperity of the ports spread to the farm households, where Shandong has widened the income gap between itself and other Yellow River provinces. It did this in no small part by using the water of the river.

Table 5.3: Average farm household income, 1990 and 1997
(Shandong = 100)

Province	1990	1997
Shandong	100	100
Henan	78	76
Shanxi	89	76
Inner Mongolia	89	78
Shaanxi	78	56
Ningxia	85	66
Gansu	63	52
Qinghai	82	58
All China	101	91

Source: *Zhongguo nongcun tongji nianjian 1998* (China Rural Statistical Yearbook 1998), Beijing: China Statistics Press, November 1998, p. 249. The per capita rural farm income in Shandong was 680.18 yuan in 1990 and 2292.12 yuan in 1997 (in current prices).

When the Water Stopped Flowing

Few things are as effective at intensifying conflict between upstream and downstream as when the water disappears. By 1990, nearly half of the annual run-off in the Yellow River was being used consumptively. This rose to over 60 per cent, to more than 30 cu km out of less than 50 cu km, during the 1990s (Huanghe Shuili Weiyuanhui, 2001: 14). A total of 122 projects in the lower reaches divert up to 4000 cu m/s from the Yellow River in eighty-five counties and municipalities to irrigate over eight million hectares of land (Zhu Erming, 1997: 10). These include some very productive cotton, tobacco, and rice fields on the north China plain.

By far, most of the downstream diversions are in Shandong province, with fifty-four gates, fourteen syphons, and twenty-eight pumping plants with a designed capacity of about 3,000 cu m/s (Cheng et al., 1997: 51). Downstream diversions began in 1957, but led to widespread secondary salinization of the soil and were suspended for five years in the early 1960s. After 1966, the water of the Yellow River was once again used, almost entirely for irrigation. Average annual diversions were 2,330 million cu m during the initial 'Great Leap Forward' years (1957–61). It fell to 1,630 million cu m during the restoration years (1966–69), increased dramatically to 4,820 million cu m during the 1970s, and jumped to 7,640 million cu m during the drought-prone 1980s. In the particularly dry year of 1989, Shandong's agricultural withdrawals leaped to 12,346 million cu m (Cheng et al., 1997: 52), well above the quota set only two years previously, which will be discussed later.

In addition, water-short cities such as Qingdao on the Shandong peninsula have come to rely on diversions from within Shandong, as have cities on the plains north of the river, sometimes including the distant great metropolis of Tianjin. The largest non-agricultural destination of Yellow River water, however, has been the delta area, including the Shengli oilfield. As can be seen in Table 5.4, though, non-agricultural uses of the river's water are relatively minor contributors to the shortage. Of these uses, industry is not the most significant, perhaps in part because of reliance on groundwater. Even as late as in 1993, 94.4 per cent of the diverted water went to irrigation (Cheng et al., 1997: 51).[8] Since they tend to be at the end of the line,

---

[8] Agriculture took 92 per cent of water used for the entire river in 1998 (Wang Jinzhong, 1999: 10).

Table 5.4: Yellow River diversions in Shandong province

Purpose or destination of diversion	Period	Total diverted (million cu m)	Annual average	Industrial use (entire period)
Irrigation	1980–9	52,700	5,270	nil
Estuary, including Shengli oil field	1983–3	8,040	731	2,533
Qingdao	1989–94	1,160	194	260
Tianjin	1981–93	893	69	200
Other cities	1984–94	–	–	351

Source: Derived from Cheng et al. (1997: 54–55).

non-agricultural users are among the most affected when the water fails to flow into the Bohai Gulf.

Virtually unknown before, flow stoppages in the mainstream of the Yellow River began to appear in 1972. The severity of stoppages worsened over the years, especially during the 1990s. Water flowed to the sea for only 143 days in 1997, an extremely dry year and the worst year on record for stoppage. At one point, the flow trickled out 703 km inland at the outskirts of Kaifeng (*Huanghe queshui* 1997: 1–2, Ke, 1997: 34; Nickum, 1998: 76–7).

Are the Losses Real?

The economic losses to Shandong, the province by far most affected,[9] are given as follows (Zhong Bin et al., 1998: 45):

'In the nineteen years, according to analyses by experts, flow stoppages in the Yellow River between 1972 and 1996 led to direct economic losses in industrial and agricultural value of 23,400 million yuan in Shandong, an annual average of 1,230 million yuan. Out of this, losses amounting to 18,500 million yuan occurred during the 1990s, an annual average of 3,090 million yuan. Direct economic losses from stoppages in 1997 reached 13,500 million yuan, 4,000 million of which was to industry, 7,000 million to agriculture, and 2,500 million to other sectors. In the nineteen years from 1972 to 1996, direct economic losses

---

[9] As noted, in 1997, the flow stopped upstream as far as Henan, where Kaifeng is located. Also, some Yellow River water diverted in Shandong is provided to Hebei and Tianjin. Stoppages in these flows probably constitute a small portion of total losses, however.

to industry from Yellow River flow stoppages were 12,400 million yuan, an average of 650 million yuan per year.'

Is this a lot or is it a little? Much is unanswered in the data provided by Zhong, including how economic losses are computed, whether they are in value added GDP or material product (gross value of output) terms, whether the yuan is in constant or current values, and whether output reductions due to flow stoppages have been separated out from total losses due to drought. Some indication of significance may be gained, however, from noting that the GDP of Shandong in 1995, in current prices, was 500,234 million yuan (*China Statistical Yearbook 1996*: 43). This implies that the cited direct losses amount to less than 2 per cent per annum in an economy that is growing at a much faster rate. Indirect losses can be much higher, of course, depending on the multiplier effect.

Furthermore, drought losses occur every year. A calculation of indirect losses (estimated at about seven times direct losses) in industrial and agricultural value in the Yellow River basin due to drought indicates an average annual loss of 15 per cent from 1950 to 1990, but with no clear intertemporal trend (Figure 5.1).[10]

Hence, while they may be locally traumatic, it is possible that the bulk of economic losses, using the current measure, should be considered a normal kind of slack or customary loss below an unattainable ideal (like structural unemployment is to full employment), unless they grow over time as a percentage of economic output. Even then, they might reflect an economy that is growing so fast that it is outpacing infrastructure and policy responses. Thus, for example, eighteen of the thirty-three cities in the basin deemed water-deficient in 1994 were so because of inadequate infrastructure, not a lack of source (Chen et al., 1997: 63). It should also be noted that pricing distortions, especially the lack of full costing for the economic value of water, might overstate the level of loss.

[10] The estimate of indirect losses used a simple exponential function to derive $y = 10^{-6} \times 9.548x^{2.3757}$, where y is gross value of industry and agriculture in the Yellow River basin (in 100 million yuan) and x is grain output (in 10,000 tonnes). This function makes the dubious assumption of a direct relationship between grain output and total output value, with uncertain bias over the period. For the period as a whole, a linear regression of annual indirect losses yields a slight positive coefficient ($y = 0.001x$), with an insignificant R^2 of 0.029—in other words, no statistically significant trend up or down over time.

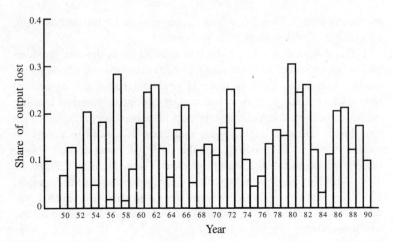

Figure 5.1: Estimated indirect loss due to droughts in the
Yellow River basin, 1950–90

Source: Huanghe Liuyu, pp. 379–81, 1996.

It should also be remembered that the largest user of Yellow River water by far in Shandong is agriculture, which is a low-value use relative to industry and municipal water.[11] Thus temporary shortages creating cessations of flow could in principle be addressed within Shandong, for example, by cutting back on agricultural uses or by tightening controls over those who, fearing shortage, draw more water than they need, thereby exacerbating shortages for others (Guojia Huanjing Baohu Ju, 1997). It is generally felt, however, that Shandong's farmers are already relatively efficient in their water use, and that the greatest gains can come from the large irrigation districts upstream, especially in the second-largest user of the Yellow River, Inner Mongolia.

An example of the political clout of the delta area with its oilfields came in 1992, when there was what to that date was a record flow stoppage. On 8 July, Shandong province sent an urgent cable to the

---

[11] Cheng et al. (1997) estimate that value added to irrigation in the 1980s was 0.53 yuan/cu m, while that for industry in roughly the same period was about 60 per cent higher, at 0.82 yuan/cu m. Most studies indicate that the marginal difference between agricultural and industrial uses anywhere is much larger, by several orders of magnitude.

State Council in Beijing asking for immediate remedial action. Such action was forthcoming, with cabled orders to Henan and Shandong to adopt remedial actions, including strictly forbidding withdrawals for irrigation. The Central government also ordered the upstream and midstream provinces of Gansu, Inner Mongolia, Shaanxi, and Shanxi to stop using water for agricultural use temporarily, and ordered the Liujiaxia Reservoir to increase its water releases. In the crunch, agriculture was squeezed (Yang Chaofei, 1997: 2).

Environment

Most likely, a more serious problem than the economic aspects of the flow stoppages are their environmental effects, notably increased siltation of the channel, and the lowered capacity of the river to dilute an ever-growing amount of industrial and agricultural pollutants. Here, there are intertemporal trends in recent years, much for the worse.

As shown in Table 5.5, flood losses in the Yellow River basin have historically been of a comparable order of magnitude to those from drought, especially in Henan province. Although one would think that a reduction of flow, which occurs during the short high-flood season as well as the dry season, would reduce the danger of flood, this is not the case in the highly silty Yellow River. Silt deposits have increased markedly in the channel, increasing the likelihood of a catastrophic

Table 5.5: Total economic losses attributed to flood, drought, and urban shortage, 1950–90 (in million yuan Renminbi at 1990 prices)

Province	Flood	Drought	Urban shortage
Qinghai	737	623	
Gansu	1,970	5,611	
Ningxia	189	1,444	
Inner Mongolia	1,324	2,968	131
Shaanxi	4,111	8,821	9,629
Shanxi	5,111	6,094	
Henan	15,508	4,524	
Shandong	2,598	2,205	9,175
Total	31,548	32,290	18,935

Note: Drought losses are to agriculture, except in Gansu, Ningxia, and Inner Mongolia, where pasture losses are included. Only in Inner Mongolia are these significant, at 1,411 million yuan.
Source: Huanghe liuyu, p. 44, 1996.

breach of the banks. This is of much greater concern to the experts than are the direct effects of the flow stoppage.

Pollution of the water of the Yellow River has also become a major environmental concern in recent years. Rising concentrations because of declines in flow have combined with increased discharge of untreated wastewater by industry, growing cities, and ever more agrochemicals. It is considered to be the second-most heavily polluted of China's seven major rivers (Wang Jianzhong et al., 1999: 11). Over 70 per cent of its length is considered to be highly polluted. In some cases, especially near the mines in the water-desperate Shanxi province, the water is rendered useless for other purposes (Ongley, 2000: 230).

Allocations Among Provinces

In September 1987, China's State Council issued State Document Number 61 (1987), transmitting the 'Report on a Scheme for the Allocation of Water that can be Supplied from the Yellow River' of the State Planning Commission and the Ministry of Water Resources and Electric Power as a non-binding set of allocations among the eleven provinces and municipalities using the river. Listed in Table 5.6, these quotas were set on the basis of average flow, hence in absolute, not proportional amounts. In effect, this could favour upstream users in a time of drought.

As in the case of the Colorado River, setting quotas did not go too far towards solving conflicts when water was short, especially chronically short, and still necessitated after-the-fact negotiations among affected provinces in case of shortage (World Bank, 2001: xi–xii). Also similar to the Colorado River, average flow was set at a relatively abundant time, and run-off has subsequently fallen below the figures used to set the quotas. Rainfall in the period 1990–97 was more than 10 per cent below average. This, combined with increased withdrawals, led average annual run-off at Huayuankou (at the debouchment into the plains) to fall to 26,800 million cu m during the same period, vastly below the 48,200 million cu m average during the wetter 1950s (Wang Jianzhong et al., 1999: 10).

Eventually, a revised set of more mandatory quotas were imposed in 1999. For the time being at least, these restrictions, in combination with more plentiful rainfall, appear to have succeeded in stopping the flow stoppages.

Table 5.6: Allocation of consumptive use of water (in million
m³ per annum) from the Yellow River prior to the completion
of the south to north transfer project(s)

Province	Quota	Actual use 1989	Actual use 1990	Actual use (late 1990s)	Actual use (after 1999)
Qinghai	1,410	983	999		
Sichuan	400				
Gansu	3,040	2,326	2,362		
Ningxia	4,000	3,411	3,542		
Inner Mongolia	5,860	6,056	6,464		
Shaanxi	3,800	1,958	1,851		
Shanxi	4,310	1,440	1,230		
Henan	5,540	3,723	3,295	3,500	2,800
Shandong	7,000	13,479	8,092	7,500	
Hebei and Tianjin	2,000				
Subtotal	34,960	33,376	27,835		
Total	37,360				

Note: Subtotal is without Sichuan, Hebei, or Tianjin.
Source: Chen Yongqi et al. (1997: 12); Zhang Ren (1997: 67) (1989, 1990);
Ka Shizhong et al. (2001: 83) (Henan in 1990s); Zhang Guangdou (1997: 1)
(Shandong before 1999).

The quotas would appear to establish the basis for a crude sort of property right assignment among provinces. If a real-time monitoring system can be installed at the provincial borders, as is now being discussed, it would be possible to assign rights in a way that might even allow a market solution, whereby water-rich provinces could sell water to cash-rich ones. Not long ago, this approach would have seemed rather far-fetched, but it is not at all outside the realm of consideration these days (for example, by Qian and Wang, 2001: 61). Problems remain of who pays whom, especially within the provinces, but merely having these allocations sets the basis for solving one of the biggest problems of setting up a water market, which is establishing initial entitlements (Randall, 1981).

Inner Mongolia

Inner Mongolia is dominated by the giant (1,160,000 ha.) Hetao Irrigation District. Although it has ancient origins, in the harsh desert environment with strong winds, the old systems fell into collapse, often buried under the sand (Chaolunbagen, 1998: 2). Hetao's modern history begins with a rehabilitation around the year 1830, after which irrigation relied on a run-of-the-river diversion, subject to the vagaries of the level of the river, until a control structure was built in 1961. Irrigated area expanded rapidly thereafter, but a lack of adequate drainage on the extremely flat land (with a slope of 1/3,000 to 1/5,000) led to a rise in groundwater levels that in turn created widespread secondary soil salinization. In 1981, a major drainage system was built to address this problem (Wang et al., 1993: 1). Water applications remain heavy in Hetao, however, in part to flush out the remaining salts. This may address the salinization problem in the farm land, but at the cost of disrupting the ecology of the Wuliangsu Lake at the end of the system through overloading it with salts and nutrients washed off the flushed fields (field observations, 1999). Allocated 4,000 million cu m of the region's quota of 5,860 million cu m, Hetao actually uses about 5,200 million cu m, contributing to Inner Mongolia's distinction as the only province besides Shandong that regularly exceeds its 1987 quota (Tian, 2001: 40).

Grains produced are primarily for local consumption by people and livestock. It is said to be difficult to diversify into higher value crops because of the remoteness of major markets and price instability—the latter perhaps more a reflection of continuing government price support for grain than of intrinsic instability in non-grain crops. Nonetheless, it is clear that even in its smaller cities such as Linhe, Inner Mongolia has shared in China's recent prosperity and reforms, boasting convenience stores, bowling alleys, and real estate agents. At least in the cities, and probably in the irrigated areas, the lagging behind downstream is in relative prosperity, not absolute impoverization. At the same time, the vast grasslands that are the traditional basis of the Mongolian herding economy have degraded into Nevada-like deserts, dotted here and there with cashmere on the hoof (personal observations in and around Hetao, 1999).

Since the tightening of regulations over the Yellow River allocations in 1999, pressure has been placed in particular on the Hetao Irrigation District to reduce its water use. Approaches adopted include regulation of water intake, a relatively simple matter to deal with centrally as it is controlled by a very small number of structures; greater exploitation of shallow groundwater, often saline and declining with the reduction in surface flow; lining of critical stretches of canals (mains and field ditches, the two levels where seepage is highest); changes in irrigation practices; reduction in irrigated area (Liu Huizhong, 1999; and Zheng Tonghan, 2002); improved pricing; and the development of a decentralized system of management, based on water users' organizations. Both as part of international discourse and perhaps a bit out of desperation, economic pricing, user-based management, transparency of operations, and even water markets are widely being accepted by institutional reformers. Nonetheless, putting them in practice can be difficult.

Thus, for example, many agencies at many levels are involved in setting water prices, in a highly politically charged process. Administrative decentralization has complicated matters. The Price Bureau and the Water Resources Bureau of Inner Mongolia both set a cost of 0.053 yuan per cubic meter for water in the Hetao Irrigation District, but the price bureau of the local government, Bayan Naoer League, which includes Hetao, was able to block this increase and hold it down to 0.04 yuan, below operations and maintenance costs, on the grounds of keeping farmer burdens low. Yet water fees have fallen steadily as a share of costs of growing wheat in the league, from 9.6 per cent in 1989 to about 6 per cent in 1999, while chemical fertilizer costs, which are not so political, constituted 28 per cent of costs in the latter year. A further complication is that conservation measures are often successful. Especially with price ceilings set, this can lead to a fall in revenue to management agencies (Zheng Tonghan, 2002; and field notes, 1999).

Conclusion

Currently, the upstream-downstream conflict in the Yellow River basin is largely one where considerations of equity and efficiency collide. The efficiency considerations tend to win, especially when linked to economic power. Hence, Shandong can exceed its quota and still in effect call upon poorer upstream provinces to release more water, largely without recompense, because its value added is higher downstream. At the same time, it must be noted that doing so only partially displaces the difficult problems of making more rational use of water intra-provincially. Shandong is also called upon to use its water more efficiently, and within its quota.

Enforcing lowered water use and higher efficiency in upstream provinces such as Inner Mongolia may prove beneficial for the latter in the end, especially if it succeeds in introducing more progressive forms of pricing and cost recovery and user-organization-based management. As the basin's capacity for real-time flow measurement improves, moreover, the intriguing possibility arises that market solutions can effect a more mutually agreeable trade of funds for water.

References

Chaolunbagen, 'Neimenggu Zizhiqu guan'gai paishui gaiquang' (Conditions of irrigation and drainage in Inner Mongolia Autonomous Region), mimeo 1999.

Chen Fong-ching and Jin Guantao, *From Youthful Manuscripts to River Elegy*, The Chinese University Press, Hong Kong, 1997.

Chen Yongqi, Wang Tiemin and Qiao Xixian (eds.), *Huanghe liuyupian queshui chengshi shui ziyuan gongxu yuce* (Supply and demand projections for water resources in water-short cities in the Yellow River basin tract), Huanghe Shuili Chubanshe, Zhengzhou, 1997.

Cheng Jinhao, Li Zuzheng and Wang Xuejin, 'Shandong Huanghe shui ziyuan liyong xiaoyi fenxi yu zhanwang' (An analysis and future prospects of the benefits of utilization of the water resources of the Yellow River in Shandong), in Guojia Huanjing Baohu Ju Ziran Baohu Si, *Huanghe duanliu yu liuyu kechixu fazhan*, pp. 51–9, Zhongguo Huanjing Kexue Chubanshe, Beijing.

China Statistical Yearbook 1996 (*Zhongguo tongji nianjian* 1996), China Beijing Statistical Publishing House, Beijing, 1996.

Gleick, Peter H. (ed.), *Water in Crisis*, Oxford University Press, New York, 1993.

Guojia Huanjing Baohu Ju Ziran Baohu Si (Nature Preservation Office, State Environmental Protection Bureau), *Huanghe duanliu yu liuyu kechixu fazhan* (Flow stoppage in the Yellow River and the sustainable development of the river basin), Zhongguo Huanjing Kexue Chubanshe, Beijing, 1997.

Huanghe liuyu ji xibei pian shuihan zaigai bianweihui (Editorial committee on flood and drought disasters in the Yellow River basin and the Northwest stretch [*pian*]), (ed.), *Huanghe liuyu shuihan zaigai* (Flood and drought disasters in the Yellow River basin), Huanghe shuili chubanshe, Zhengzhou, 1996.

Huanghe Shuili Weiyuanhui (Yellow River Water Commission), 'Liuyu shuiliang tongyi diaodu di shijian yu renshi' (The practice and understanding of the unified control of water quantities in river basins), *Zhongguo shuili* (China Water Resources), February 2001, pp. 14–17.

Ka Shizhong, Guo Fang, Huang Shaoxia, and Zhao Jianpeng, 'Henan Huanghe shui ziyuan kaifa liyong cunzai di wenti yu duice' (Problems and policies in the development and use of Yellow River water resources in Henan), *Zhongguo shuili* (China Water Resources), November 2001, p. 83.

Ke Lidan, 'Huanghe xiayu duanliu yuanyin fenxi ji duice yanjiu' (An analysis of the causes of flow stoppage in the lower reaches of the Yellow River and a study of countermeasures), in Shuilibu Shuizheng Shui Ziyuan Si (Water Policy and Resources Office, Ministry of Water Resources), *Huanghe duanliu ji qi duice* (Flow stoppage in the Yellow River and countermeasures), pp. 34–7, Zhongguo Shuili Dianli Chubanshe, Beijing, 1997.

Li Guoying, 'Huanghe di zhongda wenti jichi duice' (The greatest problems in the Yellow River and policies to address them), *Zhongguo shuili* (China Water Resources), January 2002, pp. 21–3.

Liu Huizhong, 'Hetao Guanqu yinhuang zhibiao shuiliang jianxiao hou, guanqu jieshui di chulu chutan' (A tentative exploration of the path forward for saving water in the Hetao Irrigation District after Yellow diversion targets reduce the quantity of water), mimeo, 1999.

McCaffrey, Stephen C., 'Water, Politics, and International Law' in Peter H. Gleick (ed.), *Water in Crisis*, Oxford University Press, New York, pp. 92–104, 1993.

Nickum, James E., 'Dam Lies and Other Statistics: Taking the Measure of Irrigation in China, 1931–91', East-West Centre Occasional Papers, Environment Series No. 18, Honolulu, 1995.

—— 'China's Water Resources Facing the Millennium', *Constraints on Development—Focus on China and India*, Population and Development Series No. 23, The Asian Population and Development Association, March 1999, pp. 71–85.

Ongley, Edwin D., 'The Yellow River: Managing the Unmanageable', *Water International*, 25(2), June 2000, pp. 227–31.

Qian Xi and Wang Lei, 'Guanyu jianli Huanghe shui shichang di zhengce jianyi' (A proposal for establishing a water market policy for the Yellow River), *Zhongguo shuili* (China Water Resources), May 2001, p. 61.

Qian Zhengying, 'Jinkuai zhiding guihua, zonghe zhili Weihe' (Draw up a plan for the comprehensive control of the Wei River as soon as possible), *Zhongguo shuili* (China Water Resources), January 2002, pp. 6–7.

Randall, Alan, 'Property Entitlements and Pricing Policies for a Maturing Water Economy', *Australian Journal of Agricultural Economics*, 25(3), 1981, pp. 195–220.

Shuilibu, (Ministry of Water Resources), *Huanghe queshui xingshi ji qi duice* (The Situation of Water Shortage in the Yellow River and countermeasures), Shuilibu Shuili Xinxi Zhongxin, Beijing, 1997.

Shuilibu Huanghe Shuili Weiyuanhui Huanghe Shuilishi Shuyao Bianxiezu (Compilation Group for Huanghe Shuilishi Shuyao of the Yellow River Water Commission, Ministry of Water Resources), *Huanghe shuilishi shuyao* (A history of water in the Yellow River), Shuili chubanshe, Beijing, 1982.

Shuilibu Shuizheng Shuiziyuan Si (Water Policy and Water Resources Office of the Ministry of Water Resources), (ed.), *Huanghe duanliu jichi duice* (Flow stoppage in the Yellow River and countermeasures), Shuili Shuidian Chubanshe, Beijing, 1997.

Tian Kejun, 'Hetao Guanqu guan'gai guanli tizhi gaige tantao' (Exploring the reform of the irrigation management system of Hetao Irrigation District), *Zhongguo shuili* (China Water Resources), March 2001, pp. 40–1.

Wang Jianzhong et al., 'Huanghe duanliu qingkuang ji duice' (The situation of flow stoppage in the Yellow River and countermeasures), *Zhongguo shuili* (China Water Resources), April 1999, pp. 10–13.

Wang Lunping, Chen Yaxin, and Zeng Guofang (eds), *Neimenggu Hetao Guanqu guan'gai paishui yu yanjianhua fangzhi* (Irrigation, Drainage and Salinization Control in Neimenggu Hetao Irrigation Area), Shuili Dianli Chubanshe, Beijing, 1993.

World Bank, *China: Agenda for Water Sector Strategy for North China*, Volume 1: Summary Report (Report No. 22040-CHA), 2 April 2001 draft.

Yang Chaofei, 'Huanghe duanliudi shengtai sikao' (Ecological considerations from the Yellow River flow stoppage), in Guojia Huanjing Baohu Ju Ziran Baohu Si (Nature Presentation Office, State Environmental Protection Bureau), *Huaghe duanliu yu liuyu kechixu fazhan*, pp. 1–9, Zhongguo Huanjing Kexue Chubanshe, Beijing, 1997.

Yang Jin, 'Huanghe liuyu di youguan shuju' (Relevant data on the Yellow River basin), *Zhongguo Shuili* (China Water Resources), February 2000, 57.

Zhang Guangdou, 'Huanghe duanliu wenti' (The problem of flow stoppage in the Yellow River) in Shuilibu Shuizheng Shuiziyuan Si (ed.), *Huanghe dualiu jichi duice*, pp. 1–5, Shuili Shuidian Chubansha, Beijing, 1997.

Zhang Ren, 'Huanghe xiayu duanliu wenti jiqi duice' (The problem of flow stoppage in the lower reaches of the Yellow River and countermeasures), in *Guojia Huanjing Baohu Ju Ziran Baohu Si*, 1997, pp. 66–9.

Zheng Tonghan, 'Zhidu jili yu guanqu di kechixu yunxing' (Institutional drivers and the sustainable operation of irrigation districts), *Zhongguo shuili* (China Water Resources), January 2002, pp. 33–5, 40.

Zhong Bin and Zhang Xuejian (eds), *Huanghe duanliu wanli tanyuan* (A long-distance search for the sources of the Yellow River flow stoppage), Jingji Ribao Chubanshe, Beijing, 1998.

Zhu Erming, 'Huanghe xiayu duanliu duice cuoshi tantao' (An inquiry into measures to counter the flow stoppage in the lower reaches of the Yellow River), in Shuilibu Shuizheng Shuiziyuan Si (Water Policy and Water Resources Office of the Ministry of Water Resources), (ed.), *Huanghe duanliu jichi duice* (Flow stoppage in the Yellow River and countermeasures), pp. 9–13, Shuili Shuidian Chubanshe, Beijing, 1997.

6

Current Situation and Future Prospects of Water and River Management in Japan

Yutaka Takahasi

Current Situation of River and Water Management

Japanese river engineering and administration are now increasingly considering environmental requirements. However, flood control works are always important, as the Japanese Archipelago is located in the Asian monsoon region, which suffers from frequent floods due to intense rainfalls. The important issue thus is how best to harmonize flood control and water resources development practices with a new river management policy which emphasizes environmental issues.

One of the most important problems related to water management is the integration of several kinds of water administrations. Although the water in rivers is managed by river administration under the River Law, once withdrawn from the river channel, water is managed differently under different laws. The many issues relating to water management such as flood control, hydroelectric development, agricultural and industrial water use, water supply to cities and towns, water quality, sewage and drainage, sediment control such as erosion control works for mountainside and torrential course, forest conservation, prevention of coastal erosion, maintenance of the ecosystems in rivers and lakes, and river landscape and the like are managed by different administrators.

Issues such as the prevention of land subsidence resulting from the over-pumping of groundwater, flexible and efficient applications of water rights, and the protection of the quality of river water and groundwater and the like cannot be addressed by individual laws or administrators alone. The problems that need to be solved are discussed next.

Increasing Potential of Damage by Floods and Sediment Flows

Active flood control projects and measures to protect against disasters have decreased the number of flood victims over the years, but the level of safety remains low. Between 1986 to 1995, about 90 per cent of all municipalities in Japan suffered a flood or a disaster arising out of excessive sedimentation. Urbanization and industrialization in Japan after the Second World War took place mainly on alluvial plains where rivers are prone to floods. Since the beginning of this century, and in particular during the high economic growth period of the 1960s and 1970s, urbanization has been concentrated along the coastal zones, which are prone to storms. With the progress of urbanization and industrialization, the retention and detention functions of river basins and the wetlands along the rivers and coastal zones have been gradually lost, while at the same time flood discharges in river channels have increased, thereby increasing the risks of flood, high-tide damage, and also the destruction of the ecosystem. The damage potential of sediment flows including debris flows, landslides, and steep slope failures has increased rapidly with the development of hillsides and mountainsides.

Increased safety from floods has led to the further concentration of city functions and housing on floodplains, resulting in a higher damage potential. Floods still cause economic damage. The potential of damage that can be caused by floods and sediment flows is an ever increasing one. As Japan relies on intensive urban and industrial activities, it is necessary to minimize the damages caused by floods and debris flows, especially in large cities and their surrounding areas. With an increasing population of senior citizens, foreigners, and tourists, the number of people particularly vulnerable to disasters is expected to grow. It is therefore necessary to take new measures to protect the people against possible disasters.

Increasing Frequency of Droughts

Precipitation has been on the decrease in recent years in Japan, causing frequent droughts. The drought of 1994 in western Japan had a far-reaching and serious impact. An oil company had to buy water from Korea, Hong Kong, and even Vietnam during this period. The conventional approach of meeting water requirements by constructing dams and other structures on rivers is expected to become increasingly difficult because of the scarcity of suitable dam sites, the considerable cost and time required for relocation, the declining economic efficiency of water resources development projects, and more recently, the growing public concern for the natural environment.

As water use has increased, Japan's social and economic activities have become increasingly vulnerable to water shortages. The establishment of lifestyles dependent on the intensive use of water and the growing number of people who are vulnerable to water shortages have made cities particularly susceptible to droughts. Droughts in the future are likely to have serious impacts on Japan's social and economic activities.

Deteriorating River Environment

Rapidly changing socio-economic conditions have reduced water availability and greenery in river basins. It is also an undeniable fact that flood control and water development projects in the past, which gave priority to early completion with main focus on economic efficiency, failed to give adequate consideration to the environment and the riparian landscape. These factors have given rise to a number of environmental problems such as decreasing biodiversity and shrinking habitat, depletion of springs and decreases in usual stream flow, and water pollution in rivers, channels, and lakes.

One of the results of urbanization and modern river works has been that the urban people have become less aware of issues relating to rivers. However, now, relationships among communities living along rivers, the administration and the river itself are becoming increasingly important. Conflicts between the administration and environmentalists are occuring over new river projects. It is necessary to find methods to obtain consensus on issues relating to rivers.

Historical Review of River Management Since the End of the Nineteenth Century

The old River Law was enacted in 1896, just one year after the end of the Sino-Japanese War. The Imperial Diet was set up in 1890, and so the old River Law was a constitutional policy. On the basis of this Law, a number of large-scale flood control projects began on several important rivers. Following the severe flood of 1910, and the damages it caused, the number of projects was increased to cover more than twenty rivers. Almost all the flood control projects were completed between 1930–1.

As a result of the construction of river projects to protect against floods, continuous embankments were built in areas that were not protected earlier. The river projects were designed on the basis of the maximum recorded flood. However, in the case of several rivers such as the Tone River, the largest river basin area in Japan, floods exceeding the design flood discharge occurred successively. Consequently, the plan was revised each time the design flood discharge was exceeded.

In order to meet the growing demand for power generation, municipal water supply, and the like, the River Water Control Scheme for flood control and water utilization, which focused on dams and weirs, was initiated in 1937.

The history of Japan after the Second World War is an extraordinary story of a country transcending economic poverty to become a world economic giant. A radical transformation has been brought about not only in the socio-economic field but also in the field of water-related issues.

Drastic changes mark the post World War II era in Japan: an increase in population, gross national product (GNP), water demand, flood damage, and the change in the industrial structure from agriculture, forestry and fishery to industry and service. To cope with these rapid changes and the increased demand for water and power, several measures have been introduced, for example, the construction of multipurpose reservoirs, hydroelectric dams, saline barriers, and river improvement works.

The period of over half a century following the Second World War may be divided into four periods based on socio-economic and related hydrological considerations: (i) the severe flood damage period from 1945 to 1959; (ii) the water shortage period from 1960 to 1972; (iii) the post-high economic growth period from 1973 to 1989; (iv) the environment and amenity period from 1990 to date.

Unfortunately, severe and frequent storms and rainfalls were caused by extraordinarily strong typhoons during the rainy seasons from June to July almost every year from 1945 to 1959. One of the worst was the Ise-Bay typhoon of 1959, which is often compared with the 1953 storm-surge in the Netherlands. This typhoon lashed mainly at the coast of Ise-Bay, the Pacific side of middle Japan, killing more than 5100 inhabitants.

Almost all of the important rivers in Japan experienced the breaking of a levee at some time, affecting more than a thousand people annually between 1945 and 1959. The river works focused on flood damage repair and flood control during this period. The reasons why such severe flood damages occurred during this period were record-breaking precipitation, and the change of the river regime. Since the end of the last century, continuous levee systems have been built, especially in the middle and downstream reaches of important rivers where they flow through large alluvial plains with floodways and cut-offs in several cases, in order to protect rice fields and urbanized areas. As a result of these works, the productivity of farm lands has increased rapidly, and the extent of regularly flooded areas has markedly diminished. On the other hand, flood flows have become concentrated within the river channels between continuous levees on almost all the rivers in the country, thereby intensifying peak discharges, raising the level of river stages, and increasing the velocity of flood waves. This implies that the flood regime is decisively influenced by human activities, including the flood control works themselves and economic development in the river basins. This is why the increase in flood discharge for the same amount of storm-rainfall has become one of the most important reasons for successive severe flood damages. Rapid changes in the factors affecting the condition of river basins are clearly visible in Japan, a country that achieved industrialization in a relatively short span of time.

Urbanization and industrialization since the latter half of the 1950s led to a rapid increase in water demand in the urbanized and industrialized areas, especially in big cities such as Tokyo, Nagoya, and Osaka, and the industrial zones along the Pacific coast. This rapid increase in the demand for water resulted in water shortages in big cities and industrial zones everywhere in Japan.

Though the Water Resources Development Acceleration Law was formulated in 1961, it was not implemented in time to tackle Tokyo's water shortage in 1964, just before the Tokyo Olympic Games. A number of dams were constructed between 1950 and 1970 to meet the increasing demands for municipal and industrial water supply, and also for flood control and hydroelectric power. For example, the number of dams higher than 15m, constructed between 1950 and 1978, was 304 for irrigation purposes, 222 for hydroelectric power, and 226 for multiple purposes emphasizing flood control. As a result, there are now few major rivers without dams, and their appearance at upper river reaches have changed strikingly since the 1950s.

The oil crisis of 1973 deeply influenced not only the Japanese economy but also the water problems. The trend towards increasing water demand by the municipal and industrial sectors slowed down or stopped after the oil crisis. The cost of dam construction escalated due to the paucity of suitable dam sites and the increase in compensation costs, including the so-called 'social cost' of removing the local population living near the dam sites. Water, thus became a 'precious' and irreplaceable resource.

In 1964, the River Law was revised to meet the changing needs of the river administration, which resulted from socio-economic progress and administrative reform.

The Special Measures Law concerning upstream area development was enacted in 1973 to stimulate the economy of dam-site areas which are generally less attractive to settlers as compared to urban areas further downstream. The law recognized the social value of water resources development in the upstream areas. It also recognized the necessity of finding a solution to the conflicting interests of upstream and downstream areas, arising from the construction of dams.

In the latter half of the 1970s, citizens began to make new demands for recovering amenities from and accessibility to the rivers. The main topics of interest in water-related sciences have changed since then, with socio-economic requirements receiving increasing attention. Issues related to water and the environment such as pollution, aesthetic design on the river front, and ecological balance have gradually become important issues for river projects.

In the 1980s, environmental problems affected many of Japan's rivers. In 1990, the River Bureau, Ministry of Construction (MOC) initiated a river restoration project. So-called natural diverse river improvement methods were applied to use natural materials such as stone, rock, and vegetation instead of concrete lining at revetment work on levees, in order to maintain and conserve the ecosystem and create sound river environments.

In 1995, the River Council produced a report on 'How River Environment Must Be'. The River Law was amended in 1997.

Amendment of the River Law in 1997

One of the most important amendments to the River Law is the addition of environmental issues to Article 1.

Article 1.   The purpose of this Law is to contribute to land conservation and the development of the country, and thereby maintain public security and promote public welfare, by administering rivers comprehensively to prevent occurrence of damage due to floods, high tides, etc., utilize rivers properly, maintain the normal functions of the river water by maintaining and conserving the fluvial environment.

'Maintaining and conserving the fluvial environment' are a new addition to 'the purpose of the River Law'. Before the amendment, the purpose of the River Law, that is, the purpose of river works, were flood control and river water development and utilization. There were no environmental items in the River Law before 1997.

The next important amendment is the addition to reflect the opinion of the people.

Article 16-2-3.   When river administrators intend to draft a river improvement plan, they shall consider opinions from persons with experience or academic background when necessary.

Article 16-2-4.   In connection with the previous paragraph, river administrators shall take necessary measures, such as public hearings etc., to reflect the opinion of the people concerned whenever necessary.

Article 16-2-5.   When river administrations intend to establish a river improvement plan, they shall consider opinions from concerned prefectural governors and mayors in advance as provided in the Government Ordinance.

Article 16-2-6.   When river administrators establish a river improvement plan, they shall make a public notification to that effect without delay.

The river improvement plan requires concrete measures to be taken in accordance with the fundamental river management policy, which is authorized by the River Council. It must include matters related to the objective of the river improvement plan, the execution of the plan, purpose, type, and location of the river works and a description of the functions of the river administration facilities to be provided as a result of the execution of the river works.

The next amendment concerns the speedy countermeasures to be implemented in case of unusual drought.

Article 53.   In case of an unusual drought which makes it difficult to adequately use the river water for the permitted utilization purpose, or when such a situation is expected, the persons who have obtained permissions to use the water shall make efforts to consult with one another.

By this amendment, more efficient and speedy reaction is expected in case of an unusual drought.

Eighty-five per cent of water use in Japan depends on river water, and the remaining 15 per cent on groundwater. In the event of low precipitation and a drought more severe than the design drought, it is necessary to conciliate between water uses in different areas and for different purposes. In Japan, just as in the other countries of the Asian monsoon area, older water rights have traditionally been deemed superior to recent water rights. Older water rights consist primarily of irrigation water rights. As water utilization associated with drinking water supply, which is closely connected to everyday life, has increased, giving priority to older water rights over drinking water rights cannot necessarily be deemed rational. The River Law stipulates that in the event of a drought, which usually does not require action as urgently as a flood, the river users concerned first should try voluntarily to adjust their water uses. However, in view of the present conditions of the river basins, with increasing water utilization, and the fact that construction of dams is becoming increasingly difficult, the law seeks to make river administrators play a primary role in water use adjustments from the early stages of a drought by requiring them to provide the necessary information to the users.

'Unusual drought' refers to a drought that is more severe than a drought whose recurrence interval is used as a basis for water resources development planning and the granting of water rights. In Japan, for droughts a recurrence interval of about ten years is usually used as a basis of planning. But recently there has been a tendency for droughts to occur more often due to abnormal climatic conditions.

It is also worth nothing that 'fluvial woods' (trees planted along levees or reservoirs in a long narrow strip for flood mitigation and/or water conservation purposes due to the MOC Ordinance) are recognized as an integrated component of river management facilities like dams, weirs, sluices, levees, revetments, and the like, under Article 3. Here, the river administration facility has the function of increasing public benefits from the water of a river.

Trees must be planted on the following types of lands:

a) In the case of a group of trees located along a levee, land within approximately 20 m from the toe of the back slope of the levee;

b) In the case of a group of trees along a reservoir, land within approximately 50 m from the water surface–land boundary, during the highest level of river water stored in the reservoir.

It is also essential and important that the activities of the two administrative bodies, the MOC and the Forest Agency, be properly coordinated.

Adoption of these suggestions are likely to contribute to the engineering techniques used for flood control and water utilization which meet the requirements of environment and nature conservation. Such integration is likely to enhance both human welfare and environmental conservation.

Furthermore, under the amendment of 1997, a river administrator could cover all or part of the expenses for river maintenance such as control of water pollution. Also, unlawful activities of pleasure boats can be controlled by law.

Thus, the amendment of the River Law in 1997 focused not only on environmental matters, but also provided for ways to cope with socio-economic changes.

Toward a Sound Management of Catchment Areas

The Hydrological Cycle Sub-committee, River Council, Ministry of Construction, submitted a report to the Minister of Construction in July 1998 titled 'What the hydrological cycle should be within catchment areas'. The main issues considered in this report are given next.

The hydrological cycle plays an important role in the movement of various substances in the earth and the atmosphere. Human civilization has generally developed on the basis of consideration of the implications of the hydrological cycle.

Inadequate cooperation between the various institutions entrusted with the management of rivers, lakes, marshes, waterways, sewerage systems, municipal water supplies and wastewater systems, and agricultural and hydropower developments has led to formulation of ad hoc, piecemeal and unintegrated policies, with no consideration for comprehensive and integrated measures. These have resulted in adverse impacts on different components of the hydrological cycle. Furthermore, land use planning in the country did not consider properly the integrity of the hydrological cycle. In other words, the administrative measures and practices for urban and rural areas, forests and agricultural lands have often been not adequate.

Therefore, a new framework is necessary, which should be oriented towards overall development, conservation, and management of national land resources, which respects, and is in conformity with, the needs of the hydrological cycle. The importance of assessing the impacts on the hydrological cycle of all development activities needs to be further emphasized in the future.

In Japan, a new framework has been formulated for flood control and water resources management, whose components are as follows: Flood control system must include not only river improvement projects but also flood plain management, including land use considerations. Water resources development and management must consider not only dams, weirs, and other hydraulic structures, but also reuse of water, rainwater harvesting, water rights, desalination, and other similar measures.

The sudden and immense changes in social and economic activities have had numerous impacts on the hydrological cycle of each catchment area. In certain areas, new types of problems have sprung up. These are considered below.

a) Changes in the hydrological cycle in forests and farmlands: During the twentieth century, Japan developed what were non-arable lands earlier. However, water retention and movement functions have changed significantly due to the development of farm lands, and hydraulic infrastructures in many catchment areas. The ways in which agricultural water use has changed have also had an influence on the hydrological cycle.

b) Changes in flood patterns and increased flood damage potential due to urbanization: Floods occur shortly after heavy rainfall in catchment areas which include urban districts. This is because of the expansion of impervious areas caused by urbanization which contributes to the reduction in water retention capacities of the soil. The nature of flood hydrographs have changed because of these developments. Peak discharges have increased. Because of these changes, new types of flood damages started to occur, which were first noted in the western part of Tokyo in 1958. Thereafter, flood control projects had to be reviewed in the newly urbanized areas. All these developments made implementation of effective flood control and flood management practices an increasingly complex and difficult task.

In addition, underground spaces are increasingly being put to use in a number of ways without adequate measures against inundations. For example, in 1999, two persons died in underground rooms due to severe storms in Tokyo and Fukuoka respectively. Also, the potential for flood damages has increased very significantly because of the extensive use of sophisticated equipment which are susceptible to flood damages.

c) Increase in water shortages: As discussed earlier, there has been an increase in water shortages, one of the severest being the water shortage faced by western Japan in 1994.

d) Drop in the normal river discharge: The expansion of impervious areas due to urbanization, and the development of sewerage systems have decreased the infiltration into the sub-surface, and have diverted the flow of water, which used to earlier flow into the rivers. The channel-type power generation system uses water, which used to flow earlier into rivers.

These have caused the flow of the river to decrease. Also, due to the diversion of the water in certain areas, rivers now receive less water than they used to. This has damaged the river environment to some extent not only in big cities but also in the suburban areas.

e) Water shortages due to disaster prevention measures: The Great Hanshin–Awaji Earthquake in 1995 exposed the vulnerability of overpopulated urban communities. During the earthquake, numerous fires started simultaneously. This incident made it clear that water in rivers, waterway, and ponds near overpopulated cities was a must to help extinguish fires at an early stage and prevent their spread. Water is also required to meet the demand for domestic use. However, despite this requirement, rivers, and waterways, are frequently reclaimed or diverted through conduits.

f) Water pollution: Recently, the quality of groundwater, which was believed to be good in Japan, was found to be widely contaminated with chemical substances such as trichloroethylene near factories and laundries. Once contamination takes place, improving the quality of water is difficult. Groundwater contamination by nitrogen, nitrate and nitrite is also widely observed.

g) Drop in groundwater levels and land subsidence: In the eastern part of Tokyo, and the other cities where the problem of land subsidence was particularly acute, administrative controls have been placed for pumping groundwater. However, in regions where control does not exist, such as the northern prefectures of Tokyo, land subsidence and other problems occur due to the pumping of excessive amounts of groundwater during periods of water shortage.

h) 'Heat island' phenomena: The increase in reclaimed land, the development of closed sewerage system conduits, and urban activities have caused the so-called 'heat island' phenomenon in which temperatures rise in cities.

i) Change in the ecosystem: Environmental changes taking place in the hydrological cycle in rivers and in catchment areas have contributed to changes in the ecosystem.

j) Loss of water-related culture: In ancient times, Japan was called a land of 'purple hills and crystal streams' and a culture which could never be detached from water was formed. However, with major changes in human activities which have affected the hydrological cycle, the continuation and nurturing of the 'water culture' is doubtful and it is further feared that this could lead to the loss of Japan's identity.

Nagara River Estuary Barrage Project

Severe conflict arose between the river administration and environmentalists in the 1990s over the Nagara River Estuary Barrage Project.

The construction of the barrage began in March 1988 and was completed in 1995 amidst much heated discussion both in parliament and in the media. The objectives of the barrage were flood control and development of water resources. The barrage is located 5.4 km from the estuary near Nagoya City in central Japan. With a total length of 661m, it consists of ten main gates and two types of fishways.

The Nobi Plain, the largest plain below sea level in southern Japan through which the Nagara River flows, has suffered many floods during its long history. The Ise-Bay Typhoon that hit the region in 1959 left over 5,000 people dead and the flood of 1976 severely damaged about 3,500 houses.

The downstream area along the Nagara River is so densely populated that it is practically impossible to expand the river width. As such dredging the river channel would be essential for expanding the cross-sectional area of the river. Rising of levees is perilous because it would increase the potential for damages by flood in such low-lying areas.

However, river-bed dredging would have allowed sea water to flow into the river channel. This would make the river water saline and limit the availability of fresh water for agriculture and industry. Without any preventive measure, salinization of groundwater and soil along the river would be certain and agricultural activities would have been seriously damaged. To prevent the intrusion of saline water, the Nagara River Estuary Barrage was built. The gates of the barrage are kept closed during normal periods to block saline water, but are opened during floods to allow the flood water to flow safely into the sea.

According to the administrators, a stable water supply—essential to maintaining uninterrupted social and economic activities during periods of drought—is provided by the project. The barrage also helps in development of water resources.

The Ministry of Construction, which is responsible for the project, has noted that a comprehensive study of the environmental impacts due to the construction of the Barrage had been carried out during the planning stage of the project, and also during the construction phase of the Barrage, in close consultation with the Environment Agency. These studies considered issues like potential impacts of the proposed structure on anadromous fishes, aquatic and terrestrial fauna and flora, and water quality.

It should be noted that many environmental organizations were strongly opposed to the construction of the Barrage. For example, the Fish Protection Association of Japan provided the following arguments against the project to *The London Times* on 22 May 1992.

a) The water resources development for the areas is not necessary, because the current water supply is quite sufficient. The plan was based on the socio-economic conditions that existed during the high economic growth era of Japan. Even through the economic conditions have changed radically, the Ministry did not make any modification to the proposed plan to reflect the new conditions.

b) The MOC stressed the importance of the Barrage for flood prevention, but it is likely to exacerbate damages once severe flooding occurs.

c) The quality of water upstream of the barrage is likely to become very poor due to stagnation. The natural influx of sea water into the estuary had a purifying effect on the river water in the past. This would cease with the damming of the estuary.

d) The MOC claims that the barrage would prevent salt water contamination in agricultural fields. Although low-lying fields were damaged in this way once in the past, advances in irrigation techniques make a recurrence of salt water damage most unlikely.

e) The MOC planned to construct special fishways to assist the Satsukimasu Salmon to migrate from the sea to upstream of the river. However, even if the fish could successfully negotiate the fishways, it is highly unlikely that they would survive the long swim through the stagnant waters of the 25-km barrage-reservoir to the river further upstream. If they reach the river, the Satsukimasu Salmon and many other species of fish would find survival almost impossible in its poor quality water.

The Freshwater Fish Protection Association emphasized that it was important to protect the migration route of the Satsukimasu Salmon indigenous to Japan, which lies between the sea to the south of the Japanese archipelago and the Nagara River on Japan's west coast. It was claimed that this species of salmon would be in danger of extinction if the barrage was constructed.

Among the supporters to save the Nagara River were the Ecological Society of Japan, the Ichthyological Society of Japan, the Japanese Society of Limnology, the World Wildlife Fund of Japan, the Nature Conservation Society of Japan, and the Wild Birds Society of Japan.

Almost all forms of journalism, including influential newspapers, television broadcastings, and various magazines supported the agenda of the environmental groups. They asked the MOC to stop the construction at least till a more detailed survey on the environmental impacts of the construction of the barrage was completed. But the MOC continued and completed the barrage in 1995, with the support of every assembly of cities and towns in the Nagara River basin.

Though the barrage has been constructed, the discussion is still continuing on the survey of the environmental impacts of the barrage. Both the environmental groups and the engineers of the MOC are talking around the same table and exchanging data. The dialogue between the MOC and the environmental groups may lead to a consensus towards better river improvement works in harmony with the ecosystem. The road to reach a consensus is not easy but the situation is gradually becoming better.

Taking a lesson from the conflict of the Nagara River estuary barrage, the MOC started the Aqua Restoration Research Centre (ARRC) on the Kiso River area in November 1998. The objective of the ARRC is to analyse basic phenomena such as the relation between the change of the discharge and the biological habitat, and to establish and disseminate the methods for environmental conservation and restoration. The research facility consists of three experimental rivers, (each 800 m long straight channel with revetment, and two biotope channels).

Table 6.1: History of river management in Japan

Year	Event
1896	First River Law was enacted.
1910	Large-scale flood control works started on many rivers.
1923	Great Kanto earthquake.
1930	Almost all flood control work accomplished.
1945	End of the Second World War.
1945–59	Severe storm-rainfall almost every year leads to the big flood damage period.
1959	Ise-Bay typhoon. The largest flood and storm-surge damage in the twentieth century.
1960–72	Water shortage period results from a rapid increase in water demand due to urbanization and industrialization.
1961	Formulation of the Water Resources Development Acceleration Law.
1964	Severe water shortage in Tokyo. The new River Law was enacted.
1973–89	Post-high economical growth period.
1973	Special Measures Law concerning upstream area development was enacted.
1977	New policies for comprehensive flood control measures in urbanized river basins were enacted.
1991	Promotion of river restoration projects. The beginning of the 'Natural Diverse Improvement Method'.
1994	Construction of the estuary barrage on the Nagara River accomplished inspite of a severe opposition movement and heated discussion.
1995	River Council presents its report on 'How River Environment Must Be'.
1997	Amendment of the River Law.

Concluding Remarks

The high level of economic growth during the latter half of 1950s and 1960s in Japan, enabled construction of numerous major infrastructures, including dams and river improvement works. At the initial stage of a dam project, the most important social issue was generally the compensation cost.

But since the 1970s, the resistance movement against dam construction has become increasingly more intense with progressive urbanization. In 1973, the Special Measures Law concerning upstream area development was enacted to stimulate the economy of upstream regions. The law is applicable to towns where the number of submerged houses is more than thirty, or where agricultural lands submerged are over 30 ha, revised later to twenty houses or 20 ha. The objective of the law is to improve the natural environment in the surrounding areas, ensure the welfare of the communities in such areas, and to promote the construction of dams.

According to this law, a reservoir area improvement plan has to be formulated, with appropriate public works, such as the construction of roads, land improvement, small-scale water supply and sewerage. The Central and regional governments are to take the necessary financial measures to assure the livelihood of residents where properties are submerged as a consequence of the construction of dams.

The law ensures that social costs of water development in the upstream areas are recognized, and also that it is necessary to find a solution to the conflicting interests between the upstream and the downstream regions due to the construction of dams.

Furthermore, after 1975, a fund for reservoir area development has been established to improve the economies of dam-site areas for several important river basins where many dams have been planned. This will contribute simultaneously to regional development and environmental conservation.

In view of the increasing importance of water development on the environment, various environmental policies are now applicable to water development projects. Among these policies are the following:

a) Implementation of basic studies of natural and social environments of reservoirs, such as regular and continuous surveys of the habitats of fish, birds, plants, and other organisms living in and around dam reservoirs.
b) Formulation basic policies for conservation, development, and management of water system environments. Adoption of the master plans, which are determined taking into account the opinions of local residents and academic experts.
c) Implementation of environmental impact assessment, in which surveys are conducted prior to the implementation of dam projects to predict and evaluate the impacts of the projects.

d) Implementation of measures to maintain water in a good condition and to conserve the ecological system through the creation of recreational facilities such as parks and sports grounds, ensuring their harmony with the rich natural environment around dams and reservoirs.

Since the latter half of the 1990s, the dam projects have to formulate and implement regional development activities in areas surrounding the dam site. For example, the Miyagase Dam, which was completed in 2000, primarily to supply 15.5 m^3/s of water to Yokohama city and the surrounding areas, is one of the latest representative example. This emphasizes regional development and environmental conservation measures, including provision of new recreational spaces, biotopes, etc. The total cost of this dam was around US$ 3.3 billion. Of these, the compensation cost, including regional development, amounts to US$ 1.1 billion.

References

River Bureau, Ministry of Construction, *The River Law* (with commentary by article, compiled and commented by Toshikatsu Omachi), Infrastructure Development Institute, Japan, 1999.

River Bureau, Ministry of Construction, Japan, *What the Hydrological Cycle should be within Catchment Areas*, Japan, 1998.

Takahasi, Y., 'New Concept of Water Management in Japan', Paper presented in the World Water Symposium, Montevideo, 1999.

Tamura, H. and K. Niki, *The Nagara River Estuary Barrage Project*, Civil Engineering in Japan, JSCE, Tokyo, 1992.

The Freshwater Fish Protection Association, 'Save the Nagara River', *The London Times, Infotech Times*, 22 May 1992.

7

Participative Water-based Regional Development in the South-Eastern Anatolia Project (GAP): A Pioneering Model

Olcay Ünver • Rajiv K. Gupta

South-Eastern Anatolia Project

The South-eastern Anatolia Project (GAP)[1] aims to develop water resources for poverty alleviation, socio-economic and balanced regional development. The GAP, in its historical context, was formulated as a package of water and land resources development project in the 1970s, which was later on transformed, in the early 1980s, to a multi-sectoral, socio-economic regional development programme, and then into a sustainable human development project in the 1990s. This massive launch for development has special emphasis and priority for the economic, social and cultural advancement and well-being of the whole country in general, and of the people of the South-eastern Anatolia region—the Upper Mesopotamia or fertile crescent (as historically known) in particular (Figure 7.1). The basic objectives of the GAP are: to remove inter-regional disparities in the country by alleviating conditions of abject poverty and raising the income levels and living standards in the region; to enhance productivity and employment opportunities in rural areas; and to improve the population-absorbing capacity of larger cities. Therefore, we shall first examine the social and economic conditions prevailing in the GAP region and then analyse as to how the water resources development and integrated socio-economic development strategy have been used to remove regional backwardness

[1] GAP (in Turkish)—Güneydoğu Anadolu Projesi.

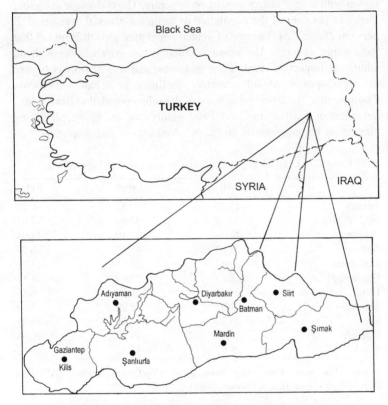

Figure 7.1: Map of the GAP Region in Turkey

Source: Based on the records of Geographical Information System Unit of GAP Administration, 2002.

and empower people by employing a participative planning and development paradigm.

Demographic Features

As of the year 2000, about 10 per cent of the total population of Turkey lived in the region which had the highest population growth rate (Table 7.1), fertility rate, infant mortality rate, and lower life expectancy at birth (Table 7.2) than the national average. Significantly, it is also a region with a very young population structure. The 0–14 year age group forms 41 per cent of the population as against a national average of 29 per cent (Table 7.3). The rate of annual population growth is higher than the country average. The urban population has grown faster with the additional impact of rural–urban migration and this growth is beyond the development of infrastructure facilities in urban settlements. Consequently, the living conditions of the urban population have further deteriorated. Furthermore, the large number of young people in the population has aggravated problems of education and employment.

Table 7.1: Decadal population growth rates in 2000 (percentage)

	Total	Urban	Rural
Turkey	15.20	27.84	–8.72
GAP region	26.64	42.54	–7.65
Adıyaman	40.01	73.10	–12.01
Batman	21.63	43.31	–30.27
Diyarbakır	22.41	42.28	–17.78
Gaziantep	15.69	22.67	–6.14
Kilis	–24.20	–41.01	0.50
Mardin	21.03	48.29	–17.29
Siirt	10.70	42.84	–45.55
Şanlıurfa	37.67	49.40	19.08
Şırnak	28.82	53.92	–11.12

Source: Hacettepe University, Institute of Population Studies (HÜNEE), *Demographic and Health Survey*, 2000.

Table 7.2: Vital statistics and age groups in 2000

	Unit	Turkey	GAP
Annual population growth rate	Thousand	16	25
Crude birth rate	Thousand	22.6	35.9
Crude death rate	Thousand	6.3	7.0
Infant mortality rate	Thousand	38.9	56.1
Life expectancy at birth		Turkey	GAP
Total	Years	69.4	63.0
Male	Years	67.1	60.9
Female	Years	71.7	65.1

Source: Hacettepe University, Institute of Population Studies (HÜNEE), *Demographic and Health Survey*, 2000.

Table 7.3: Distribution of population in 2000 by three
major age groups (per cent)

	Age group 0–14	Age group 15–34	Age group 34–64	Age group 65+
GAP region	41.2	36.2	18.5	4.1
Adıyaman	36.6	36.8	21.6	5.1
Batman	45.2	35.0	15.9	3.9
Diyarbakır	42.2	36.1	17.9	3.8
Gaziantep	36.0	37.3	22.0	4.7
Kilis	35.8	35.2	22.6	6.4
Mardin	41.2	36.6	18.0	4.3
Siirt	42.6	36.6	16.6	4.2
Şanlıurfa	44.8	34.5	17.2	3.5
Şırnak	46.2	39.1	11.7	3.0

Source: Hacettepe University, Institute of Population Studies (HÜNEE), *Demographic and Health Survey*, 2000.

In this region, the family is an economic unit as much as it is a social one. It is the micro unit of production in rural areas. The size of rural families is larger. Table 7.4 gives figures for the year 1995 on the share of families having more then ten members for the GAP provinces. Since the national average for the prevalence of this large family size is 18.4 per cent, all GAP provinces except Adıyaman and Gaziantep are above

Table 7.4: Share of households with ten or
more members (1995), GAP region

Province	%
Adıyaman	15.0
Diyarbakır	21.5
Gaziantep	11.5
Mardin	28.9
Siirt	30.2
Şanlıurfa	18.8
Batman	29.5
Şırnak	31.4
GAP	18.3
Turkey	18.4

Source: Hacettepe University, Institute of Population Studies (HÜNEE), *Demographic and Health Survey*, 2000.

the national average. This means that in these provinces fertility patterns peculiar to rural life still prevail. This is of course related to the process of rural–urban migration which has taken place within the last ten years. In the normal course of development, family size gets smaller not only through a spatial movement (that is, movement from rural to urban areas) but also with the participation of family members in various economic activities. Nevertheless, the extraordinarily poor state of affairs in this region, including terrorist activities taking place mostly in rural areas, make it necessary to interpret demographic indicators and urban family types in a somewhat unique manner.

Income Distribution, Poverty, Unemployment, and Backwardness

As can be seen in Table 7.5, the region of south-eastern Anatolia remains at the bottom of the list in terms of average household income (AHI). In other words, the AHI in this region is far below other regions, rural sector, and national averages.

Table 7.5: Average household income (1994)

('000 Turkish Lira)

| | Total | Household percentiles (%) | | | | |
		1st 20%	2nd 20%	3rd 20%	4th 20%	5th 20%
Turkey	165,089	40,095	71,221	104,089	157,057	452,984
Urban	202,471	48,915	82,878	120,133	181,196	579,231
Rural	117,203	32,668	59,430	86,709	127,687	279,523
Marmara region	239,787	51,968	89,508	128,181	196,046	733,229
Aegean region	146,085	39,349	70,786	102,936	152,620	364,737
Mediterranean region	146,279	38,382	66,211	96,775	143,261	386,766
Central Anatolia	142,171	36,015	64,743	97,909	155,616	356,573
Black Sea region	140,000	35,876	66,255	96,372	141,094	360,402
Eastern Anatolia	132,151	41,657	74,585	107,045	152,792	284,673
South-eastern Anatolia	99,579	35,268	55,493	74,192	103,237	229,703
Gaziantep	101,991	40,673	59,067	80,821	115,120	214,273
Diyarbakır	107,073	36,999	53,810	70,927	102,654	270,977

Source: DIE (State Institute of Statistic), *Household Income Survey*, 1994.

Considering that household size is on average larger in south-eastern Anatolia, it becomes apparent that the situation in per capita terms is even worse. Another indicator that supports this view is per capita gross domestic product (GDP). The per capita GDP in the GAP region increased at the rate of 15.86 per cent, slightly higher than the national average, at fixed prices in the period 1995–98. The 1994 Household Income Survey shows that Diyarbakır are one of the provinces where income distribution pattern is most distorted. In Diyarbakır, the average household income was $3,657 at the exchange rate of the year 2000 (the national figure was $5,503) and the highest 20 per cent of the population had a share of 51 per cent in the total income of the province. In short, the per capita GDP of eastern and south-eastern Anatolia at fixed prices was far below the country average in both the 1995 and 1998 surveys (Table 7.6).

According to UNICEF (2000), the group of provinces represented by Diyarbakır are the lowest and most slowly growing of all the fourteen sub-regions. Depending upon the province concerned, 21.8 to 44.7 per cent of households in the region live under the line of poverty.

Table 7.6: Per capita GDP in GAP provinces according to
DPT (State Planning Organization) and DIE
(State Institute of Statistics) sources (at 1998 prices)

	1995	1998	Change 1995–98 (%)
Turkey	1,587,953	1,829,754	15.23
Marmara region	2,355,568	2,667,003	13.22
Aegean region	1,990,537	2,280,039	14.54
Mediterranean region	1,506,426	1,712,882	13.71
Central Anatolia	1,484,672	1,708,810	15.10
Black Sea region	1,099,269	1,366,704	24.33
Eastern Anatolia	626,339	673,339	7.50
South eastern Anatolia	854,164	989,641	15.86
Adıyaman	783,250	808,715	3.25
Diyarbakır	923,766	997,531	7.98
Gaziantep	1,223,659	1,447,224	18.27
Mardin	690,868	827,362	19.76
Siirt	655,126	666,821	1.79
Şanlıurfa	717,702	877,410	22.25
Batman	817,617	1,091,508	33.50
Şırnak	345,591	456,412	32.07
Kilis	–	1,718,495	

Source: GAP, *Report of the Social Working Group, New Master Plan*, 2001.

A comparative analysis shows that the most disadvantaged region of Turkey is south-eastern Anatolia in terms of per capita and per household minimum food expenditures and per capita cost of basic needs (Erdoğan, 1997).

The 1994 DIE Income Surveys pointed out that among all regions south-eastern Anatolia is the one where food security is lowest. According to another survey conducted by the State Planning Organization (1996), the ranks of GAP provinces (altogether 76 provinces) are: Gaziantep (25), Diyarbakır (57), Şanlıurfa (59), Adıyaman (61), Batman (65), Mardin (66), Siirt (68) and Şırnak (75). Kilis is out of this ranking since it was given provincial status after this survey. What is more important than this data is that there are some significant disparities among the provinces of the region as well (see Table 7.7).

As can be clearly seen in Table 7.8, the GAP region does not display a uniform character in socio-economic development. There is some differentiation on the basis of provinces having 'irrigable' and 'non-irrigable' land.

However, these sorts of developments are not enough to solve the problem of employment. Still, the human capital of the region is either underemployed or unemployed. The rate of unemployment is still very high in the region. Parallel to this, a factor that hides the underemployment of the labour force is the involvement of a large portion of the labour force in agricultural activities. Most of the labour force is not paid duly for their agricultural activities and the real disadvantageous group in this respect is the women since their contribution to agriculture is higher than the male population (Table 7.9).

As Table 7.9 shows, Turkey's ratio of the waged and employer status is higher than the GAP region. A reverse tendency is observed in the ratio of self-employed and unpaid family worker since in these two categories the GAP region has a higher ratio than the rest of Turkey. In these four categories, the last category shows the underemployment of the labour force. In this category, women's disadvantageous stance as well as invisibility are highlighted since nearly two-third of the women not paid for their labour.

From Table 7.9 one can also conclude that while underemployment affects largely the female population, unemployment is high among the male population. Moving from the fact that the youth between ages 14–39 constitute a large proportion of the total population of the GAP region, the conclusion that a considerable amount of the unemployed

Table 7.7: Daily poverty lines and poor households in terms of per capita and per household minimum food expenditures and cost of basic needs

	1	Average no. of HH members	Ratio of poor HHs %	2	Ratio of poor individuals %	3	Average no. of HH members	Ratio of poor HHs %
Turkey	4.6	4.66	11	1.0	15	6.6	4.46	31
Marmara	4.4	4.15	5	1.1	7	7.3	4.15	29
Aegean	3.5	4.82	3	0.9	4	6.9	4.82	24
Mediterranean region	4.2	4.52	7	0.9	11	8.5	4.52	29
Central	3.6	4.28	10	0.8	12	5.7	4.28	30
Black Sea	4.8	4.69	13	1.0	19	6.5	4.69	34
East Anatolia	4.8	5.56	18	0.9	25	6.3	5.56	33
South-eastern Anatolia	4.4	5.78	18	0.8	24	6.5	5.78	37

Notes: 1) Poverty line by the cost of minimum food expenditure per HH, daily, in dollars.
2) Poverty line by the cost of per capita minimum food expenditure, daily, in dollars.
3) Poverty line by cost of basic needs per HH, daily, in dollars.
Source: DIE, *Household Income Surveys,* 1994.

Table 7.8: Ranking of GAP provinces in terms of their
socio-economic development status

Province	District	Ranking
Adıyaman	Centre	170
	Gölbaşi	299
	Sincik	854
Batman	Centre	155
	Kozluk	711
	Sason	837
Diyarbakır	Centre	62
	Ergani	513
	Kocaköy	849
Gaziantep	Metropolitan area	13
	Kilis	246
	Nurdaği	748
Mardin	Centre	179
	Kiziltepe	576
	Savur	834
Siirt	Centre	186
	Aydinlar	547
	Pervari	853
Şanlıurfa	Centre	105
	Birecik	458
	Harran	830
Şırnak	Centre	582
	Cizre	386
	Güçlükonak	857

Source: Prime Ministry-DPT, *Ranking of Provinces in Terms of Socio-economic Development Indices*, 1996.

in the GAP region are young people is not misleading. In South-eastern Anatolia, school enrolment rates are well below the country averages (Table 7.10).

The region is well below the country averages in terms of the number of schools and of student–teacher and student–classroom ratios (one teacher for thirty-one students in the rest of Turkey versus one for forty-two students in this region; one classroom for forty-two students in the rest of Turkey versus one classroom for fifty-eight in the GAP region). Given a societal structure which still holds on to its feudal past and a bias against women, female children in the region do not adequately enjoy educational opportunities at all levels. All the

Table 7.9: Labour force status in the GAP region, 2000

Provinces	Waged			Employed			Self-Employed			Unpaid Family Worker		
	Total	Men	Women	Total	Men	Women	Total	Men	Women	Total	Men	Women
Turkey	47.05	59.01	24.82	2.61	3.73	0.54	22.15	25.39	16.13	28.19	11.86	58.51
GAP	35.74	49.04	11.88	1.20	1.85	0.03	23.21	29.40	12.10	39.85	19.71	75.99
Adıyaman	24.20	34.33	11.9	0.39	0.72	0.00	27.37	43.24	8.09	47.81	21.30	80.01
Diyarbakır	34.22	46.83	14.05	0.95	1.33	0.36	23.73	29.65	14.27	41.10	22.20	71.33
Gaziantep	49.60	59.85	23.00	5.22	6.92	0.8	21.54	23.34	16.87	23.64	9.89	59.33
Mardin	26.11	35.74	11.99	0.64	0.65	0.61	20.68	31.44	4.92	52.57	32.17	82.48
Siirt	27.81	45.01	2.93	0.56	0.56	0.57	28.41	30.57	25.29	43.21	23.86	71.21
Şanlıurfa	29.72	40.27	11.48	0.3	0.46	0.03	26.89	36.24	10.64	43.10	23.01	77.84
Batman	15.02	24.12	3.13	0.06	0.09	0.01	32.64	55.45	2.82	52.28	20.34	94.03
Şırnak	33.70	48.78	4.68	0.82	0.31	1.80	25.11	27.91	19.73	40.37	22.99	73.80

Source: GAP, Report of the Social Working Group, New Master Plan, 2001.

163

Table 7.10: Turkey—GAP region school enrolment

Categories	2001	
	Turkey	GAP
Pre-school education (3–5)	9.80	2.12
Primary education (6–13)	97.60	82.39
General secondary education (14–16)	36.60	18.40
Vocational and technical secondary education (14–16)	22.80	6.82
Higher education (17–20)	27.80	4.30

Source: DIE, *Statistical Yearbook of Turkey*, 2001.

provinces in the GAP region have lower literacy rates than the national average (96.5 per cent, male 53.3 per cent, female 43.35 per cent). The low level of female literacy in the region contributes to the greater marginalization of women (Figure 7.2).

The gender empowerment measure based on the UNDP Human Development Report of 1997 proves the degree of women's non-participation in decision-making processes at the community level

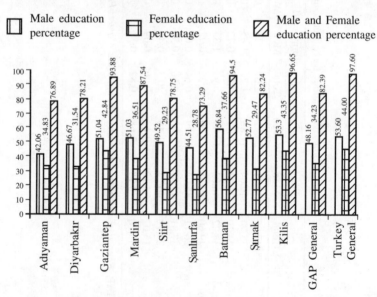

Figure 7.2: DIE, Education ratios in the GAP region according to cities and gender (%)

Source: Statistical Yearbook of Turkey, 2001.

(Table 7.11). This report very clearly reveals the 'invisibility' and non-participation of women in decision-making processes in the GAP region.

Water for drinking and other purposes is not available in sufficient quantities in this region. It is also not purified or chlorinated in many urban settlements. As Table 7.12 shows, around 28 per cent of the village population do not have drinking water, and 12 per cent have only limited quantities. Most of the water distribution systems available cannot be operated in an efficient manner, due to both lack of finance and qualified personnel, and the situation is worsened by the frequent breakdowns in the water networks. The latter factor increases the possibility of water contamination, along with the risk of increase in the incidences of contagious diseases in the summer months.

Drainage and water treatment facilities are not sufficient in the GAP region in general. In some provinces, wastewater is used for irrigating vegetable gardens. Serious problems are posed on human health and on the environment by the disorganized process of solid wastes collection, accumulation, storage, and disposal. Excessive migration from rural areas to towns contribute to the growth of these problems. The lack of these facilities has direct effects on women's lives since they are the ones who carry water from the nearest water source in the village or the quarter. Without exception, fetching water is accepted as women's work, a routine labour burden on which women have to spend nearly three hours daily depending on the size of the family. Another direct and negative effect of this shortage of water is related not solely to women but also to the household. The absence of safe water in houses affects household sanitation negatively as most of the women in this region do not pay much attention to this. Besides this, the education of female children also suffers since they are supposed to help their mothers in daily drudgery of carrying water, which takes away time which could be spent on studying.

Migration

Seasonal or permanent migration based on 'job seeking and subsistence' or 'security' and 'terrorism' as well as other horizontal population movements constitute the most pronounced demographic phenomena in the GAP region. Earlier studies (Nipon Koei Co. Ltd. and Yuksel Proje A.S., 1990) have reported that the GAP region has consistently lost migrants since the early 1960s and the median emigration rates for the region have been higher than Turkey as a whole.

Table 7.11: Rank of the provinces in the GAP Region in gender empowerment measurement (GEM) index

HDE rank	Rank in GEM	No. of chairs in local assemblies (women %)	Administrators and managers (women %)	Professional and technical occupations (women %)	Share from the earned income (per person $)	GEM value
35 Gaziantep	36	0.6	1.8	28.3	44.8	0.194
59 Diyarbakır	39	0.0	2.7	28.7	44.5	0.189
61 Adıyaman	59	0.0	2.3	24.4	45.2	0.161
63 Batman	72	0.0	1.2	18.3	43.5	0.134
64 Mardin	65	0.0	2.2	19.9	45.2	0.147
65 Şanlıurfa	63	0.0	1.5	21.4	43.1	0.150
66 Siirt	67	0.0	1.8	19.4	44.9	0.143
73 Şırnak	73	0.0	1.2	14.6	40.4	0.113

Source: UNDP, Human Development Report, UNDP, 1997.

166

Table 7.12: Breakdown of drinking water facilities in villages of the GAP Region

Province	Drinking water available		Insufficient quantity		Not available		Total		Out of service	
	Village	Population	Village	Population	Village	Population	Village	Population	Village	Population
Adıyaman	615	183,848	61	17,530	283	27,629	959	229,007	83	59,632
Diyarbakır	715	178,889	387	82,525	887	102,380	1,989	363,794	60	95,092
Gaziantep	664	208,872	9	3,752	8	1,245	681	213,869	24	58,976
Mardin	586	161,435	104	20,230	166	10,980	856	192,645	33	97,416
Siirt	330	66,102	52	7,296	120	9,242	502	82,640	9	22,314
Şanlıurfa	1,840	383,610	193	22,846	658	56,022	2,691	462,478	66	56,788
Batman	219	67,346	123	30,646	222	19,390	564	117,382	7	12,827
Şırnak	218	54,476	132	19,761	141	8,620	491	82,857	14	38,754
Kilis	141	29,753	49	10,727	9	940	199	41,420	1	2,153
Total	5,328	1,334,331	1,110	215,313	2,494	236,448	8,932	1,786,092	297	443,952

Source: DSI, General Directorate of State Hydraulic Works, *Report on Drinking Water Facilities*, 2000.

According to some field studies, difficult conditions of subsistence and lack of land have a share of 49.5 per cent as push factors explaining migration patterns in the GAP region. Massive migration from rural to urban areas put a real pressure on many services including housing, infrastructure, and social services and have further aggravated the problems of unemployment and poverty.

Water as an Engine for Sustainable Development

A large part of the GAP region lies in the Euphrates–Tigris basin. The total catchment area of the Euphrates in Turkey, upstream of the Syrian border, is 103,000 sq km, of which 22 per cent lies within the GAP region. The upper Euphrates basin is separated from the GAP region by the south-eastern Taurus mountains. The mean annual run-off near the Syrian border is estimated to be 31 billion cu m. The Tigris River, which flows east of the GAP region, drains a catchment area of 38,000 sq km north of the Syrian border; 30,000 sq km of this basin lies within the GAP region. The mean annual discharge in Turkey is estimated to be around 17 billion cu m. The water quality of both rivers is good and suitable for irrigation with electrical conductivity values below 0.6 mhos/cm and SAR below 0.4.

In order to utilize the available water resources, the GAP envisages the construction of twenty-two storage dams and nineteen hydropower plants, of which twelve dams are complete and six hydropower plants are in operation.

Of the total area of the region, around 3.9 per cent was irrigated in 1985 (Table 7.13). After the completion of GAP projects this would increase to 22.6 per cent.

As of June 2000, a total of 215,080 ha had already been brought under irrigation (Table 7.14).

Turkey produced about 17.6 billion kWh of hydropower in the first six month of 2000. In this total, GAP has a share of 39.7 per cent with its production of 7.8 billion kWh (the energy production target of GAP is 27 billion kWh). In the same period, the aggregate (thermal and hydraulic) energy production of the country was 61.2 billion kWh, of which the share of GAP was 11.4 per cent (Table 7.15).

Table 7.13: Land use in 1985

Area	Total area (ha)	% in total	Percentage in each category (%)
1. Area cultivated	3,081,170	42.2	100.0
Rain-fed agr. land	2,628,703	–	85.3
Irrigated area	120,740		3.9
Horticulture			
Fruits-vegetables	251,627		8.2
Special crops	80,074		2.6
2. Pasture and meadows	2,427,229	33.2	100.0
Meadows (grassland)	587		0.0
Pastures	2,246,642		100.0
3. Forest and bush	1,493,327	20.5	100.0
Forest	60,401		4.0
Bush	1,432,926		96.0
4. Settlement areas	25,561	0.4	100.0
5. Others	268,334	3.7	100.0
Total	7,295,624	100.0	

Source: Ex TOPRAKSU General Directorate, *Provincial Soil Sources Inventory Report*, 1985.

Table 7.14: Present irrigation in the GAP Region

Irrigation	Area under irrigation (ha)
1. Hancağız Dam and irrigation	7330
2. Şanlıurfa-Harran plains (partly)	120,000
3. Hacıhıdır Dam and irrigation	2080
4. Derik-Dumluca	1860
5. Silvan Part 1 and 2	8790
6. Nusaybin	7500
7. Silopi Nerdüş	2740
8. Akçakale Groundwater	15,000
9. Ceylanpınar Groundwater	27,000
10. Devegeçidi Dam and irrigation	7500
11. Suruç Groundwater	7000
12. Çinar-Göksu Dam and irrigation	3580
13. Garzan-Kozluk	3700
14. Adıyaman-Çamgazi (partly)	1000
Total	215,080

Source: GAP-RDA, *Status Report, South-eastern Anatolia Project*, 2000.

Table 7.15: Electricity production in the GAP Region of Turkey

GWh = million kWh

Years	Electricity Output of Turkey			GAP Hydro GWh*	GAP/Turkey	
	Thermal GWh	Hydro GWh	Total GWh		Hydro (%)	Total (%)
1995	52,548	31,973	84,521	16,114	50.0	19.0
1996 (first six months)	23,520	21,805	45,325	10,211	47.0	22.5
1996	54,448	40,423	94,871	19,314	48.0	20.0
1997 (first six months)	29,224	20,456	49,680	10,362	51.0	21.0
1997	63,299	39,764	103,063	19,385	48.7	18.8
1998 (first six months)	32,571	21,135	53,706	9,959	47.1	18.5
1998	68,677	42,224	110,901	20,053	47.5	18.0
1999 (first six months)	37,860	18,493	56,353	6,967	37.7	12.4
1999	81,800	34,629	116,429	14,781	42.7	12.7
2000 (first six months)	43,531	17,632	61,163	6,992	39.7	11.4

Source: GAP-RDA, Status Report, South-eastern Anatolia Project, 2000.

170

Impact of Irrigation on Regional Development

Effect on Small and Marginal Farmers

The ratio of farmers within the project area who did not possess land was around 40 per cent in 1990. Although not proven definitely yet, due to changes in the possession of irrigation areas and cultivation areas in the last ten years, this percentage has dropped to 23.5 per cent, according to some studies (Turkish Chamber of Architects and Engineers, 2001).

The group of farmers having only small lands or no lands at all carry out agriculture by sharing the lands of others, a practice called 'ortakçılık' (sharing the crop) or 'kiracılık' (cultivation of the leased land) when cotton farming is carried on. The ratio of crop-sharing families carrying out agriculture on the lands owned by others was 27 per cent in 1994, on the Şanlıurfa-Harran Plain which was opened up to irrigation. This ratio now stands at 57 per cent. New irrigation possibilities naturally offer new advantages to establishments possessing large lands, but this process also yields benefit to the small farmers cultivating the lands under possession of others as discussed earlier. This benefit is either in the form of a rise of income, or job opportunities in the case of families possessing no land. Parallel to growing irrigation possibilities, the number of landowners possessing large lands who move to cities grows too. A landlord thus becomes a 'businessman', dealing with trade in the city, who manages his land via agents in towns or cities, who are crop sharers most of the time.

Changes in Migratory Movements

With the introduction of intensive irrigated farming in some areas in 1995, migratory movements have started to display some changes. For example, the percentage of those moving from GAP region to Hatay and to Adana for seasonal works dropped from 70 to 12 per cent. According to some field surveys, rural settlements in Şanlıurfa started to receive population, contrary to the situation in other rural areas, after irrigation was phased-in. This presents a significant case in which the possibility of irrigation is directly related to the demand for agricultural labour force.

According to analyses made by the GAP Regional Development Plan, Agriculture and Rural Development Group, in case of 'full irrigation', which means a land of 1.8 million ha eventually opening up to irrigation, the provinces of the region will have differing needs for labour within a period of nine years. Table 7.16 shows the demand for labour in the various GAP provinces.

As can be seen in Table 7.16, 'full irrigation' will create employment for 834,000 people and Şanlıurfa and Diyarbakır will have a share of 60 per cent of this employment.

Effect on Agricultural Productivity

The total agricultural production prior to irrigation amounted to $31.5 million with a crop pattern of mainly wheat, barley, and some vegetables.

Table 7.16: Labour force need in the GAP region with full irrigation

Provinces	Labour force (Man/Year)
Adıyaman	52.70
Batman	37.08
Diyarbakır	200.35
Gaziantep	66.53
Kilis	25.75
Mardin	81.35
Siirt	21.18
Şırnak	44.34
Şanlıurfa	304.63
GAP	833.91

Source: *Final Master Plan Report, The South-eastern Anatolia Project*, GAP-RDA, 1990.

The value added was $60 per decar (0.1 ha). With irrigation, cotton became the main crop in addition to wheat and barley, and secondary crops were also introduced. The most striking change has been in the land used for cotton, which has increased from 21 per cent to 45 per cent. With irrigation, production values rose to $121 million and the value added per decar to approximately $182, both showing significant improvement in only one year. Value added per decar increased three times and the annual per capita income increased from $1,034 in 1994 to $3,963 in 1995. This area has been closely monitored to assess the impact of irrigation on different facets of life and economy.

Irrigation in the Harran Plain (in 103,000 ha of land), which started in 1995 yielded an output valued at $98,262,000 at the end of 1998 (Table 7.17). The total value of full irrigation in the GAP region is estimated as $3 billion.

Table 7.17: Gross agricultural output value (GAOV) and value added
from irrigation in the Harran Plain as a sample area

Year	Area (ha)	GAOV*			Value added**		
		(million $/year)	($/ha)	($/per)	(million $/year)	($/ha)	($/per)
Prior to Irrign.	30,000	31.5	1,050	1,044	18.0	600	596
1995	30,000	65.4	2,180	2,168	49.8	1,661	1,652
1996	40,000	87.5	2,187	2,229	67.4	1,685	1,717
1997	60,000	125.8	2,097	1,896	100.6	1,667	1,516
1998	90,000	178.8	1,987	2,388	148.8	1,653	1,987
1999	103,000	99.0	1,932	2,185	160.1	1,554	1,748

Notes: * GAOV: Production quantity × Yield price
** Value added: Value production – Production cost
Source: GAP-RDA, *Status Report, South-eastern Anatolia Project*, 2000.

The most significant change has been in terms of cotton production
in the region. Around half of Turkey's cotton is being produced here
with only partial irrigation coming into force (Figure 7.3). The increase
in cotton production has been responsible for increasing income levels
of farmers.

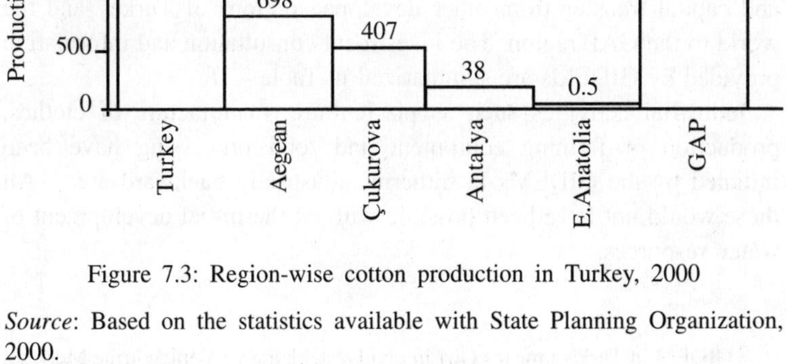

Figure 7.3: Region-wise cotton production in Turkey, 2000

Source: Based on the statistics available with State Planning Organization,
2000.

Between 1990 and 1998, the growth rate of the gross regional product (GRP) in Şanlıurfa was 9.15 per cent. When the added value created by the Atatürk Dam and Hydroelectric Power Plant is deducted from the provincial product value, the annual growth rate is about 6.75 per cent. This rate is same as the growth rate of 6.8 per cent envisaged under Alternative C in the master plan for the region. In the 1990s, both the GRP and the population increased at a rapid rate in Şanlıurfa. As the regional population had an annual rise of 2.5 per cent in 1990–2000, Şanlıurfa enjoyed an increase rate of 3.6 per cent per annum. This represents an increase rate that is some 25 per cent higher than the one envisaged by the master plan for the region. The developments in Şanlıurfa confirms the hypothesis that water resources development by provision of irrigation is the first and maybe the absolute condition to make any major progress in the context of regional development.

Effect on Industrial Development

The increasing prosperity in agricultural income has also stimulated increase in small and medium industrial enterprises. In selected province of the region (Figure 7.4) the number of industrial units grew considerably after part of GAP had been commissioned. Many of these are agro-based industries like cotton processing, spinning, textile, food processing, etc. This development has been accelerated by the establishment of the Entrepreneurship Support and Guidance Centres (GIDEMs)[2] by the GAP administration. The original idea behind the creation of GIDEMs has been to ensure that funds received in the form of compensation for land expropriation and extra returns from irrigated farming find their way to productive capital investment rather than being used for speculating. Another aim has been to help in technology and capital transfer from other developed regions of Turkey and the world to the GAP region. The investment consultation and information provided by GIDEMs are summarized in Table 7.18.

Industrial activities such as pisciculture, manufacture of clothes, production of farming equipment and food processing have been initiated by the GIDEMs in hitherto industrially backward areas. All these would not have been possible without the initial development of water resources.

[2] GIDEM in Turkish means Girişimciyi Destekleme ve Yönlendirme Merkezi.

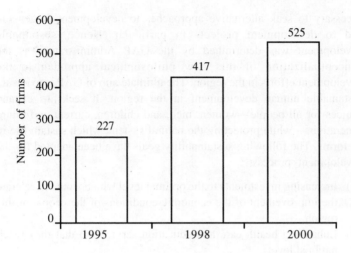

Figure 7.4: Manufacturing firms (with more than ten employees) in Şanlıurfa, Adıyaman, Diyarbakır, and Mardin

Source: Based on the statistics available with State Planning Organization, 2000.

Table 7.18: GAP-GIDEM services (15 September 1997–31 May 2000)

	Areas of Invest.	Mac. and Equ.	Incentives	Financing	Other	Total
Adıyaman	143	106	102	91	106	548
Diyarbakır	88	163	111	126	146	634
Gaziantep	185	48	70	102	100	505
Mardin	142	98	75	68	132	515
Şanlıurfa	75	58	61	59	144	397
Total	633	473	419	446	628	2,599

Source: GAP–RDA, *Status Report, South-eastern Anatolia Project*, June 2000.

GAP—A Paradigm Shift

It was felt that worldwide efforts over the last five decades, while resulting in new methods, techniques, pioneering technologies, rational use of resources, and hence output growth, had failed to prevent greater problems in social equality, environmental destruction, and the general disruption of ecological equilibrium. These conditions make it necessary to seek alternative approaches to development in general, and to development projects in particular. Hence, sustainable development was determined by the GAP Administration as the conceptualization of this new philosophical approach to the development efforts in the region. The ultimate aim of GAP is to ensure sustainable human development in the region. It seeks to expand choices for all people—women, men, and children, current and future generations—while protecting the natural system, which sustain life in all forms. The following sustainability goals have been adapted for the development process:

a) Increasing investments to the optimal level which would accelerate the improvement of the economic conditions of the people of this region.
b) Enhancing health care and education services so that they reach national levels.
c) Creating new employment opportunities.
d) Improving the quality of life in the cities and improving urban and social infrastructure so as to create healthier urban environments.
e) Providing the necessary infrastructure to rural areas in order to aid the optimal development of irrigation facilities.
f) Increasing inter- and intra-regional accessibility.
g) Meeting the infrastructural needs of existing and new industries.
h) Protecting water, soil, and air, and the associated ecosystems as a priority consideration.
i) Enhancing community participation in decision making and project implementation.

The main components of sustainability for GAP are: social sustainability, physical and spatial sustainability, environmental sustainability, economic viability, and sustainable agriculture and irrigation. In accordance with the sustainable development approach and strategies of GAP, special programmes and projects have been initiated to emphasize the human dimension of development through project implementations concerned with basic social services (education, health, housing), gender equity, urban management, irrigation facilities, agricultural and environmental sustainability, institutional and community capacity-building, and public participation.

As the GAP project has progressed, principles of sustainable development have been applied to a variety of activities ranging from marketing studies for new agricultural products to establishing women's community centres in poor neighbourhoods. Some brief case studies are available, and these will be used as a springboard to the issues of integration at the local, national, and international levels. In addition to these, there are dozens of other programmes and activities in sectors like agriculture, industries, health, etc., which are being implemented for the development of the GAP region.

Gender Empowerment

The multipurpose community centres (ÇATOMs) in the GAP region are result of the search for integrated and participatory alternatives for social change for gender-balanced development, a need which was identified by the survey 'Women's Status in the GAP Region and Their Integration into the Process of Development'. This survey was conducted by the Development Foundation of Turkey (TKV) in 1994.

ÇATOMs are community-based centres established in poor urban neighbourhoods and in rural communities and provide a gathering place for women (ages 14–50 years) who otherwise would remain isolated.

A flexible and modular programme is used with social interactions and training to achieve the following short-and medium-term objectives:

a) Raise the literacy level of women.
b) Promote awareness and provide information related to health issues.
c) Increase their knowledge about balanced nutrition.
d) Train women in childcare.
e) Improve women's income-generating skills.
f) Enable women to better express themselves.
g) Promote skills and understanding of cooperative work.
h) Help women become more aware of problems in their communities.
i) Enhance their self-confidence.
j) Improve women's access to public services.

The ÇATOMs only lay the groundwork and serve as catalysts for the process of change. The basic principle here is to show women what they can do under certain conditions, rather than dictating to them what they should do. It is the women themselves who decide what each local ÇATOM will become. Women who participate in income-generation programmes are encouraged to organize their work together. One outcome of this has been the project 'Improvement of Income-generating Activities and Management Capabilities for Women in

South-eastern Anatolia' in the year 2001, with the cooperation of the International Labour Organization (ILO) in the provinces of Adıyaman and Kilis. Another project, 'Start Your Own Business', aims to instil in women the idea of starting their own businesses, developing their management skills, improving employment and raising income levels. This project was started in 1999 with thirty-one women, who attended courses in entrepreneurship, business plans, marketing, legal procedures, insurance, enterprise visits, cost accounting, use of resource persons and capital funds. As a result of this project, twenty-four of the thirty-one women developed concrete business plans and decided to start their own businesses in fields such as sewing, embroidery, raising sheep, daycare centres, and snack bars. The first enterprises that followed were 'SYB Dowry' started by two women in the Adıyaman and an embroidery and textile workshop started on credit by two women in Kilis. Now the present target is to replicate the SYB project in all other ÇATOMs.

Under the social programme, activities and meetings are organized to promote the sense of community and an excursion through the region as per the request of participants. Since the objective is participatory community development, there are also meetings in neighbourhoods and villages where ÇATOM participants get together and discuss matters with the rest of their community members.

In addition to programmes conducted at ÇATOM centres and workshops, field staff are involved in home visits and consulting services to provide information, mainly on health and gender issues. During these home visits, field staff also have the opportunity to identify common problems, interests, and needs of women for the purpose of developing further projects.

Although ÇATOM is basically a project focusing on women's development, it was decided in 1999 to diversify its programmes and include men. The first experiments along this line include saj (musical instrument) courses for men in the Şırnak ÇATOM, and a 'Training of Fathers' course in the Adıyaman ÇATOM. Psychological counselling and guidance services are also provided to fathers through the ÇATOMs.

The activities of ÇATOMs are guided by field personnel who have been recruited from among local women who have graduated from high school and who receive periodic training. This model contributes to local capacity building. Since 1997 there have been efforts to form ÇATOM committees elected from among its members. These committees are envisaged as a means for ÇATOM women to take part in the management of relevant activities and so contribute to local capacity building and to help ÇATOMs become fully self-sustaining organizations. In some ÇATOMs, the members of these committees have already started to function as development volunteers. As of May 2001, ÇATOMs services reached around 45,000 persons. In 2001 alone, 4,512 people (3,843 women, 371 men, and 298 children) directly benefitted from ÇATOM programmes (GAP Report, 2001). Around 377 women benefitted from income-generating programmes such as kilim weaving, stone working, machine knitting, hair dressing, and the like, and health training was given to 1,576 people (including 120 men) in 2001.

A social impact assessment was conducted in 1998 (Turkish Development Foundation, 1998) to evaluate the relevance, approach and methods of programmes, and to assess the impact of ÇATOM activities on various sections of local communities. It was found that ÇATOMs reached the poor at the bottom of the social pyramid. One of the most meaningful outcomes of ÇATOM activities has been that the participants have started functioning as community health volunteers. For example, participants in the ÇATOMs in Batman and Şırnak were assigned by the directorates of health to take part in their immunization campaigns. It is a point of specific concern to ensure that all ÇATOM participants are capable of working as health volunteers and encourage others in their communities to develop awareness of health, sanitation, and environmental protection issues. Present trends are promising in the sense that ÇATOM participants have the potential to act as local community leaders and help others utilize available health services.

Producing More Food and Using Water More Productively

Traditional farming methods do not make the best use of irrigation. Therefore, the GAP Administration has coordinated a project for training local farmers and organizing them into water user groups who would be responsible for planning among themselves the use of the available water. The main objective of the model on management, operation and maintenance (MOM) of irrigation systems is to provide an institutional and organizational framework within which the proposed management model can be replicated with the following objectives:

a) To maximize net benefits derived from irrigated agriculture.
b) To ensure the financial and physical sustainability of irrigated agriculture.

The crucial work of formulating the MOM model for the project covered many subjects and disciplines. These include:

a) Environmental problems and the controls to minimize the negative effects on water and soil resources and human health.
b) Water distribution organization and management methods of different water supply systems.
c) Regulatory and legal aspects related with water supply management, water use, land ownership, and establishing farmers' organizations.
d) Issues such as social and family structures, variety of work, farming applications, cultural preferences and differences, level of irrigated agriculture training needs and character of water uses.
e) Technical aspects such as using canal and piped systems, irrigation methods, crop pattern and intensity, soils and topography, drainage needs, evaluation of water resources, operating large drainage and distribution systems.
f) Microeconomic issues for irrigation management organizations such as water charge structure and policy, the cost of finance and possible economies of scale and their impact on operation and maintenance costs.

The resulting model is based on bottom-up, participatory approach. It provides a framework for water distribution and water use efficiency leading to sustainable production techniques that ensure the protection of the land and water resources. It also allows independence in matters related to finance and in the decision-making process of the management organization while maximizing the responsibility of the individual. Twenty-four on-farm demonstrations have been established to test and show different irrigation systems and methods to farmers. Thanks to project activities, 11 per cent of the total irrigation water in this district was saved and cropping intensity has been 177.5 per cent in demonstration fields. Ongoing training courses for technical staff of the irrigation district and other related organizations are conducted periodically. Formation of water user groups is under way and intensive farmer training is the major activity in that area.

Role of Participation in Planning for Regional Development

As we have seen, participation is one of the main issues that is given ultimate importance in all integrated regional development programmes in the GAP region. It means participation of all stakeholders from public institutions to the private sector, from organized social groups to disadvantaged social groups in all the stages of the project cycle, from project preparation to implementation, and monitoring and evaluation.

The 1989 GAP Master Plan for Regional Development emphasized issues ranging from regional physical infrastructure development, to the neglect of social and sustainability issues. It was partially a conventional regional investment plan with a multi-sectoral approach. Concepts such as environmental, economic and social sustainability, gender issues, participatory planning and implementation, and the inclusion of the private sector as an active participant were either missing or not given much importance. The changing needs and conditions in the Harran plains in the GAP region and paradigm discussed here are the reasons for preparing a new participative water-based regional development plan.

Components of New Plan Participation

The new plan has become not only the basis for dialogue and development strategies but has also helped set up national, regional, and inter-regional stakeholders' networks, which are concerned with the region's problems and which help solve these problems in addition to guiding and propelling local initiatives from conception to reality. The premise of the dialogue is that local people and organizations best determine which regional and sub-regional problems need urgent attention and that local solutions to local problems have a better chance of succeeding.

Structure of Dialogue

The plan's structure was designed to get the most from continuous dialogue and participation at two levels (national and local) with five participatory methods: information, consultation, working together, delegation, and governance. All stakeholders were informed about planning procedure and every step of the new plan via workshops, small meetings, regional meetings, newsletters, web page, advertisements in the press, national private television, and local provincial television. The New Regional Development Plan was prepared with related agencies, non-governmental organizations (NGOs), universities, both national and local chamber of commerce and industry, the Union of Chambers and Stock Exchange (TOBB), the National Businessman Association (TÜSİAD), Chambers of Engineers, and the like. Some subjects or development sector studies were transferred to the related NGOs, universities, associations, or chambers for their analysis without working with the planning team, but while working with the stakeholders.

The critical issue learned from the new plan procedure is that people's participation in development is concerned with four variables:

a) educational levels,
b) structural relationships,
c) people's capacity and skills, and
d) cultural entities.

We need to make a stakeholder analysis to understand people's interests, needs, and wishes. People's knowledge and skills should be seen as a positive contribution and as a network mechanism. People's participation is a time-consuming mechanism but is useful for understanding the needs and limitations for development in the local perspective. The exercise also required different attitudes/perspectives from governmental official than the normal top-down approach.

Regional Development Scenario under the New Plan

Three different development alternatives have been formulated for the region in the framework of the New Regional Development Plan and they have been analysed from different angles. These are: Alternatives A, B, and C.

Alternative A projects the results that might be obtained in case the development trends in the region during the last decade should continue until 2010. Alternative B envisages that all the investments and projects identified by the 1989 master plan and the subsequent studies will be realized by 2010 and takes to the forefront principles such as participation, sustainability, humanitarian and social development, which constituted the starting points for the new development plan. In a sense, Alternative B projects the results that can emerge in case adequate financing can be secured and if technical conditions allow. Alternative C is a more realistic version of Alternative B, where public financing restrictions and the technical realization periods of the physical investment projects, particularly irrigation and energy projects, in other words, technological restrictions, are of a decisive nature.

The population, employment, and GRP figures, and their growth rates projected by all of the three alternatives are outlined under

Table 7.19. The public sector investment requirements under Alternatives A and B are TL 1.5 quadrillion and TL 6.2 quadrillion at 1998 prices, respectively, with the overall investment spending amounting to TL 3.07 quadrillion for Alternative A and TL 11.0 quadrillion for Alternative B. Alternative C calls for a public investment spending of TL 3.12 quadrillion, with the total investment spending amounting to TL 6.08 quadrillion. The gross additional area for which irrigation is to be provided by the public sector during the planned period is 355.6 thousand ha, 1,726.3 thousand ha, and 1,115.8 ha under Alternatives A, B, and C (see Table 7.20).

Much attention has not been paid to Alternative A because it is an option which just assumes that the past trends will continue to prevail and does not take into account the need for the formation of a plan within the framework of the new strategy. The realization of Alternative B has been considered difficult because of public financing restrictions and limited private sector investment capabilities. Focus has therefore been concentrated on Alternative C. The applicability of Alternative C, along with the available means for that, is composed of more realistic projections. This alternative corresponds to the people's expectations expressed at consultative meetings. In addition to its easier application, it is more likely to be adopted by the people.

It changes the inter-regional investment breakdown in favour of the region to some extent. As discussed later in this chapter, Alternative C does not carry this to unacceptable levels nationwide in the context of other underdeveloped regions and areas in particular, as in the case of Alternative B. Besides this, Alternative C is more compatible with the national development plan compared with the other alternatives.

Financing the New Regional Development Plan

As noted for Alternative C, the total financing requirement is TL 6,082 trillion at 1998 prices. This implies a fixed capital product of about 2 per cent. This ratio, which expresses the effective use of capital, is also a natural result of infrastructure investments such as dams, highways, airports, etc., which have been constructed in the region but have not yet yielded dividends.

The plan has a fund-generating potential which is self-consistent and dynamic. According to the findings of the new plan, the ratio of tax collection to the GRP in the region is 0.045. The minimum tax revenues calculated for the plan period (when it is assumed that the

Table 7.19: Projections as per the development plan alternatives

	1998*	2010			Annual growth rates (1998–2010) (%)		
		A	B	C	A	B	C
I. Population (thousand)	6,225.0	8,591.6	9,865.3	8,601.8	2.72	3.91	2.73
i) Urban (thousand)	4,026.0	6,674.2	5,893.3	5,754.5	4.30	3.23	3.02
ii) Rural (thousand)	2,199.0	1,917.4	3,972.0	2,847.3	-1.14	5.05	2.18
II. Employment (thousand)	2,105.7	2,926.2	4,079.1	3,305.1	2.78	5.66	3.83
i) Agriculture	1,295.0	1,770.3	2,550.0	1,929.2	2.64	5.81	3.38
ii) Industry	184.0	266.3	378.3	308.8	3.12	6.18	4.40
iii) Services	626.0	889.6	1,150.8	1,067.1	2.97	5.20	4.54
III. Gross product (At 1998 prices, TL trillion)	2,718	3,667	7,682	6,226	2.53	9.04	7.15
i) Agriculture	817	1,003	2,387	1,954	1.73	9.35	7.54
ii) Industry	408	804	1,420	1,043	5.81	10.94	8.13
iii) Services	1,493	1,860	3,875	3,229	1.84	8.27	6.64
IV. Per capita product (At 1998 prices, TL million)	437	427	779	724	-0.19	4.94	4.30

Note: * The 1998 population figures are the mid-year population figures by the DIE (State Institute of Statistics).
Source: GAP-RDA, *Draft Summary of the New GAP Regional Development Plan*, 2002.

Table 7.20: Investment and areas to be irrigated under the
development alternative plans

	Alternative A	Alternative B	Alternative C
I. Total investments (1998 prices, TL Trillion)	3,068	11,013	6,082
II. Public investments (1998 prices, TL Trillion)[1]	1,534	6,205	3,120
III. Agriculture	331	2,902	1,692
IV. New area to be irrigated (public, gross thousand ha)[2]	355.6	1,726.3	1,115.8
V. Total irrigation area (public, gross thousand ha)[3]	645.0	2,015.7	1,405.3
VI. Irrigation by people (additional, gross, thousand, ha)[4]	29	29	29
VII. Irrigation by people (total, gross, thousand, ha)	100	100	100
VIII. Total irrigation area (gross, thousand, ha/ V + VII)	745.0	2,115.7	1,505.3
IX. Total irrigation area (net, thousand, ha)	633.3	1,798.2	1,276.2

Notes: [1] Including investments by the local administrations.

[2] New area for which irrigation will be provided by DSİ.

[3] Including 289.4 thousand ha envisaged to be provided with irrigation
by the public agencies as of the end of 2001.

[4] New area to be irrigated by people during the plan period.

Source: GAP-RDA, *Draft Summary of the New GAP Regional Development
Plan*, Draft Summary, 2002.

current tax infrastructure and collection/assessment ratios will still be
effective) is TL 1,900 trillion at 1998 prices. It is envisaged that the
funds necessary for the public sector fixed capital investment of TL
3,120 trillion, as required by the plan, can be generated within the plan
period.

However, this will not be so easy in an environment where the ratio
of total public sector investments to the gross national product (GNP)
falls to 3.7 per cent and the ratio of the consolidated budget investment
allocations to the GNP is below 2 per cent. In this case, it is of
particular importance to attract local and foreign private sector into
those fields where services can be charged.

Certain energy and irrigation projects may need to be funded through private sector participation. Macroeconomic stability and predictability on a nationwide basis in general must be provided to ensure direct investment by the local and foreign private sector. This situation makes it essential to realize the public sector spending and revenue reform in general and the consequent conditions such as efficient and good management, transparency, etc. The creation of these conditions parallel to the Eighth Five-Year Development Plan can relieve the public sector.

The following points must be given attention in the development of policies as part of the regional development approaches in order to facilitate the financing of private sector investments and private sector involvement in infrastructure investments:

a) Elimination of public financing imbalances and management problems by means of a comprehensive public reform. Public sector domination in the financial markets should be minimized through reform programmes and more efficient use of scarce resources should be ensured.

b) Establishment of a sound and stable macroeconomic environment which will facilitate the sophistication and expansion of the financial markets and which will give a strong investment impetus to the real sector.

c) Implementation of supportive policies which aim at strengthening the financial infrastructure, improving the services of the financial system, and facilitating compliance with international norms.

d) Attracting foreign resources, particularly in the form of direct foreign investment and long-term investment, to the country and the GAP region.

Intra-regional Development Strategies

The new plan aims at eliminating intra-regional development differences to the largest extent possible by economic and social development with a human focus. In this context, the strategies envisaged in the framework of the sub-regional means and restrictions are outlined as follows:

a) Strategy Set 1 (improvement measures): This applies to sub-regions to which irrigation has been made available, namely, Gaziantep, Kilis, Adıyaman and Şanlıurfa provinces in the west, and Diyarbakır, Batman, and Mardin provinces in the central GAP region.

These are sub-regions where irrigation will form the real driving force for development. The components of the strategy are: provision of infrastructure for development (irrigation, transportation, urban and rural amenities, and infrastructure); increasing the industrial and service capacities of the major cities to make use of economies of scale and concentration, and increasing the level of organization.

b) Strategy Set 2 (mixed measures): This applies to sub-regions which cannot be provided with irrigation, yet enjoy positional superiority. This includes part of the west and part of the central provinces of this region. These areas which cannot be provided with irrigation, have limited land resources but enjoy superiority in view of their geographic positions which can benefit sectors such as transport and tourism. Different measures are adopted for the different areas and sectors which fall in this region. Improvement measures are carried out in sectors that are doing well and in those areas which already enjoy superiority in terms of having a viable tourism industry, transport industry, mining industry, etc., while rehabilitation measures are carried out in the areas not covered by irrigation.

c) Strategy Set 3 (improvement and repair measures): This covers sub-regions which cannot be provided with irrigation, face many natural restrictions, and lack any positional superiority. This largely comprises the eastern provinces of Siirt and Şırnak and parts of the western and central part of the region. Improvement and repair measures are measures which must be implemented in those areas which generally have more natural restrictions and display underdevelopment in view of different indicators. The fundamental elements of this set of measures are: strengthening the economic structure, encouragement of investments in those sectors enjoying relative superiority, and improvement of the standard of living by adopting all kinds of improvement measures with regard to the social structure. The components of this strategy are: economic improvement and genetic improvement of the animal stock in particular, diversification of agricultural production (poultry, pisciculture apiculture, etc.), implementation of programmes for higher employment in the management and afforestation of the upper basins of dams, encouragement of industries based on natural resources, development of handicrafts and home economy, and improvement of the standard of living through urban and rural infrastructure investment.

Conclusion

As noted, the GAP has been conceived and implemented as a means of integrating water resources development with overall human development in the poorest and most backward regions of Turkey. The strategy aims at bridging the gap between physical, spatial development and human-centred development. Despite constraints and challenges initial results have justified it not only as a tool for bringing about economic changes but also as an instrument for social engineering. It has sought to resolve the dichotomy between the dominant interventionist paradigm of development and the participative and decentralized model, by using a balanced sustainable strategy of integrating water resources development with the overall socio-economic development of the region, in a just and equitable way.

The new regional development plan with comprehensive strategies for different sectors is expected to lead to effective water-based governance for balanced regional development in the south-eastern Anatolia region. However, in view of the shortfalls in achievements of regional development strategies in the past, there is a need to realize more efficiency of investments in various sectors by implementing broad-based supportive public policies.

References

Devlet İstatistik Enstitüsü (DIE), (State Institute of Statistics-SIS), *Hanehalkı Gelir İstatistikleri* (Household Income Survey), Ankara, Turkey, 1994.

Devlet İstatistik Enstitüsü (DIE), (State Institute of Statistics-SIS), *Statistical Yearbook of Turkey*, Ankara, Turkey, 2000.

Devlet Planlama Teşkilatı-(DPT), (State Planning Organization-SPO), *Ranking of Provinces in Terms of their Socio-Economic Development Levels (1996)*, Ankara, Turkey, 1996.

Devlet Su İsleri-DSI (State Hydraulic Works), *Report on Drinking Water Facilities in GAP Region*, Ankara, Turkey, 2000.

Erdoğan G., 'Yoksulluğun neresindeyiz?' (Where are We in Terms of Poverty?), *Ekonomik Forum* (Economic Forum), 4(4), pp. 26–8, 1997.

Gorelik, M., G. Bechar, J. Margulies, G. Krayn and Y. Porat, *Planning Project, Harran Plain-Şanliurfa Province Southeastern Anatolia, Turkey, Integrated Rural Development, David Publications Series*, Development Study Center, Rehovot, Israel, 1999.

Hacettepe Üniversitesi, Nüfus Etüdleri İdaresi-HUNEE (Hacettepe University, Institute of Population Studies), *Nüfus ve Sağlik Araştirmasi* (Demographic and Health Survey), Ankara, Turkey, 2000.

Nipon Koei. Co. Ltd. and Yuksel Project A.S. Joint Venture, *Final Master Plan Report, The South-eastern Anatolia Project, Master Plan Study*, GAP-RDA, Ankara, Turkey, 1990.

Özbilen V., 'A Participatory Approach in the South-eastern Anatolia Region Development Plan (2002–2010)', International Workshop on South-eastern Anatolia Development Plan, Antalya, Turkey (unpublished), 2001.

Report of Social Working Group, New Master Plan for GAP 2001, Ankara, Turkey (unpublished).

Republic of Turkey, South-eastern Anatolia Project—Regional Development Administration, GAP-RDA, *Status Report South-eastern Anatolia Project*, GAP, Ankara, Turkey, June 2000.

—— Status Report on Performance of ÇATOMs, GAP Ankara, Turkey, 2001.

—— South-eastern Anatolia Project New Regional Development Plan, Draft Executive Summary, Ankara, Turkey (unpublished), 2002.

Topraksu Genel Müdürlüğü (Ex General Directorate of Soil and Water), *Provincial Soil Sources Inventory Report*, Ankara, Turkey, 1785.

Turkish Chamber of Architects and Engineers, the Chamber of Agronomists, TARGEV, Addendum 3, *Report on Farmers Surveys*, Ankara, Turkey, January 2003.

Türkiye Cumhuriyeti, Güneydoğu Anadolu Projesi-Bölge Kalkınma İdaresi GAP-BKİ (Republic of Turkey, South-eastern Anatolia Project—Regional Development Administration, GAP-RDA), *A Report on the Activities of ÇATOMs in 2001*, GAP-RDA, Ankara, Turkey (Unpublished), 2001.

Türkiye Kalkınma Vakfı (Turkish Development Foundation), *Social Impact Assessment for Multi Purpose Community Center*, Ankara, Turkey, 1998.

Türkiye Kalkınma Vakfı (Turkish Development Foundation), *Women's Status in the GAP Region and their Integration in the Process of Development*, Executive Summary, Ankara, Turkey, 1994.

UNDP, *Human Development Report*, UNDP, Ankara, Turkey, 1997.

UNDP, *Human Development Report*, UNDP, Ankara, Turkey, 2000.

UNICEF, *Türkiye'de Bölgelerin Gelişimi* (Development of Regions in Turkey), Ankara, Turkey, 2000.

8

South-Eastern Anatolia Project: Impacts of the Atatürk Dam

Cecilia Tortajada[1]

Introduction

The South-eastern Anatolia Project (GAP), as it is presently conceived, is a \$32 billion, multi-sectoral, integrated regional development programme. Its main objective is to strengthen the economic, social, and institutional aspects of human development in this economically disadvantaged region by raising the existing living standards and quality of life of its people. With proper planning and management, water is expected to be the engine for the sustainable development of this region in the coming decades.

The GAP region covers the provinces of Adıyaman, Batman, Diyarbakır, Gaziantep, Kilis, Mardin, Siirt, Şanlıurfa, and Şırnak which represent approximately 10 per cent of the area of Turkey. The region also accounts for nearly 10 per cent of the country's total population (6.1 million inhabitants according to the 1997 census). According to the present plan, by the year 2010, the GAP is expected to generate 27 billion kilowatt-hours (kWh) of hydroelectric energy annually, and irrigate 1.7 million hectares (ha) of land, accounting for nearly one-fifth of the irrigable land of Turkey. This would be accomplished through the

[1] The support of Dr Olcay Ünver and Mr Kaya Yasinok, President and Vice President of the GAP Administration, and Prof. Dogan Altinbilek, Director General of the DSI, are gratefully acknowledged. Ms. Berrin Basak of the GAP Administration provided much assistance in the data collection process. The study was carried out for the GAP Administration and UNDP. The opinions expressed are those of the author and not necessarily those of GAP and UNDP.

construction of twenty-two dams, nineteen hydropower plants with a total installed capacity of 7500 megawatts (MW), and extensive irrigation and drainage networks. The project is expected to almost double the existing area of artificial lakes to 228,136 ha in the country. The irrigated land would increase from 2.9 per cent to 22.8 per cent of the total area of the region, and concurrently rain-fed agriculture would decrease from 34.3 per cent to 10.7 per cent (Biswas and Tortajada, 1997).

On the basis of currently available anecdotal evidence, the disparities between the GAP region and Turkey as a whole are slowly reducing in terms of several socio-economic indicators. Some indicators of the improvement in the lifestyle of the population in the area are literacy (which increased from 55 per cent in 1985 to 67 per cent in 1997); infant mortality (which decreased from 111 per 1,000 in 1985 to 62 per 1,000 in 1995); landless population (which fell from 40 per cent in 1985 to 25 per cent in 1997); rural and urban water supply (which rose from 57 per cent and 15 per cent to 67 per cent and 57 per cent, respectively), decrease in emigration and very significant improvement in the regional economy (Unver, 2000a). While these are all encouraging statistics, further detailed studies have to be conducted to determine to what extent the GAP Project per se has been responsible for improving these socio-economic indicators, especially as only part of this project has been implemented.

Regarding infrastructural development, one of the seven GAP schemes on the Euphrates River is the Lower Euphrates Project. It consists of projects like the Atatürk Dam and Hydroelectric Power Plant (HEPP), Birecik Dam and HEPP, Şanlıurfa Tunnels, Şanlıurfa–Harran irrigation, Mardin–Ceylanpınar irrigation, Siverek–Hilvan pumped irrigation and Bozova pumped irrigation (DSI, 2000; Unver, 1997). In 2000, the Birecik Dam and HEPP was completed through a private sector build, operate, and transfer (BOT) project. The magnitude and complexity of the GAP can be seen from Table 8.1.

The construction of the Atatürk Dam was started in 1983, and it was completed in 1992. Its height from foundation is 169 metres (m), with a maximum water elevation of 542.0 m above mean sea level (msl). The volume of the reservoir is 48.7 billion m^3, and it has an area of 817 km^2. The construction of the diversion tunnels was initiated in November 1983 and was completed in January 1986. The diversion of water through the tunnels started in June 1986. The impoundment of the reservoir was started in January 1990 and was completed in August of the same year. With the construction of the Atatürk Dam, some 81,700 ha of land were inundated (GAP Administration, 1999a).

Table 8.1: Water and land resources development projects in the GAP Region

	Euphrates River	Tigris River	Total
Installed capacity:	5,304 MW	2,172 MW	7,476 MW
Energy production:	20,098 GWh	7,247 GWh	27,345 GW
Irrigated area:	1,091,203 Ha	601,824 Ha	1,693,027 Ha
Number of dams:			22
Number of HEPPs:			19

Project	Installed capacity (MW)	Energy production (GWh)	Irrigated area (Ha)	Status
I. Karakaya Project				
Karakaya Dam and HEPP	1,800	7,354		OP
II. Lower Euphrates Projects	2,450	9,024	706,281	
* Atatürk Dam and HEPP	1,800	8,900		OP
* Şanlıurfa HEPP	50	124		OP+U/C
a) Şanlıurfa-Harran Irrigation			140,000	OP
b) Mardin-Ceylanpınar Grav. Irrigation			185,639	M/P+U/C
c) Mardin-Ceylanpınar Pump Irrigation			149,000	M/P
* Siverek-Hilvan Pump Irrigation			160,105	Rec.
* Bozova Pump Irrigation			69,702	Rec.
III. Border Euphrates Project	852	3,168		
* Birecik Dam and HEPP	672	2,516		U/C
* Karkamış Dam and HEPP	180	652		U/C

(Contd.)

Table 8.1 Contd.

Project	Installed capacity (MW)	Energy production (GWh)	Irrigated area (Ha)	Status
IV. Suruç-Yaylak Project				
* Yaylak Plain Irrigation			146,500	U/C
* Suruç Plain Irrigation		509	18,322	Rec.
V. Adıyaman-Kahta Project	195		128,128	U/C
* Çamgazi Dam and Irrigation			77,824	U/C
* Gömikan Dam and Irrigation			6,536	M/P
* Koçali Dam and HEPP	40	120	7,762	M/P
* Sırımtaş Dam and HEPP	28	87	21,605	M/P
* Fatopaşa HEPP	22	47		M/P
* Büyükçay Dam, HEPP and Irrigation	30	84	12,322	M/P
* Kahta Dam and HEPP	75	171		M/P
* Pumped Irrigation from Atatürk Reservoir			29,599	M/P+U/C
VI. Adıyaman-Göksu	7	43	71,598	M/P
* Çataltepe Dam Irrigation				M/P
* Erkenek HEPP	7	43		M/P
VII. Gaziantep Project			89,000	O/P
* Hancağız Dam and Irrigation			7,330	U/C
* Kayacık Dam and Irrigation			13,680	U/C
* Kemlin Dam and Irrigation			1,969	M/P

(Contd.)

193

Table 8.1 Contd.

Project	Installed capacity (MW)	Energy production (GWh)	Irrigated area (Ha)	Status
* Pumped Irrigation from Birecik Reservoir			53,415	M/P
* Belkis-Nizip Irrigation			11,925	U/C
VIII. Tigris-Kralkızı Project			126,080	
* Kralkızı Dam and HEPP	204	444		U/C
* Tigris Dam and HEPP	94	146		U/C
* Tigris Rights Bank Grav. Irrigation	110	298		U/C
* Tigris Rights Bank Pump Irrigation		54,279		U/C+D/D
		71,801		
IX. Batman Project		483	37,744	U/C
* Batman Dam and HEPP	198	483		U/C
* Batman Left Bank Irrigation	198		18,758	U/C
* Batman Right Bank Grav. Irrigation			18,758	U/C
X. Batman-Silvan Project		964	257,000	Rec.
* Silvan Dam and HEPP	240	623		Rec.
* Kayser Dam and HEPP	150	341		Rec.
* Tigris Left Bank Grav. Irrigation	90		200,000	Rec.
* Tigris Left Bank Pump Irrigation			57,000	Rec.
XI. Garzan Project		315	60,000	Rec.
* Garzan Dam and HEPP	90	315		Rec.
* Garzan Irrigation	90		60,000	Rec.

(Contd.)

194

Table 8.1 Contd.

Project	Installed capacity (MW)	Energy production (GWh)	Irrigated area (Ha)	Status
XII. Ilisu Project	1,200	3,833		
* Ilisu Dam and HEPP	1,200	3,833		Imp.
XIII. Cizre Project	240	1,208	121,000	
* Cizre Dam and HEPP	240	1,208		Imp.
* Nusaybin-Cizre Irrigation			89,000	Rec.
* Silopi Plain Irrigation			32,000	Rec.
Individual Projects on the Euphrates River				
OP	14.4	42	60,440	
M/P			6,353	
Nusaybin Irrigation			7,500	OP
Çağ Çağ HEPP	14.4	42		OP
Akçakale, groundwater			15,000	OP
Ceylanpınar, groundwater			27,000	OP
Hacıhıdır Project			2,080	OP
Dumluca Project			1,860	OP
Suruç groundwater			7,000	OP
Besni Dam and Irrigation			2,820	M/P
Ardıl Dam and Irrigation			3,535	M/P

(Contd.)

195

Table 8.1 Contd.

Project	Installed capacity (MW)	Energy production (GWh)	Irrigated area (Ha)	Status
Individual Projects on the Tigris River				
Devegeçidi Project			7,500	OP
Silvan I and II Irrigation			790	OP
Nerdüs Irrigation			2,740	OP
Çinar-Göksu Project			3,582	OP
Garzan-Kozluk Irrigation			3,700	OP

Legends: OP: In operation
  U/C: Under construction
  D/D: Detailed design completed
  M/P: Master Plan
  Rec.: Reconnaissance
  Imp.: On Implementation Programme

Note: Individual projects are not included in grand total.
Source: General Directorate of State Hydraulic Works, *DSI in Brief*, Ministry of Energy, 2000.

The Atatürk Dam generates 8.9 billion kWh of energy annually. By September 1999, the cumulative energy production in the region (Karakaya and Atatürk Dams) was 155 billion kWh, representing a revenue of $9.3 billion. Expressed in terms of alternative sources of commercial energy, it corresponds to the import of about 39 million tons of fuel oil or 30 billion m^3 cubic metres of natural gas (GAP Administration, 1999b).

Water reaches the Şanlıurfa–Harran plain through the Şanlıurfa tunnel system, which consists of two parallel tunnels, each 26.4 km long and 7.62 m in diameter (DSI, 2000). One of these two tunnels was completed in 1995, and the other by the end of 1998. Irrigation in the Harran Plain covered 90,000 ha of land with economic returns of around $85 million (GAP Administration, 1999b). The Şanlıurfa main irrigation canal is expected to irrigate 43,000 ha of land by gravity and 5,000 ha by pumping. The Harran main irrigation canal is expected to irrigate 98,500 ha of land by gravity (Unver, 1997).

In addition to the benefits of such large infrastructural development projects at the national level in terms of electricity, and to the region in terms of increased agricultural production through irrigation, it is of fundamental importance to identify the benefits of the projects at the local level, and also review the nature of the beneficiaries. Assessment of the social, economic, and environmental impacts of different water projects, both positive and negative, should be undertaken so that specific policy decisions can be taken in a timely manner, which will maximize the positive benefits and minimize the adverse social and environmental costs. This knowledge, based on an objective assessment, is especially necessary to ensure that expected benefits from the project accrue as planned. It can also be used to improve the planning, construction, and management of similar water development projects in the future in the regions concerned, as well as elsewhere.

GAP Regional Environmental Management

Since the GAP is a large-scale, multi-sectoral regional development project, an integrated approach is essential to achieve its targets and objectives. It is evident that the irrigation and other infrastructures that have already been constructed, and would be built in the future, would contribute significantly to rapid economic growth and social changes. All these changes, in turn, would unquestionably have discernable environmental impacts, both positive and negative, which need to be carefully managed to ensure the sustainability of the project and to improve the quality of life of the local people.

In Turkey, the legislation for environmental impact assessment (EIA) was enacted in 1993. Since then, EIA studies have been prepared for all water projects, including dams. However, it was in 1992, even before this legislation, that the EIA guidelines for water development projects in Turkey were developed by the DSI, with financial support from the United Nations Food and Agriculture Organization (FAO) (Akkaya, 2001).

Due to the construction of the Atatürk Dam, and the ensuing advantages in economic activities, some urban areas (for example, Şanlıurfa) face an influx of population, with the attendant need for more housing, water, education, health services, employment opportunities, and more efficient transportation systems. If these changes cannot be properly managed, environmental degradation in terms of water, soil, and air pollution could become a serious problem. With such a multiplicity of problems, it is important to identify specific priorities, policies, and actions needed to address the most immediate issues related with sustainable urban development, including the necessary social and technological infrastructures.

A number of environmental studies have been carried out for some of the projects in the GAP region. These include potential environmental impacts of irrigation schemes: hydrology, pollution, seismic, sedimentation, ecology, human health, socio-economic conditions, and cultural heritage (Chamber of Agricultural Engineers of Turkey, 1993; Development Foundation of Turkey, 1994; GAP Administration, 1995, 1998; Harmancioglu et al., 2001; METU 1994a, 1994b, 1993; Sociology Association, 1994). The main environmental benefits of the GAP that have been identified are the control and use of flood waters mainly for energy and agricultural purposes, availability of a regular supply of high-quality water for human and industrial needs, preservation of the natural flora, increase in the aquatic fauna, creation of recreation areas, etc. On the other hand, some of the adverse impacts are considered to be inundation of natural and cultural areas; modification of hydrological patterns; changes in land use; contamination of soil and surface and groundwater; salinity, waterlogging, sedimentation, and erosion; increase in the use of fertilizers and pesticides; increase in the incidence of water-borne diseases, etc. (Harmancioglu et al., 2001).

In the specific case of irrigation activities, some of the mitigation measures proposed include the increased efficiency of irrigation systems, including drainage canals, improvement in irrigation techniques, and reuse of the irrigation return flow and urban water in irrigation (Harmancioglu et al., 2001).

Environmental considerations need to be integrated within the physical planning projects of the GAP region for the protection of the natural resources of the region. Experiences from other parts of the world indicate that big rural and urban public investment projects, when not properly implemented, could adversely affect the ecosystems, and thus contribute to increased environmental degradation. Accordingly, the formulation of regional strategies to manage water, land, and biotic resources need the integration and implementation of land use policies and practices for both urban and rural areas that could be affected by the development projects.

The existing environmental conditions in the GAP region, as well as the potential positive and negative impacts which could result from the different developmental activities, would have to be carefully analysed and managed within an integrated sustainable framework in order to maximize the total benefits and minimize the overall costs that could accrue to the society. A regional environmental management study is thus an essential prerequisite for sustainable development. Accordingly, adequate baseline information is necessary, with which future changes could be compared, and then appropriate steps could be taken for their management. A methodology is currently being developed for preparing reliable environmental profiles, that would be most appropriate for the region. The results could be effectively used for planning, decision making, and overall management of the region.

Atatürk Dam: Assessment of the Economic, Social, and Environmental Impacts

In order to understand and appreciate the changes that the construction and operation of the Atatürk Dam would bring to the people of the region through economic and social development, a study was carried out some eight years after the construction was completed. The study proposed to objectively determine the extent and magnitude of the actual social, economic, and environmental impacts of the dam and the reservoir on the region. The emphasis of the study was on economic, social, and environmental issues, both direct and indirect, over the short to medium terms, which could be objectively estimated and evaluated with reasonable accuracy. It included the evaluation of the direct impacts (positive and negative) on the people living in the two provinces affected directly— Adıyaman and Şanlıurfa—as well as on the region as a whole. The evaluation of the impacts of the dam at the national level was not included in this analysis (Tortajada, 2000), primarily because of the absence of data.

The assessment carried out included extensive fieldwork and intensive discussions, both in Ankara and the project area, with the staff members from the GAP Administration and from other different planning and implementing institutions, especially the DSI, the State Planning Organization, the General Directorate of Rural Affairs, the Middle East Technical University, etc. The interviews were carried out with senior members of national and international institutions from within and outside the region, numerous representatives of the affected population at different locations, concerned private sector institutions, and non-governmental organizations (NGOs). The members of the local population were selected at random.

After the initial set of discussions, it was decided to focus on issues like new economic activities and employment generation during the construction of the dam, the reservoir, and the associated hydraulic structures. Studies were conducted on farms using pumped irrigation directly from the reservoir, and how their agricultural yields, and thus incomes, may have changed. The resettlement process due to inundation caused by the reservoir was reviewed vis-à-vis their impacts on health and education and overall changes in the quality of life of the population living in the project area.

During the process of carrying out the studies, it was evident that the social, economic, and environmental impacts of the construction of the Atatürk Dam and its reservoir on the region were substantial through a variety of pathways. Both the dam and the reservoir have acted as the engine for economic growth and integrated regional development in a historically underdeveloped area. The benefits that are now accruing to the country by the increased electricity generation alone are very substantial. Equally, for the population living in the project area, and especially for the majority of people living near and around the reservoir, the benefits can be counted mainly in terms of infrastructural development (health and educational infrastructures), transportation and communication networks, capacity building, etc.

The areas around the dam were primarily rural, with limited infrastructural facilities. Before the dam was constructed, the transportation and communication networks between the various population centres near and around the dam site were limited. When the construction of the dam started, a good road network was constructed, which significantly improved the communication in the area. It became much easier and less time consuming for people and goods to move from one place to another. Commercial activities increased. For example, before the dam was constructed, there was not even a drugstore in the area. People had to travel to larger urban centres to obtain drugs and receive medical attention. Shortly after the construction started, one of the engineers working in the dam site opened a drugstore, which ensured that the local people could obtain drugs without having to travel long distances. The transportation network constructed also ensured that schools became more accessible to boys and girls. All these new facilities improved the social milieu of the local people.

According to the studies carried out on the impact of the construction of the Atatürk Dam, the way of living of the local population has changed. Employment opportunities have increased and working conditions have improved significantly. Expanded economic activities have encouraged migration from the rural to the urban areas within the region in many cases. In other words, rural migration to major urban centres in Turkey, like Istanbul or Ankara, were reduced. People migrated from the rural areas to the centres of the region, which started to grow significantly because of the new economic activities and employment opportunities. Thus, new urban growth poles are developing, like in Şanlıurfa, which has helped somewhat in balancing the national urbanization process.

The following aspects were identified as the main changes generated by the construction of the Atatürk Dam in the project area and beyond.

Employment Generation

Construction of the Atatürk Dam, Hydropower Plant, and Tunnels

Employment opportunities in the GAP region have been historically limited because of the limited economic opportunities. High population growth, economic stagnation, limited agricultural activities, untrained labour force, and political instability have all contributed to the underdevelopment of the region during the pre-1990 period. This situation started to change with the beginning of the construction of the Atatürk Dam and the associated hydraulic infrastructures by the

DSI. The area became a magnet for people seeking employment in the construction activities from both within and outside the region. The private sector companies, which carried out all the construction suddenly opened up a new vista for employment for skilled and non-skilled personnel, the extent and magnitude of which had never been witnessed in the region earlier.

For the construction of the Atatürk Dam and the hydropower plant, the main contractor was ATA Insaat Sanayi ve Ticaret A.S. In August 1983, the contract value of the dam was TL 102,842,062,500, which was later increased to TL 171,085,000,000 because of additional costs. The contract was signed on 28 October 1983, and the construction of the dam started immediately thereafter, on 4 November 1983. Although the duration of the work was initially estimated to be 108 months, the actual construction period had to be extended up to December 1997, a total period of 169 months. The payment was TL 214,293,000,000,000 at 1997 prices, including value added tax (VAT) equivalent of $1,916,437,700,000 (at January 1997 exchange rate of $1.00 = TL 111,818.40).

On the basis of information collected for the present analysis from the ATA Construction Company, the construction of the dam started in November 1983 with only eighty-nine workers (Figure 8.1). The total number of workers employed during the construction of the dam was 16,431, of which 466 were technical staff, and the rest were skilled and unskilled workers (nearly 1,000 of them were sub-contracted). Between November 1983 and May 1996, there was an average of 3,100 person-months of work.

The technical and skilled staff came to the region from other parts of Turkey, since local people neither had the knowledge nor the skill to construct such a large and complex structure. Most of the skilled workers who migrated to the dam site had gained their knowledge and experience during the construction of other similar structures like the Keban and Karakaya dams earlier. Local people represented 95 per cent of all the workers employed, but all of them were initially recruited as unskilled labourers. During the course of the construction of the Atatürk Dam, many unskilled personnel were trained. They later worked as drivers, machinery operators (light, medium, and heavy), carpenters, turners, metal workers, etc. Following the completion of the hands-on training period, the unskilled workers gradually became skilled workers.

As can be seen from Figure 8.1, the number of people working on the construction of the dam steadily increased with time. As noted earlier, the number of employees in November 1983, when the

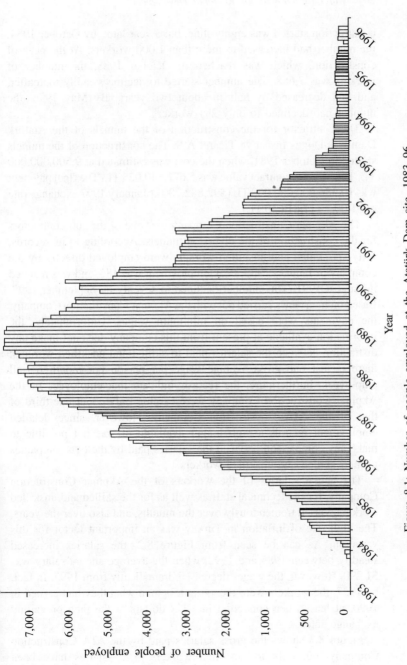

Figure 8.1: Number of people employed at the Atatürk Dam site, 1983–96

Source: ATA Construction Company (1996), Number of People Employed at Atatürk Dam Site, 1983–1996, Turkey.

construction started was eighty-nine, but a year later, by October 1984, the number had increased to more than 1,000 workers. At the peak of construction, which was reached in October 1988, the number of workers was 7,688. The number started to decline steadily thereafter, and had decreased by half in about two years. By May 1996, the number had declined to only 281 workers.

The contractor for the construction of the tunnels of the Atatürk Dam was Dogus Insaat ve Ticaret A.S. The construction of the tunnels started in October 1981, when the cost was estimated at 9,500,000,000 TL. The original contract value was 5,671,849,025 TL. The final payment was 10,386,000,000,000 TL ($92,882,700 at January 1997 exchange rate of $1.00 = 111,818.40 TL).

The Akpinar Construction Company was one of the sub-contractors for the construction of the diversion tunnels. According to its records, sixty-six skilled and unskilled workers were employed directly by the company in January 1985. Peak employment was 186, and was reached in August 1991. This number declined to seventy by September 1997. Based on interviews carried out with the Akpinar Construction Company, the payments were based on minimum wages. Unfortunately, the information available at present from this company does not make any distinction between the salaries paid to skilled and unskilled workers; there is just one gross average wage per person per month for all employees. Furthermore, the records indicate that employees of the Akpinar Construction Company were earning only about one-third of that of the ATA Construction Company. However, since detailed employment records are no longer available, it was not possible to make any comparison between the salaries paid by the two companies to their skilled and unskilled workers.

The actual salaries of the workers of the Akpinar Construction Company, for the technical staff as well as for the skilled and unskilled workers, varied tremendously over the months, and also over the years. The high rate of inflation in Turkey was an important factor for this variation. As can be seen from Figure 8.2, the salaries increased steadily between 1985 and 1992, when the average annual salary was $1,290. However, the wages decreased dramatically from 1993. In fact, by 1997, the workers were receiving less than what they had earned in 1990, at least when converted into US dollars at the then prevailing exchange rates.

Figure 8.3 shows the gross salaries paid by the ATA Construction Company, but only to unskilled workers. The salaries have been

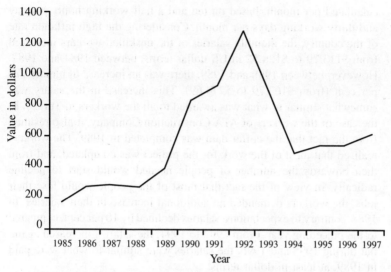

Figure 8.2: Annual gross salaries for personnel working for the
Akpinar Construction Company, 1985–97

Source: Akpinar Construction Company (1997), Annual Gross Salaries for
Personnel Working at Akpinar Construction Company, Turkey.

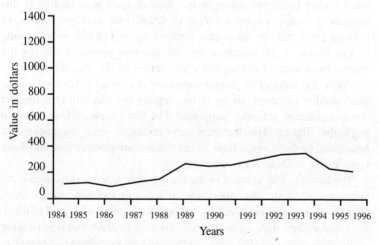

Figure 8.3: Monthly gross salaries of unskilled workers at the Atatürk
Dam, ATA Construction Company, 1984–96

Source: ATA Construction Company (1996), Monthly Gross Salaries of
Unskilled Workers at the Atatürk Dam, 1984–96, Turkey.

calculated per month, based on ten and a half working hours per day and thirty working days per month. Considering the high inflation rate of the country, the monthly salaries of the unskilled workers increased from $113.79 to $138.52 in US dollar terms, between 1984 and 1987. However, between 1988 and 1989, there was an increase of almost 170 per cent (from $163.79 to $274.69). This increase in the salary was somewhat similar to what was awarded to all the workers in Turkey. In the case of the workers of ATA Construction Company, it also resulted from the fact that the coffer dam was completed in 1989. The workers realised that most of the work for the project was completed, and from then onwards the number of people needed would start to decline radically. In view of the fact that most of the people would lose their jobs, the workers demanded an additional increase in their salaries. In 1988, contrary to expectations, salaries declined by 10 per cent, compared to the preceding year. From 1992 to 1994, the salaries increased again, but during 1995 and 1996, the salaries were similar to what were paid in 1989, at least in dollar terms.

It should be noted that nearly 100 per cent of the unskilled workers employed by the different companies during the construction of the dam and the associated hydraulic structures were recruited from among the people living in the GAP region. The estimated number of people hired during the construction of the Atatürk Dam was 16,400. If this number is multiplied by a factor of seven (the average number of persons per family in south-east Turkey), some 114,800 people living in the region were sustained by the income generated due to the employment created during the construction of the Atatürk Dam.

While the number of people supported is a small percentage of the total number of people living in this region, the fact still remains that the construction activities supported 114,800 people, which is not a negligible figure. The incomes they obtained were unquestionably beneficial to them since their living conditions generally would have been worse without such incomes.

The benefits that accrued to the local people who worked during the construction of the Atatürk Dam, at least in the case of the ATA Construction Company, went far beyond the high salaries paid to them. As noted earlier, thousands of local unskilled workers received training in different activities, thus enabling them to gain knowledge, experience, and skills in different areas. As a result of this training, many workers who were unskilled and mostly unemployed and unemployable earlier received marketable skills, and thus an opportunity to get both permanent and seasonal jobs in various construction companies after the work on the Atatürk Dam was completed. Furthermore, since the construction of the dam continued over several years, many employees were entitled to retirement pensions. The workers received additional social benefits as well during the time they worked for the ATA Construction Company, including social insurance for the employees and their families, as well as health services. The ATA Construction Company employed four doctors, and five nurses, and had four ambulances to provide medical services to the workers and their families, as well as to the local people. These were important social and medical benefits which were basically unavailable to the local people before the construction began. Clearly, these developments had beneficial impacts on the lifestyles of the local population.

The ATA Construction Company noted that of the hundreds of students who were trained at the Şanlıurfa Vocational School of Industry, many were hired by the company after their graduation. The manager of the ATA Construction Company estimated during an interview that approximately 25 per cent of the skilled people working on the Atatürk Dam were hired later for construction projects in other parts of the country.

New Economic Activities Due to the Construction of the Atatürk Dam and Reservoir

It was natural that several new economic activities were generated during the construction and operation of the Atatürk Dam and the associated hydraulic infrastructures. Among these activities were fishing and fishing-related industry (boat building, fish-net construction and repair, and fish processing and marketing), agricultural production through pumped irrigation directly from the reservoir, tourism, developments in the agro-industrial and industrial sectors, and the like.

Fishing and Fishing-related Industries

The south-east of Turkey is a semi-arid region. Accordingly, most of the agriculture practised was rain-fed, and fishing and fishing-related activities were basically unknown to most people before the construction of the dam and the reservoir. It is important to note that the Atatürk Dam lake is the longest reservoir in Turkey, and is one of the largest man-made lakes of the world. Thus, properly planned and managed, it could have significant fishing potential.

The GAP Administration and the DSI are aware of the economic, social, and environmental impacts, both positive and negative, that may directly occur due to the construction of water projects in general. Thus, both institutions are carrying out activities whose objectives are to enhance the positive impacts and mitigate the negative ones.

One of the important tasks of the DSI as a planning and implementing agency is to advise the local populations as to how best to take advantage of the newly available water resources for their own benefits as well as those of other communities concerned. The DSI is also the institution responsible for carrying out activities related to water conservation and maximizing the economic benefits that could accrue from the water projects. These activities include income-generating activities, improvements in water supply and sanitation facilities, crop diversification and increase in agricultural yields, promotion of fishery, and the use of new varieties of food that were not locally available earlier (Safak et al., 1999).

The DSI and the General Directorate of Rural Affairs, have signed a protocol governing the water products-related activities in the reservoirs of the DSI. This protocol defines the responsibilities of each of the two parties. It also establishes that once the studies on fisheries are completed, the estimated leasing prices and fishing periods have to be determined. Afterwards, the reservoirs are to be leased to local cooperatives and to the private sector by the Ministry of Finance (Safak et al., 1999).

Fishery-related activities in reservoirs developed by the DSI have been carried out since 1959. In general, these include limnological research, breeding and feeding activities in hatcheries, stocking of reservoirs, stock assessment and cage-culture activities. In reservoirs which are under operation, fish production from commercial fishing, cage-culture project application, and sport fishing are encouraged by the DSI. By the end of 1998, the DSI had carried out limnological studies in 180 reservoirs and restocking studies in 160 reservoirs and 191 ponds (Safak et al., 1999).

The financial contribution of fishery production in Turkey from reservoirs is estimated at approximately $6 million per year. Additionally, the fishing activities in the reservoirs have provided new jobs to approximately 20,000 local people. However, some of the experiences of the DSI are that the reservoirs in the country generally have fishes that are either not marketable or have low economic value. Thus, the DSI notes that in order to improve the economic production from the reservoirs, the existing hatcheries need to be improved significantly.

Additionally, the fry production capacity for restocking and cage-culture projects should be increased with species with higher economic value (Safak et al., 1999).

The fisheries activities carried out by the DSI in the country up to 1998 are shown in Table 8.2.

In the case of the Atatürk Dam, the department of operations and maintenance, water products branch of the DSI, prepared an 'Assessment of Water Products and Fishing Ground in the Atatürk Dam Lake' (Safak et al., 1994). Among the main objectives of this assessment were the definition of the characteristics of the lake, study of the flora as well as any structure that would be covered by the water in the reservoir, limnological studies, estimation of the fish production potential of the reservoir (including stock assessment and feeding requirements), establishment of a water products station, and provision of necessary support to establish a cooperative for fishermen. The fieldwork on which the assessment was based was carried out between May 1992 and March 1993. Laboratory work was conducted between July and November 1993, and the report was completed in 1994.

Water Products. Limnological studies concluded that fishes of economic importance were not present in the reservoir. The stock of the existing species was very low, and there were considerable problems in terms of hatching. Accordingly, it was decided to introduce large fish populations (especially carps) in the reservoir, using 5–6 cm fingerlings from the Elazig-Keban Water Products Centre. In 1991, 200,000 carp fingerlings were released into the lake. The number of fingerlings released increased subsequently to 600,000 in 1992, and then to 2,000,000 each year in 1993 and 1994. Table 8.3 shows fish species that existed in the lake, as well as their percentage compositions.

Species like *Carasobarbus luteus, Tor grypus, Silirus triostegus, Vimba vimba,* and *Cyprinion tanuiradius* also are present in the reservoirs of the lower Euphrates river system, in both Syria and Iraq.

Based on the data, it appears that the density of fish in the Atatürk Reservoir is less than what have been observed in Karakaya and Keban. This is to be expected and is primarily due to the fact that the Atatürk Reservoir is new and thus the amount of nutrients available for fish production is low. According to the investigations carried out by the DSI, the fish stock in the Atatürk Reservoir was about 850 tons/year when the assessment referred to earlier was carried out. This stock comprised mainly varieties like biyikli, bizir, in, sis, cultured carp, fresh water scud, and bass.

Table 8.2: Fishery activities conducted by the DSI

| Years | Limnological studies | Restocked reservoirs | | | Fish Fry produced (1000 nos.) | Restocked fish (1000 nos.) | Reservoirs in operation | Annual production (tons) |
		Dam lakes	Regulated natural lakes	Ponds				
Before 1980	55	34	3	14	1,274	1,203	14	4,700
1980	58	40	3	18	2,124	1,903	19	5,000
1985	78	56	3	39	8,294	8,253	51	5,700
1990	116	98	6	48	16,584	16,253	73	4,400
1995	153	133	8	130	59,634	59,067	97	4,900
1996	162	145	8	145	75,634	74,367	103	5,300
1997	169	151	8	162	93,634	91,556	105	5,320
1998	180	160	8	191	110,134	108,056	108	5,350

Source: Safak et al., DSI, Turkey, 'Reservoir Fisheries in Turkey', in M. Turfan (ed.) Benefits of and Concerns about Dams, Case studies, 1999.

Table 8.3: Existing fish species in the Atatürk Dam lake

Fish species	Turkish name	Percentage
Barbus raganorum	Bizir	27
Carasobarbus luteus	Egrez baligi	26
Vimba vimba	Sis baligi	21
Aspius vorax	Kultur sazani	8
Cypinius carpio	Tatli su kefali	8
Tor grypus	Sabut	5
Silirus triostegus	Mezopotamya yayini	3
Alburnus alburnus	In baligi	2
Capoeta sp.	Siraz	*
Chondrostoma regium	Kababurun	*
Mastacembelus simack	Firat Yilanbaligi	*
Chalcalburnus mossulensis	Musul kolyozu	*
Cyprinion tanuiradius	Biyikli balik	*

Note: *These species are present in very small percentages.
Source: Safak et al., 1994 *Assessment of the stock of water Products and Fishing Grounds in the Atatürk Dam Lake*, DSI, Turkey.

According to this report, there are many bays in Adıyaman at present in the Atatürk Reservoir. The report noted that it should be possible to successfully establish cage fishing in these bays. In fact, cage fishing is considered to be an important potential economic activity for people living near the reservoir. However, this may have some implications on the quality of water which need to be considered carefully.

Fishing Activities. The data included in the assessment of the stock of water products and fishing grounds in Atatürk Dam lake (Safak et al., 1994) is based on the discussions the DSI staff carried out with the fishermen living in the project area as well as on the basis of questionnaire surveys conducted in 1993 in the villages surrounding the lake. According to this report, there were about 900 fishermen in the districts around the reservoir, who had 153 fishing boats. The average fish catch was 2,390 kg/day in both Adıyaman and Şanlıurfa (Table 8.4). Based on information available (Safak et al., 1994), it is not clear whether the people referred to were already fishermen before the dam was constructed, or whether they decided to become fishermen once they realized the economic potential of fishing activities of the newly constructed reservoir. Nor is it possible to determine at present if the fish catch indicated was for a specific year, or if it was an average over several years. Information is also not available as to whether fish

Table 8.4: Fishing activities in the Atatürk Reservoir

Province-District	No. of fishermen	No. of boats	Average fish yield (Kg/day)
Adıyaman, Kahta	165	23	500
Adıyaman, Gerger	70	13	320
Adıyaman, Centre	333	50	1,100
Adıyaman, Samsat	45	6	90
Şanlıurfa, Hilvan	45	14	100
Şanlıurfa	244	47	280
Total	902	153	2,390

Source: Safak et al., 1994, *Assessment of the Stock of Water Products and Fishing Grounds in the Atatürk Dam Lake*, DSI, Turkey.

caught were sold, or consumed by the fishermen themselves. If they were sold, it would be desirable to know how much was sold, at which locations, and what were the market prices. Accordingly, on the basis of information available, it is not possible to estimate the economic potential of fishing-related activities for the local population living near to the lake in any definitive manner. Nor is it possible to estimate the nutritional aspects of the fish catch on the local population.

A cooperative for the fisherman has already been established to facilitate fishing activities. The General Directorate of Organization and Support of the Ministry of Agriculture is responsible for assisting the fishermen in establishing the cooperatives, and provide further assistance to its members in terms of its efficient running as and when necessary.

When the Atatürk Dam was constructed, 146 villages (43,198 ha of land) were expropriated: eighty-four in Adıyaman, forty-nine in Şanlıurfa, and thirteen in Diyarbakır. Since the reservoir affected three different provinces, and ten administrative districts, it was necessary to divide the reservoir into several fishing grounds so that the cooperatives could be properly established. The fishing grounds were identified by the DSI, General Directorates of Agricultural Production and Development, Conservation and Control, and Organization and Support, Ministry of Agriculture, on a combined basis. The fishing grounds were established based on a number of factors, which included their areas in hectares, geographical boundaries, state of expropriation, and studies carried out by the DSI which included fish stocks and ongoing production activities in various locations. The established fishing cooperatives are likely to require some support in the short and medium terms, if they are to be viable commercial operations.

Nine fishing grounds were initially established:

a) Adıyaman. Four fishing grounds covering a total of 51,200 ha: Central (15,400 ha), Samsat (16,800 ha), Kahta (15,200 ha), and Gerger (3,800 ha);

b) Şanlıurfa. Three fishing grounds covering a total of 29,400 ha: Bozova (15,000 ha), Hilvan (7,500 ha), and Siverek (6,900 ha); and

c) Diyarbakır. Two fishing grounds covering a total of 1,100 ha: Cermik (800 ha) and Cungus (300 ha).

The different ministries concerned with fishery-related activities have agreed that the Ministry of Agriculture should train its own staff working in the fishing areas and also organize the fishermen into cooperatives. The rent that the individual fishing cooperatives would pay would be determined on the basis of a variety of factors, like the nature and extent of rural settlements, poverty in the area, employment opportunities, information available on fishing activities, local consumption of fish, marketing, etc.

According to the information available from the GAP Administration on the project area, the nine original fishing grounds were subdivided into twenty-one by 1996. There were 290 fishermen in Adıyaman (compared to 613 fishermen noted by the DSI in the 1994 report, and 172 in Şanlıurfa (compared to 285 reported by the DSI). At this stage, it is not possible to identify the reasons for discrepancies between the statistics of the DSI and that of the GAP Administration. There are two possibilities. First, the DSI and/or the GAP Administration did not accurately enumerate the fishermen. Second, both the statistics are correct, but during the intervening years some fishermen decided to pursue alternate employment opportunities for certain reasons. Most probably, however, the discrepancy could be accounted for because of incorrect data collection in one of the studies.

Both the DSI report (Safak et al., 1994), which was based on information collected through interviews with the Department of Fisheries, and the report by the GAP Administration, noted that fishing cooperatives were still not properly organized in the project area. The only cooperative that was active in 1994 was the one in the Bozova district. The rest of the fishing districts were either in the process of being established, in the bidding process, or there were no viable activities at all. As of October 1997, all the fishermen who did not belong to any specific cooperative were still fishing, even though this was not possible under the existing legislation. In addition, and according to the provincial agricultural directorate, no cooperative was organized in Diyarbakır, and there were no data on either the number of fishermen or their catch.

The assessment of fishing activities carried out by the DSI in 1992–93 pointed out that no marketing of products could be discerned from the fishing activities. However, on the basis of data collected by the GAP Administration, fish products were already being packed in Adıyaman and Şanlıurfa, and sold to fish traders for consumption in Gaziantep, Adana, Izmir, and Manisa. However, neither the prices of the fish products nor the amounts marketed were noted.

Data available from the GAP Administration and the DSI, as well as the interviews carried out as part of this study with the local population, indicated that fishing boats are not being constructed in the project area.

The Directorate of Agriculture in Şanlıurfa provided information on quantities and market values of freshwater fish that were caught in Şanlıurfa and Adıyaman from 1993 to 1997. These data are shown in Tables 8.3 and 8.4, and Figures 8.4 and 8.5. There are no data from 1 April to 1 July for any year, since no fishing is legally permissible during these months.

The Directorate further advised not to consider any information available for the periods January–March 1993, 1994, and 1995, because 'it simply did not exist officially'. The implications of this advice were not very clear. It could mean that fishing activities existed during these periods even though these were not legally permissible, or that there were no reliable records in the Directorate.

Adıyaman

The Directorate of Agriculture in Adıyaman provided information on the quantities and market values of freshwater fish caught during January–March 1993. However, since the Directorate of Agriculture from Şanlıurfa had advised not to consider the statistics for these months, these were not included in the present analysis. Similarly, the information for January–March, 1993–5, were not considered as well.

Based on statistics available, and if the production for January–March is ignored as advised by the Directorate, more than 370,000 kg of fish were sold in 1994, for a total market value of $38,742.

In 1995, the revenues decreased dramatically because only 105,000 kg of fish were sold during three months (July–September) of the year. The total market value was $10,512. This represented a decrease of more than 70 per cent in revenue from the earlier year.

In contrast, in 1996, about 365,000 kg of fish were sold, similar to the amount sold in 1994. However, the fish sold in 1996 was over a nine-month period compared to a similar catch over only six months in 1994. The revenue in 1996 was 13 per cent higher than that of 1994.

In 1997, the amount of fish sold until September appears to be very low compared to the previous year. Only 13,300 kg of fish was sold for a total market value of $16,000 (Figure 8.4, Table 8.5).

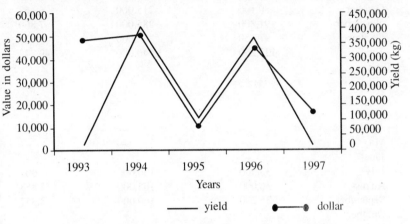

Figure 8.4: Freshwater fish yield in Adıyaman, 1993–7

Table 8.5: Freshwater fish yield in Adıyaman, 1993–7

Date	Quantity sold (Kg)	Value X1000 TL	Value Dollars
1993			
January*	5,700	130,000	14,952
February*	6,400	148,000	16,387
March*	7,300	159,000	16,957
Total	19,400	437,000	48,296
1994			
January*	8,500	65,000	4,286
February*	9,500	65,000	3,671
March*	10,000	65,000	3,157
April	—	—	—

(Contd.)

Table 8.5 Contd.

Date	Quantity sold (Kg)	Value	
		X1000 TL	Dollars
May	—	—	—
June	—	—	—
July	33,500	160,000	5,166
August	36,000	165,000	5,211
September	35,200	160,000	4,717
October	80,100	215,000	6,163
November	91,800	285,000	7,860
December	97,650	360,000	9,625
Total	402,250	1,540,000	49,742
1995			
January*	—	—	—
February*	—	—	—
March*	—	—	—
April	—	—	—
May	—	—	—
June	—	—	—
July	33,500	160,000	3,607
August	36,000	165,000	3,458
September	35,200	160,000	3,357
October	—	—	—
November	—	—	—
December	—	—	—
Total	104,700	485,000	10,512
1996			
January	27,000	400,000	6,634
February	274,000	430,000	6,730
March	28,000	460,000	6,754
April	—	—	—
May	—	—	—
June	—	—	—
July	5,500	262,500	3,627
August	6,150	339,750	4,011
September	8,000	455,000	5,134
October	5,882	282,500	3,024
November	4,382	324,750	3,305
December	6,383	490,000	4,698
Total	365,297	3,444,500	43,917

(Contd.)

Table 8.5 Contd.

Date	Quantity sold (Kg)	Value	
		X1000 TL	Dollars
1997			
January*	0	0	0
February*	100	40,000	336
March*	150	60,000	481
April	—	—	—
July	1,800	360,000	2,355
August	4,200	710,000	4,356
September	7,050	1,485,000	8,748
Total	13,300	2,655,000	16,276

Note: *Data from January, February, and March 1993–5 do not exist 'officially'.

Şanlıurfa

The market values of the fish sold varied significantly between 1993 and 1994. In 1993, 24,000 kg of fish were sold, which had a market value of about $35,000. In 1994, 20 per cent of this amount (less than 5,000 kg) were sold for 65 per cent more. In 1995, within six months, 30,000 kg of fish were sold for almost $90,000. However, nine months of fishing activities in 1996 resulted in 41 per cent more of fish being sold, but the market value increased by only 15 per cent. In January–March 1997, 5,000 kg of fish were sold for a market value of more than $30,000. Data on fish sold for January–March 1995 were not considered (Figure 8.5, Table 8.6). There were no data available from July to September 1997.

If the data available are correct, it has to be concluded that fishing activities do not appear to be an attractive business proposition. Total fish catches varied widely from one year to another. Equally, the market value of fish varied tremendously over time. The general consensus appeared to be that fishing is a relatively new economic activity in the area for most of the population, and thus, not surprisingly, people had very limited knowledge as to how best the incomes from this activity could be maximized through an appropriate management process. If this hypothesis is correct, and if fishing is to be made an attractive income-generating activity which could also be socio-economically beneficial to the region, considerable effort has to be made to develop training and educational activities in the area. Fishing could definitely be an alternative profession for some people, but the population concerned,

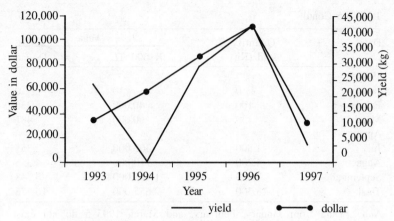

Figure 8.5: Freshwater fish yield in Şanlıurfa, 1993–7

Table 8.6: Freshwater fish yield in Şanlıurfa, 1993–7

Date	Quantity sold (Kg)	Value	
		X1000 TL	Dollars
1993			
July	1,710	155	13
August	1,500	155	13
September	1,740	155	13
October	6,370	155,000	12,416
November	7,200	155,000	11,610
December	5,790	160,000	11,400
Total	24,310	470,465	35,465
1994			
January	—	—	—
February	—	—	—
March	—	—	—
April	—	—	—
May	—	—	—
June	—	—	—
July	495	260,000	8,395
August	490	260,000	8,211
September	520	260,000	7,665
October	1,000	415,000	11,897
November	1,140	415,000	11,445
December	1,225	415,000	11,095
Total	4,870	2,025,000	58,708

(Contd.)

Table 8.6 Contd.

Date	Quantity sold (Kg)	Value	
		X1000 TL	Dollars
1995			
March	—	—	—
April	—	—	—
May	—	—	—
June	—	—	—
July	3,330	550,000	12,402
August	3,400	550,000	11,827
September	8,830	645,000	13,535
October	4,040	815,000	16,341
November	4,675	840,000	16,075
December	5,370	930,000	16,434
Total	29,645	4,330,000	86,614
1996			
January*	2,530	820,000	13,600
February*	3,180	940,000	14,712
March*	3,550	970,000	14,242
April	—	—	—
May	—	—	—
June	—	—	—
July	3,230	890,000	10,788
August	1,600	890,000	10,508
September	8,830	975,000	11,001
October	5,950	1,150,000	12,310
November	6,075	1,150,000	11,706
December	7,050	1,170,000	11,217
Total	41,995	8,955,000	110,084
1997			
January	1,740	1,210,000	10,821
February	1,435	1,330,000	11,185
March	2,160	1,230,000	9,874
April	—		—
May	—		—
June	—		—
July	No data	No data	No data
August	No data	No data	No data
September	No data	No data	No data

Note: *Data from January, February, and March 1993–1995 do not exist 'officially'.

at least during the initial years, should be supported in terms of technical knowledge, economic and marketing skills, and equipment.

Tourism and Recreational Activities

On the basis of available evidence and site visit, no significant increase in tourism and recreational activities can be noted around the Atatürk Lake. This is to be expected since this area has not been considered in the past as a tourist spot and thus infrastructure like hotels, restaurants, and holiday camps have to be developed, and events which could attract tourists have to be organized. The GAP Administration is aware of this issue, and thus has been organizing an annual water festival in recent years, in which the main attractions are various types of water sports in the reservoir. While this has been a good beginning, additional consideration needs to be given as to how to attract more people for longer periods. National and international meetings could be organized in conjunction with the water festival on a regular basis each year. One possibility could be to expand the water festival to an annual occasion for discussions on various aspects related to water as a focus for regional development and integrated water and land resources management. Such a dual process of water festival and technical meetings has considerable potential to boost economic activities in the region.

Agricultural Production Through Pumped Irrigation Directly from the Reservoir

Based on information available from the DSI and from personal interviews conducted with people living around the reservoir area, only a limited number of people are practising irrigation with water pumped directly from the reservoir. These are mostly small-scale farmers. The main advantage of this type of farming has been that it could be practised as soon as the reservoir was ready: it was not necessary to wait for irrigation to start and all canal infrastructures to be constructed.

Irrigated Agriculture

The main objective of the GAP project is to transform the region into an export base for its agricultural products. Irrigation is expected to increase crop yields and diversification very substantially, which in return would contribute to infrastructural development and increased economic activities. This is expected to further accelerate development of the agro-industry and other agricultural services.

A sustainable agricultural development requires not only conservation, reuse, minimization of the various environmental damages, but also suitable production practices, appropriate technology, profitability, enforcement of laws and regulations, incentives, and appropriate investment plans.

Before irrigation started, the main crops of the region were wheat and barley. In 1995, about 30,000 ha were irrigated with cotton as the main crop. By the end of 1998, approximately 90,000 ha were irrigated with economic returns of around $85 million. Increase in cotton production has already spurred development in cotton-related agro-industries, like cotton ginning, manufacture of cotton-seed oil, cloth mills, etc.

Additionally, rain-fed agriculture is practised to produce crops such as wheat, barley, lentils, chickpeas, and sesame as well as horticultural crops like pistachios and grapes (Altinbilek, 1997).

The production of many secondary crops has increased as well. Double and even triple cropping has been achieved due to the irrigation activities. For example, maize cultivation has increased from about 500 decars of maize in 1995, to about 4,000 decares in 1997 (10 decars = 1ha). The private sector buys this production to feed livestock.

Industrial Activities

As a direct result of developments stemming from the construction of the Atatürk Dam, industrial and commercial activities are accelerating rapidly in the region. The urban centres of the region have witnessed explosive growth in immigration because of enhanced employment potential, including self-employment like traders, vendors, providers of services like childcare, transportation, etc.

Construction activities have increased substantially, as have commercial activities in both formal and informal sectors. Transportation and communication links within the region, as well as between the region and the rest of Turkey have opened up new potentials for economic and industrial activities which simply did not exist earlier. Because of increasing demands, frequency of commercial flights between the urban centres of the GAP region and the rest of Turkey has expanded exponentially in recent years. Such increased and improved transportation and communication linkages are likely to enhance the socio-economic development of the region at an accelerated rate in the coming years.

A good example of industrial and commercial development is Şanlıurfa. The city has already established an industrial zone which is now almost full, as a result of which a second zone of about 11,000 decars is now being developed. A free zone for exports and imports will be organized in this area. The first zone is mainly occupied by agro-based industries that are cotton-related, for example, cotton ginning, manufacture of textiles, production of cotton-seed oil, etc. Prior to the construction of the Atatürk Dam, these activities were extremely limited. These agro-industries are not only generating employment, which is improving the living conditions of the people, but are also providing a major value-added service, which was not available earlier. In the process, merchants or industrial concerns are buying the raw materials from the farmers, which further boosts the economy of the region. A secondary benefit of the new agro-industrial development is that the workers employed need housing, markets, and other services which are further boosting the employment conditions of the region. The Atatürk Dam has thus directly contributed to the creation of a 'win-win' situation, whose socio-economic benefits now encompass not only the GAP region but also the rest of the country through a variety of direct and indirect linkages and pathways.

The GAP region is now at an early phase of industrial development. The levels of education and training that now exist in the region are still significantly below the national average, though in recent years the gaps between the two have been closing. While at present a very significant number of unskilled labourers can be absorbed in the labour force, the region will increasingly need more and more skilled labour, if the conditions for the generation of employment are to be sustained and employment is to be generated. Many of the unskilled labourers are progressively learning new skills, which increases their salaries and employment opportunities. However, more needs to be done if the current expectations are to be fulfilled. The potential of higher income through better-paying jobs are of course strong incentives for the people to learn new skills.

In spite of some advances, however, it is now clear that the region is already suffering from a shortage of skilled workers. As new industries are established and existing ones are modernized, demand for skilled labour is going to accelerate. This demand can be fulfilled by increasing the training facilities available in the region, and by migration of skilled workers from other parts of Turkey to the GAP area. The latter is likely to put inflationary pressures on wages for skilled labour in the country, which could reduce one of the important economic advantages the country has at present. Equally, dependence on skilled workers from outside the area will assure that the workers of the region will be increasingly restricted to low-paying and undesirable jobs. This could create social tension between the low-paid local workers and highly paid employees coming from outside the area. Thus, viewed from any direction, the best solution would be to take appropriate and timely steps that would help increase the education levels and skills of the local workers.

While increasing industrialization has ensured many benefits to the people of the region, it has also brought in its wake certain social and environmental costs. The main concern at present is the negative environmental and social impacts of wastewater management practices. Proper wastewater treatment by any industry in the region is now an exception rather than the norm. The situation is serious for the industrial zone in Şanlıurfa because of the high concentration of the industrial activities in the area. None of the industries treat their waste water, and all the wastewater is now discharged on the land adjoining the industries.

Even though the Şanlıurfa industrial estate is fifteen kilometres away from the city, the discharge of wastewater could have major social and health costs in the coming years. First, groundwater of the area may become contaminated with industrial wastewater over the years. Depending on the gradient of the flow, the groundwater of the region may become contaminated with industrial waste products, not only in Şanlıurfa but also in the different cities where similar industrial developments are being encouraged.

Second, the industrial estate is located near some villages and a stream, which serves as a source for water for many people. The people and the ecosystems of the nearby villages are likely to bear the first adverse health and environmental impacts from the existing wastewater discharges from these new industrial developments.

In the short term, monitoring the quality of the wastewater discharges and groundwater near the estate is necessary. Over the short to medium term, it is necessary to prepare a plan for treatment of waste water for the industrial zone, and then implement it. Because of the lack of data on the quantity and quantity of wastewater generated, it is difficult to make specific comments on the quality of water. However, prima facie, there appears to be a good case to construct a communal wastewater plant for treating the discharges from the industrial estate as soon as possible.

A strategy for wastewater management for the industrial estate of Şanlıurfa should now receive attention. This is because a second zone is now in the process of being established which is likely to increase the magnitude of the overall problem.

The experience with the Şanlıurfa industrial zone project indicates that any new similar project in the region must consider a wastewater and solid wastes management plan from the very beginning of the planning process. Environmental management needs to receive a higher priority than at present.

Resettlement and Rehabilitation

In Turkey, the designated executing agency for water and land development projects is the DSI. The General Directorate of Rural Affairs is responsible for resettlement and rehabilitation. Like in most other countries, expropriation of land and the subsequent resettlement and rehabilitation activities are regulated by law (Altinbilek et al., 1999a, 1996b).

The levels of compensation for the people who have to be resettled depend on several factors like the nature and size of the properties, elements which could increase the value of the properties, taxes paid on the properties, etc. The levels of compensations for all properties are decided by an independent valuation commission composed of technical experts and representatives of the affected population. The DSI transfers the amount needed to a special resettlement fund that is then managed by the General Directorate of Rural Affairs. When the values of the expropriated properties exceed the compensations offered, the differences are paid back to the owners. However, when it is the reverse, the owners are given a five-year moratorium on the debt, followed by a twenty-year interest-free repayment period (Altinbilek, 1999a). Table 8.7 shows the population which has been resettled due to the construction of large dams in Turkey between 1990 and 1997.

People have the option to decide whether they would like to be resettled in rural or in urban areas. For those families who opt for rural areas, each household is entitled to housing, farm land, credit for animal husbandry, etc. The law further stipulates that the farmers who are to be resettled must receive training on new agricultural production

Table 8.7: Resettlement of population due to the construction of the Atatürk Dam

Year	Present location			Number of families to be resettled	Number of families already resettled	Type of Settlement		Resettlement area		
	Province	District	Village			Rural	Urban	Province	District	Village
1988	Adıyaman	Various	Various	185	185	–	185	Adıyaman	Y.Samsat	–
1988	Adıyaman	Various	Various	34	34	34	–	Aydin	Soke	Yalikoy
1989	Adıyaman	Various	Various	42	42	42	–	Aydin	Soke	Yalikoy
1989	Şanlıurfa	Various	Various	20	20	20	–	Aydin	Soke	Yalikoy
1989	Diyarbakir	Cermik	Dilekpinar	3	3	3	–	Aydin	Soke	Yalikoy
1989	Adıyaman	Kahta	Geldibuldu	1	1	–	1	Burbur	Bucak	–
1989	Adıyaman	Centre	Hacihalil	1	1	1	–	Şanlıurfa	Ceylanpınar	–
1989	Diyarbakir	Cermik	Dilekpinar	1	1	1	–	Diyarbakır	Centre	–
1989	Adıyaman	Samsat	Various	6	6	–	6	Adıyaman	Y. Samsat	–
1989	Adıyaman	Centre	Karicik	1	1	–	1	Ordu	Unye	–
1989	Adıyaman	Kahta	Adali	2	2	2	–	Aydin	Soke	Denizkoy
1990	Adıyaman	Various	Various	18	18	18	–	Hatay	Reyhanli	Vazvaza
1990	Adıyaman	Kahta	Geldibuldu	1	1	1	–	Konya	Sarayumu	Kayioren
1990	Adıyaman	Kahta	Bostanli	2	2	2	–	Cankiri	Cerkes	Akhasan
1991	Adıyaman	Various	Various	2	2	2	–	Hatay	Reyhanli	Vazvaza
1993	Adıyaman	Various	Various	18	18	–	18	Adıyaman	Y. Samsat	–
1993	Şanlıurfa	Bozova	Dikili	1	1	1	–	Hatay	Kirikhan	Karatas

(Contd.)

225

Table 8.7 Contd.

Year	Present location			Number of families to be resettled	Number of families already resettled	Type of Settlement		Resettlement area		
	Province	District	Village			Rural	Urban	Province	District	Village
1997	Adıyaman	Various	Various	6	6	6	–	Hatay	Centre	Hasanli
	Adıyaman	Various	Various	369	–	369	–	Hatay	Hassa	Gurpinar*
	Adıyaman	Various	Various	416	–	–	416	Adıyaman	Y.Samsat**	
Total				1,129	344	502	627			

Notes: * The General Directorate of Rural Affairs stopped the construction of the houses (regulations of 27.3.1997).
** The land to construct the houses was given by the government, the population did not pay for it.
Source: General Directorate of Rural Affairs.

226

methods from the government. All rural resettlement units have to be provided with a health centre, a doctor, a nurse, and a midwife. For urban resettlement, the people receive a house and needed commercial facilities, as well as credits for commercial activities.

In many cases, major landowners with large properties prefer to receive cash compensation, and then use the compensation received to establish small industries or commercial activities in the cities. It is fairly common to find that part of a family whose land was expropriated moves to the city, but the rest prefers to stay in the rural area. Experiences indicate that the second generation of these families who decided to resettle in the cities and invested their money successfully, became entrepreneurs.

Overall, the main problem in the relocation of populations is the scarcity of land and not money. Experiences from different water development projects from other parts of Turkey indicate that many people prefer to take the appropriate financial compensations and then organize their own resettlement in areas of their choice. The GAP region was no exception to this practice.

People from the rural areas, often non-skilled and not familiar with the various investment opportunities have not managed the expropriation funds received properly. Accordingly, many of them have not used the funds received wisely and have ended up with no house, no land, no job, and no money. Thus, within a limited period of time, they have ended up as destitutes, with economic and living conditions significantly worse than before because of poor investment decisions and inappropriate financial management. This is an important problem for Turkey in terms of the efficiency and social acceptability of the resettlement practices. This is a national problem, not limited to the GAP region, for which proper solutions, like intensive information and communication services, have to be provided which could enhance the social and economic conditions of the people who are to be resettled.

For the current analysis, the information on resettlement was collected from the DSI and from the directorate of rural affairs. The information that the XVI Regional Directorate of the DSI provided included the status of urban and rural resettlement as of 1993, and also the status of expropriation as of 1996. The DSI data did not include information on resettlement after 1993.

On the basis of the information collected from the DSI and the Directorate of Rural Affairs, as of 15 September 1997, the expropriation of properties up to the height of 542m has been completed. The total cost has been 13,057 trillion TL at 1995 prices. By the end of 1995, and by decision of the courts, 2,979 trillion TL (at 1995 prices) have been paid to settle disputes with the resettled population.

The Directorate of Rural Affairs has estimated that 1,129 families had to be displaced due to the construction of the Atatürk Dam Project during the period 1988–97. Out of this number, 44 per cent were to be resettled in rural areas, and the balance of 56 per cent in urban areas. By 1998, only 30 per cent of the population had been resettled (344 families), and 70 per cent still had to be resettled (369 families in rural areas, and 416 families in urban areas) (Tables 8.8 and 8.9).

However, the records of the DSI do not include the total number of families that have had to be resettled. The records only note that the number of families that were involuntarily resettled from 1988 to 1997 was 344 (133 in rural areas, and 211 in urban areas).

By 1999, 375 families affected by the dam were still living in rented houses waiting to be resettled in rural areas (Altinbilek et al., 1999a, 1999b).

According to the DSI (Bayram, 2000), by July 2000, thirty-six more families had been resettled in Ayrancilar village and six more families had decided not to wait any longer to receive the support of the government. The General Directorate of Rural Affairs is planning to resettle the remaining 333 families in the Van province, but the necessary land has still not been allocated by the Treasury. Due to the delay, several families are considering resettlement in urban areas. The Kahta district of the Adıyaman province is one possibility, according to the officers from the Rural Affairs Department. However, Rural Affairs still has not carried out any detailed study related to this resettlement.

In order to obtain a clearer picture of the efficacy of the resettlement process from the perspective of the people who were affected, extensive series of discussions were conducted with the project-affected people in several villages in Adıyaman and Şanlıurfa. Collective meetings with the people who were resettled were organized in a few selected villages. All the meetings included the *mukhtar* (head of the village) and all the heads (men) of each family living in each town. The men finally selected for detailed interviews were chosen at random. In all cases, at least 50 per cent of the heads of the households affected were interviewed in one form or another.

The main issues that were discussed during all these interviews and meetings were the effectiveness of the resettlement process, status and levels of compensations paid, quality of housing and services provided,

Table 8.8: Resettlement of population due to the construction of the Atatürk Dam

Name of the project	Location		Name of the area affected	What was affected		To what extent
	Province	District		Settlement	Land	
Atatürk dam	Şanlıurfa	Bozova	Dutluca Village		Land	Partially
			Yaslica Village		Land	Partially
			Cinarli Village	Settlement		Completely
			Iğdeli Village	Settlement		Completely
			Geçibasi Village	Settlement		Completely
			Dikili Village	Settlement		Completely
			Kasimkuyu Village		Land	Partially
			Tavuk Hamlet	Settlement		Completely
			Baglica Village		Land	Partially
			Odulalan Hamlet	Settlement		Completely
			Yukaricatak Village	Settlement	Land	Partially
			Tatar Hoyuk Village	Settlement		Completely
			Bahceli Village		Land	Partially
			Tekağac Village	Settlement	Land	Partially
			Asagicatak Village	Settlement		Completely
			Yiginak Village		Land	Partially
			Acikyel Hamlet	Settlement		Completely
			Akyatir Hamlet	Settlement		Completely
			Arikok Village		Land	Partially
			Kucuktulmen Village		Land	Partially

(Contd.)

Table 8.8 Contd.

Name of the project	Location		Name of the area affected	What was affected		To what extent
	Province	District		Settlement	Land	
	Şanlıurfa	Hilvan	Gülüşağı Village	Settlement		Completely
			Kocabey Village	Settlement	Land	Completely
			Gelenek Village		Land	Partially
			Kamışli Hamlet	Settlement	Land	Completely
			Ambar Village	Settlement	Land	Completely
			Uğra Village	Settlement	Land	Completely
			Uluyazi Village	Settlement	Land	Partially
			Nasrettin Village	Settlement		Completely
			Geçiatağzi Village	Settlement		Completely
			Kavalik Village		Land	Partially
			Kovaci Village	Settlement		Completely
			Oymaağac Village		Land	Partially
			Karaburç Village		Land	Partially
			Bağrac Village		Land	Partially
			Bahçelik Village	Settlement		Completely
			Faik Village	Settlement		Partially
			Kuçak Village	Settlement	Land	Partially
			Çatak Village		Land	Partially
			Bügür Village		Land	Partially
			Taşagil Village		Land	Partially
	Şanlıurfa	Siverek				

(Contd.)

230

Table 8.8 Contd.

Name of the project	Location		Name of the area affected	What was affected		To what extent
	Province	District		Settlement	Land	
			Azikli Village	Settlement	Land	Partially
			Kalemli Village	Settlement		Completely
			Kuslogöl Village		Land	Partially
			Kayali Village		Land	Partially
			Divan Village		Land	Partially
			Narlikaya Village		Land	Partially
			Bürüncek Village		Land	Partially
			Tasikara Village		Land	Partially
			Büyükoba Village		Land	Partially
			Bekçeri Village		Land	Partially
			Kavalik Village		Land	Partially
			Çaylarbaşi Sub-district		Land	Partially
	Adiyaman	Centre	Bebek Village		Land	Partially
			Araplar Hamlet	Settlements		Completely
			Mazilik Hamlet	Settlements		Completely
			Çobandede Village		Land	Partially
			Çat Hamlet	Settlements		Completely
			Bağpinar Sub-district		Land	Partially
			Akdere Village		Land	Partially

(Contd.)

231

Table 8.8 Contd.

Name of the project	Location		Name of the area affected	What was affected		To what extent
	Province	District		Settlement	Land	
			Karicik Village	Settlement		Completely
			Battalhöyük Village		Land	Partially
			Paşamezrasi Village		Land	Partially
			Hacihalil Village		Land	Partially
			Bozhoyuk Village		Land	Partially
			Yazica Village		Land	Partially
			Yazlik Village		Land	Partially
			Kuyvcak Village		Land	Partially
			Büyükkavalik Village		Land	Partially
			Lokman Village		Land	Partially
			Kizilcapinar Village		Land	Partially
			Dişbudak Village		Land	Partially
			Çayci Village	Settlement		Completely
			Aydinlar Village	Settlement		Completely
			Gölpinar Village		Land	Partially
			Incebag Village		Land	Partially
			Boztepe Village		Land	Partially
			Sariharman Village		Land	Partially
			Ipekli Village		Land	Partially
			Sitilce Mah.		Land	Partially
	Adiyaman	Samsat	Samsat Ilcesi	Settlement	Land	Completely
			Kumluca Village	Settlement	Land	Completely

(Contd.)

232

Table 8.8 Contd.

Name of the project	Location		Name of the area affected	What was affected		To what extent
	Province	District		Settlement	Land	
			Balcilar Village	Settlement	Land	Completely
			Sütbulak Village	Settlement	Land	Completely
			Kovanoluk Village	Settlement	Land	Completely
			Tapeönü Village	Settlement	Land	Completely
			Bayirli Village		Land	Partially
			Nalgevir Hamlet	Settlement	Land	Completely
			Bağarasi Village		Land	Partially
			Doğanca Village		Land	Partially
			Kilisik Hamlet	Settlement		Completely
			Göltarla Village		Land	Partially
			Gülpinar Village		Land	Partially
			Kirmacik Village		Land	Partially
			Bizbeyan Hamlet	Settlement		Completely
			Yarimbag Village		Land	Partially
			Örentas Village		Land	Partially
			Uzuntepe Village		Land	Partially
			Taskuyu Village		Land	Partially
			Çiçek Village		Land	Partially
			Belenli Sub-district		Land	Partially
	Adıyaman	Kahta	Çakiresme Sub-district	Settlement	Land	Completely

(Contd.)

233

Table 8.8 Contd.

Name of the project	Location		Name of the area affected	What was affected		To what extent
	Province	District		Settlement	Land	
			Dardağan Village		Land	Partially
			Gurni Mah.	Settlement	Land	Completely
			Eskitaş Village		Land	Partially
			Ulupinar Village		Land	Partially
			Yiğinak Village		Land	Partially
			Arili Village		Land	Partially
			Çaybasi Village		Land	Partially
			Belören Village		Land	Partially
			Saridana Village		Land	Partially
			Güzelçay Village		Land	Partially
			Seyhbaba Village		Land	Partially
			Erikdere Village		Land	Partially
			Gölgeli Village		Land	Partially
			Büyükbağ Village		Land	Partially
			Oluklu Village		Land	Partially
			Güdülge Village		Land	Partially
			Akincilar Sub-district		Land	Partially
			Geldibuldu Village	Settlement		Completely
			Adali Village	Settlement		Completely
			Ikizce Village		Land	Partially

(Contd.)

234

Table 8.8 Contd.

Name of the project	Location		Name of the area affected	What was affected		To what extent
	Province	District		Settlement	Land	
	Şanlıurfa	Gerger	Bostanli Village	Settlement		Completely
			Narsirti Village	Settlement		Partially
			Dut Village		Land	Partially
			Kahta Centre		Land	Partially
			Kilic Village		Land	Partially
			Besgöze Village		Land	Partially
			Korulu Village		Land	Partially
			Gumuskasik Village		Land	Partially
			Köklüce Village		Land	Partially
			Beybostan Village		Land	Partially
			Yagmurlu Village		Land	Partially
			Kesertas Village		Land	Partially
			Yayladali Village		Land	Partially
			Açma Village		Land	Partially
			Dağdeviren Village		Land	Partially
			Budakli Village		Land	Partially
			Üçkaya Village		Land	Partially
			Gönen Village		Land	Partially
			Gözpinar Village		Land	Partially

(Contd.)

235

Table 8.8 Contd.

| Name of the project | Location | | Name of the area affected | What was affected | | To what extent |
	Province	District		Settlement	Land	
	Diyarbakır	Germik	Geçitli Village		Land	Partially
			Konacik Village		Land	Partially
			Ortaca Village		Land	Partially
			Cevizpinar Village		Land	Partially
			Saltepe Village		Land	Partially
			Ceylan Village		Land	Partially
			Pamuklu Village		Land	Partially
			Karakaya Village		Land	Partially
			Çukurelma Village		Land	Partially
			Armağantasi Village		Land	Partially
			Dilekpinar Village		Land	Partially
			Konakli Village		Land	Partially
			Seyhandede Village		Land	Partially
			Adalar Village		Land	Partially
			Kömürcüler Village		Land	Partially
			Bayat Village		Land	Partially

Table 8.9: Number of families resettled due to the construction of large dams in Turkey, 1970–97

Year	Name of the dam	Original location Province	Original location District	Original location Village	Number of families resettled	Type of settlement Rural	Type of settlement Urban	New location Province	New location District	New location Village
1985	Altinkaya	Samsun	Bafra	Several	25	–	25	Bursa	Gemlik	Centre
1986	Altinkaya	Sinop	Duragan	Kaplangi	41	41	–	Hatay	Reyhanli	Horlak
1986	Altinkaya	Samsun	Several	Several	130	130	–	Hatay	Reyhanli	Horlak
1988	Altinkaya	Sinop	Duragan	Several	81	–	81	Sinop	Duragan	–
1989	Altinkaya	Sinop	Duragan	Several	6	–	6	Sinop	Duragan	–
1990	Altinkaya	Sinop	Duragan	Several	155	155	–	Hatay	Reyhanli	Vazvaza
1990	Altinkaya	Samsun	Vezirkopru	Several	11	11	–		Reyhanli	Vazvaza
1990	Altinkaya	Sinop	Several	Several	5		5		Duragan	
	Total				454	337	117			
1970–1983	Hasan-Suat Ugurlu				39	39	–	Samsun	Carsamba	Demirli
1985	Hasan-Suat Ugurlu	Ordu	Akkus	Cokek	17	–	17	Ordu	Unye	Centre
1985	Hasan-Suat Ugurlu	Samsun	Carsamba	Several	100	–	100	Samsun	Carsamba	Kirazlicay
1987	Hasan-Suat Ugurlu	Samsun	Carsamba	Several	134	–	134	Samsun	Carsamba	Kirazlicay
1987	Hasan-Suat Ugurlu	Ordu	Akkus	Several	38	–	38	Samsun	Carsamba	Kirazlicay
	Total				328	39	289			
1986	Karakaya	Malatya	Several	Several	132	132	–	Konya	Centre	Yaglibayat
1986	Karakaya	Malatya	Several	Several	62	62	–	Aydin	Soke	Kirikici
1986	Karakaya	Elazig	Several	Several	33	33	–	Aydin	Soke	Kirikici
1987	Karakaya	Malatya	Several	Several	20	20	–	Aydin	Soke	Kirikici
1989	Karakaya	Malatya	Centre	Meydanci	1	1	–	Aydin	Soke	Yalikoy

(Contd.)

Table 8.9 Contd.

Year	Name of the dam	Original location			Number of families resettled	Type of settlement		New location		
		Province	District	Village		Rural	Urban	Province	District	Village
1990	Karakaya	Malatya	Centre	Tecirli	2	2	–	Kanya	Sarayon	Kayioren
	Total				250	250				
1988	Atatürk	Adıyaman	Several	Several	34	34	–	Aydin	Soke	Yalikoy
1988	Atatürk	Adıyaman	Several	Several	185	–	185	Adıyaman	Y. Samsat	–
1989	Atatürk	Adıyaman	Several	Several	6	–	6	Adıyaman	Y. Samsat	–
1989	Atatürk	Adıyaman	Several	Several	42	42	–	Aydin	Soke	Yalikoy
1989	Atatürk	Şanlıurfa	Several	Several	20	20	–	Aydin	Soke	Yalikoy
1989	Atatürk	Diyarbakır	Cermik	Dilekpinar	3	3	–	Aydin	Soke	Yalikoy
1989	Atatürk	Adıyaman	Kahta	Geldibuldu	1	1	–	Burdur	Bucak	–
1989	Atatürk	Adıyaman	Centre	Hacihalil	1	1	–	Şanlıurfa	Ceylanpınar	–
1989	Atatürk	Diyarbakır	Cermik	Dilekpinar	1	1	–	Diyarbakır	Centre	–
1989	Atatürk	Adıyaman	Centre	Karicik	1	1	–	Ordu	Unye	–
1989	Atatürk	Adıyaman	Centre	Ovakuyu	1	1	–	Aydin	Soke	Denizcoy
1989	Atatürk	Adıyaman	Kahta	Adali	1	1	–	Aydin	Soke	Denizcoy
1990	Atatürk	Adıyaman	Several	Several	18	18	–	Hatay	Reyhanli	Vazvaza
1990	Atatürk	Adıyaman	Kahta	Geldibuldu	1	1	–	Konya	Sarayonu	Kayioren
1990	Atatürk	Adıyaman	Kahta	Bostanli	2	2	–	Cankiri	Cerkes	Akhasan
1991	Atatürk	Adıyaman	Several	Several	2	2	–	Hatay	Reyhanli	Vazvaza
1993	Atatürk	Adıyaman	Several	Several	18	–	18	Adıyaman	Y. Samsat	–
1993	Atatürk	Şanlıurfa	Bozova	Dikili	1	1	–	Hatay	Kirikhan	Karatas

(Contd.)

238

Table 8.9 Contd.

Year	Name of the dam	Original location			Number of families resettled	Type of settlement		New location		
		Province	District	Village		Rural	Urban	Province	District	Village
1997	Atatürk	Adıyaman	Several	Several	6	6	–	Hatay	Centre	Hasanli
	Total				344	133	211			
1993	Kayraktepec	Icel	Mut	Evren	1	1	–	Hatay	Kirikhan	Karatas
	Total				1	1	–			
1970–1983	Keban				159	93	–	Diyarbakır	Centre	Yolboyu
						40	–	Elazig	Centre	Golkoy
						26	–	Tunceli	Pertek	Bicmekaya
	Total				159	159	–			
1970–1983	Gokcekaya				43	–	43	Eskisehir	Centre	–
	Total				43	–	43			
	Grand Total				1,604	919	685			

Source: DSI, 1999, Dams and Hydroelectric Power Plants in Turkey, Ministry of Energy and Natural Resources, Turkey.

239

as well as the impacts of the construction of the Atatürk Dam on their lives, the lives of their families, and their villages.

The three villages studied in detail were New Samsat, Akpinar and Kizilcapinar. An overall picture of the resettlement process from the perspective of the resettlers, by the villages, is discussed next.

New Samsat

Due to the construction of the Atatürk Dam, one administrative district and its centre, twenty-seven villages and seventy-three arable fields (sub-village) were totally inundated, and eight villages and ten arable fields were partially inundated (Altinbilek et al., 1999a).

The district centre of Samsat was rebuilt above the reservoir level and most of the 5,000 people of the town, as well as some people living nearby had to be relocated to a new village called New Samsat. The new place was established about eight kilometres away from the old town. The population recognized the main differences between the old and the new towns, and between their old and new lifestyles. The people who were interviewed stated that Old Samsat was an unplanned city, their houses were small, and generally made of mud. Transportation and communication networks were not well developed. Access to health, education, and electrical facilities were also highly unsatisfactory. The main economic activity of the old village was agriculture, and the main products were cereals and cotton. In contrast, in New Samsat, infrastructure facilities like roads, energy, water, and communication were excellent. This, unquestionably, was a significant improvement compared to what the population had access to earlier. In addition, their new houses were built with cement, and the people had televisions, refrigerators, telephones, and washing machines in their houses, which were earlier considered to be luxury items. People also now have running water, electrical connections, and better sanitation facilities in their houses. On the negative side, the people felt their current income was lower than before since they now mainly cultivate tobacco because of a lack of water for irrigation.

The local population also pointed out that Old Samsat did not have any medical personnel or medical facilities. The absence of medical facilities and poor transportation facilities, made access to proper health care difficult for the sick, the aged, and pregnant women. This contributed to much suffering. The situation in the new town was totally different. New Samsat currently has three doctors in the town on a permanent basis and there is also a small hospital. There are also two primary school and one high school in New Samsat while earlier there was only one primary school. When the present study was carried out, another primary school was under construction.

When the population moved to New Samsat, the DSI encouraged fishing as a new commercial activity. Fish larvae were provided. However, people did not appear to be interested in pursuing fishing as a profession. Accordingly, there were only a few fishermen. The fish that are caught are primarily used for household consumption. Some of the reasons for the unpopularity of the fishing profession could be inadequate catch, lack of stock and/or proper fishing gear, lack of commercialization, or the fact that people are not aware of the potential of the fishing profession since it is a new activity in the area.

In Old Samsat, the rate of emigration was low, probably because people had steady jobs. At the new location, however, many people are migrating to other cities to get better jobs. Some of the inhabitants appear to have migrated to Şanlıurfa to work in the cotton fields. Tobacco grown in New Samsat is sold to the government. The farmers currently receive technical advice from the Department of Agriculture on various aspects of tobacco farming.

Overall, the people interviewed agreed that their quality of life is better in New Samsat compared to what they had before. This is primarily because the conditions under which they live are better at present than before in many significant ways. However, the people insisted that from an economic viewpoint, the conditions were better earlier. It was not possible to determine if the deterioration in their economic conditions was real, or if it was a complaint made by the people in the hope that doing this could bring them additional new economic benefits from the government.

Akpinar

Akpinar is an old village. Many people from this village worked on to the construction of the Atatürk Dam as non-skilled workers. With the incomes obtained during this construction process, people constructed new houses with better materials (cement, brick, etc.) compared to before (primarily mud), and bought consumer goods like refrigerators, television sets, etc. Some people also bought tractors, or established small shops. Many farmers supplemented their incomes by working at the dam site as a secondary job. There were some fishermen in the town, but the catch was primarily for their own consumption.

The economic and social conditions of the people have improved. This is because before the Atatürk Reservoir was built, they had to grow tobacco under rain-fed conditions. The new reservoir has proved to be a source of irrigation. The farmers currently grow mainly cotton, which has increased their incomes appreciably. Thus, not surprisingly, an overwhelming percentage of people feel that the construction of the dam has been beneficial to the community, although there are some unemployed people who are looking for jobs. One of the benefits of having a higher income has been that the children (including girls, according to the people interviewed) are now being sent to schools, even though most of the population of the town attended only primary school. Absence of higher education and training has meant that most of the people from this village who worked during the construction of the dam were non-skilled. However, since the construction of the Atatürk Dam extended over a period of several years, it was possible for many of the villagers to train as drivers, motor mechanics, etc. With new skills, some of the people are now working as skilled labourers at different infrastructure development projects all over the country, and a few even abroad.

Even though the town has access to water for agricultural activities, it did not have access to drinking water at the time of the study. Their only source of drinking water was a pipe which ran constantly. Drainage at the pipe outlet was a problem, especially in terms of health. The problem could be solved easily either by the local technicians or by the appropriate municipal authority. The villagers had requested the government for help to rectify the situation, but no help was forthcoming when the study was undertaken. The villagers also had taken no initiative to solve this simple problem themselves.

Kizilcapinar Village

This is a village with about fifteen families from Caili village who were still waiting to be resettled in rural areas. They are still living in rented houses paid for by the government. They work as sharecroppers, primarily growing tobacco, which is the same activity they used to carry out before they were displaced. Even though they had limited resources, most of the villagers interviewed mentioned that their children were attending schools in Adıyaman.

On the basis of the interviews and meetings, it appears that the vast majority of the people that had to be resettled due to the construction of the Atatürk Dam were not aware of the overall process of resettlement. Nor did they have much knowledge of the resettlement or expropriation laws or their entitlements under the laws. Some people who are yet to be resettled mentioned that they tried to obtain information several times from the Directorate of Rural Affairs, but without any success. Basically they wanted to know where they were likely to be resettled and when this might take place. They were told that they would be informed when their houses would be ready, and they could move. The government officials had informed them that they were likely to be resettled 'in the future' in Hatay, which did not appear to be the situation as of October 1997. Even though six families were resettled in Hatay in 1996, the information provided by the Directorate of Rural Affairs was that the construction of these houses in Hatay had to be stopped because of social disturbances and unrest. The inhabitants of Hatay did not want people from Kizilcapinar to be resettled there. It seems that the main constraint to resettle the people was the lack of appropriate land. The population in Kizilcapinar had not been informed of this situation, and accordingly, were still expecting to be relocated to Hatay at the time of this study.

Conclusion

During the studies carried out, it was evident that the magnitude and extent of the social and economic impacts generated by the Atatürk Dam and its reservoir have been positive not only for the region but also for the country as a whole. The benefits that are now accruing to the country by the increased electricity generation alone are substantial. Equally, lifestyles of the population living in the project area, and especially of the majority of people living near and around the reservoir have improved very significantly. In retrospect, based on data currently available, the construction of the main infrastructural project in the GAP area, the Atatürk Dam, has acted as the engine for economic growth and integrated regional development in a historically underdeveloped area. The dam and the reservoir system have changed the way of living of the local people, significantly increasing the employment opportunities and improving working conditions. Expanded economic activities have encouraged migration from the rural to the urban areas.

The region has flourished and employment has been generated both during the construction and the operation of the dam. Many of the labourers who initially were unskilled and who were trained during the construction of the dam, became skilled workers, and many became entrepreneurs after the construction was over. Many of them are now working in the construction of dams all over the country. The resulting incomes come back to the region, since they send their incomes back to their families who live in this region.

The benefits of irrigation are visible mainly in Şanlıurfa, where both formal and informal jobs have increased exponentially. However, the construction of the dam did not result only in benefits in terms of employment. The daily exposure of the villagers to the different traditions of 'outsiders' during more than a decade of construction, resulted in the population expressing their ambitions in terms of increased and better housing, transportation, food and health habits, higher education, the decision of the local population to send their children (including girl children) to school, demand for information and communication, etc. The lifestyle of the population has thus started to change in the region, which can result in a better quality of life for the people of the region.

While the construction of this dam and the associated hydraulic structures have contributed significantly to the improvement of the lifestyle of the people of the region, they have had some direct and indirect adverse social and environmental impacts as well. For example, they have led to the resettlement of a large number of people from the inundated area, impoverishment of those who did not manage their expropriation funds properly, loss of productive agricultural land, and an increase in environmental contamination due to accelerating industrialization. The population affected, however, has generally been fairly compensated for their losses.

Interestingly, but perhaps not surprisingly, the overall benefits and costs of the dam were viewed differently at the local level depending upon whether the people were from Adıyaman or Şanlıurfa. Several towns of Adıyaman were affected by the construction of the dam and the reservoir. In the villages visited, the people did not have access to water for irrigation purposes. Thus, they felt that their lands had been expropriated and they were not receiving benefits from the reservoir in terms of irrigation, even though they confirmed that their lifestyles had improved in general compared to what it was before. They felt this was unfair. In contrast, the people living in Şanlıurfa appeared to have a very positive attitude towards the dam and the reservoir.

Since the socio-economic and educational levels of the people who had to be resettled varied widely, it is not possible to draw uniform and universally acceptable conclusions. However, among the positive impacts were improvements in the economic condition and lifestyles of the people who managed their expropriation funds properly. Education of children (including girls) received a significant boost, and fertility rates have started to decline. Fragmentation of families was the main adverse impact with the resettlement processes, when several members of extended families decided to settle down in different places. Impoverishment was a serious issue for those who could not manage their expropriation funds properly.

If the GAP succeeds in achieving its objectives, the development requirements of the region are likely to increase very significantly. However, if these significant and rapid social, economic, and environmental changes cannot be properly managed, environmental degradation in terms of water, soil, and air pollution could become serious problems in several urban communities, which in turn could have an adverse impact on the sustainability of the project, and on the health of the local people and the ecosystems. With such a multiplicity of opportunities and problems, it is important to identify specific priorities, policies, and actions that are needed to address the most immediate and critical issues related to sustainable urban development, including the development and facilitation of appropriate social and technological infrastructures.

Although not easy, the integration of environmental considerations within the planning processes of the region would certainly result in better living conditions for the population and in the protection of its natural resources. Experiences from other parts of the world indicate that large rural and urban public investment projects, when not properly implemented and managed, can adversely affect ecosystems and, in the long term, the socio-economic development of the region. Regional strategies to manage water, land, air, and biotic resources need the formulation, integration, and implementation of appropriate land use policies and practices for both urban and rural areas that could be affected by the development projects. While formulation of rational land use policies should be possible, global experiences indicate that generally they are very difficult to implement even under the best of circumstances. This could thus be a potential problem for the GAP region.

No large-scale infrastructure development project is possible anywhere in the world which has only positive impacts. However, even though some negative impacts are unavoidable, social and environmental impacts should be looked at in an organized way so that they can be mitigated properly. Accordingly, projects and activities must be selected such that the positive impacts significantly outweigh the negative costs.

Overall, the main problem in the relocation of populations is the scarcity of land where a group of people can be resettled, and not the availability of funds for compensation. Experiences from different water development projects from different parts of the world indicate that many people prefer to receive the appropriate financial compensation and then plan and organize their own resettlements in areas of their choice. However, this creates the problem of unskilled rural people unfamiliar with money and investment matters being unable to make proper use of the compensation funds received. They often end up in conditions worse than before. In view of the long-term economic efficiency and social acceptability of the process of resettlement, this issue needs to be seriously considered. This problem, however, is not limited to Turkey alone, but also affects nearly all developing countries, for which proper solutions have to be found. In the final analysis, people who are to be resettled must enjoy better conditions than what they had before since they are involuntarily affected by the development projects. In fact, with the increasing scarcity of land necessary for resettlement processes, governments will have to develop appropriate information and communication strategies to educate people on how to invest their expropriation funds, and make them aware of the risks of managing them unwisely.

When properly planned and implemented, resettlement programmes can become part of an overall strategy to reduce poverty in any country. Well-planned investments in new infrastructure and services (water, electricity, schools, hospitals, roads, etc.) represent an opportunity to improve the living standards of the populations directly affected by the development projects. Since the rejection of any project because of involuntary resettlement is unrealistic, it is essential to improve the knowledge base for its planning and the implementation phases in order that the entitlements of those affected are assured and their lifestyles are improved. Resettlement should be approached as a development opportunity, which it is, and not as a neglected stage of an infrastructural construction project. This, although difficult, is not impossible, especially if there are policies and resources, and also an understanding of the complexities, and the benefits and costs, of the resettlement processes from the social, cultural, economic, and ecological viewpoints.

In Turkey, there does not seem to be any negative attitude towards the construction of infrastructural projects because of their impacts on the people and the environment. The projects are understood and appreciated for what they are: essential parts of the development of the country which will improve the lifestyle of the people. There is a concern and a demand, however, that social and environmental issues be taken into consideration in the projects in order to avoid increased poverty for large number of people, unwanted migratory movements, degradation of the environment, and the like.

In Turkey, the institutional framework, as well as the practices for resettlement and rehabilitation are being improved in order to ensure better coordination between the concerned agencies during the planning and implementation stages of resettlement, as well as the rehabilitation processes. Additionally, the sustainability of each dam that is to be constructed is being objectively assessed in order to assure that it would be a win-win solution from the social, economic, environmental, and cultural viewpoints. Those projects which were to be implemented, but were later not considered to be optimal for the sustainable development of any region, have been postponed indefinitely.

The need for the reform of development policies is becoming increasingly apparent and urgent with the passage of time. The present thinking is that development policies should be formulated within a much broader and integrative framework than has been the case in the past. It should be realised that exclusive reliance on techno-economic aspects cannot provide all the necessary conditions which could ensure that the fruits of development policies are enjoyed by all the people concerned in an equitable and sustainable manner. They must go beyond techno-economic considerations and must include social and environmental factors, as well as an objective review of the capacities of the institutions that are likely to implement them.

The GAP Project, if properly planned and implemented, represents an opportunity for the development of one-tenth of the population of Turkey. In order to succeed in such a major and challenging task, development planning and implementation should explicitly consider improvements in the quality of life of the people of the region, fulfilment of their needs and expectations at the national, regional, and local levels, and preservation of the environment.

Meeting the social goals of a country has set a challenging agenda for development at the global level. The GAP Project, by meeting the economic, social, and environmental goals of the south-eastern Anatolia region, has the potential to become an excellent example of people-centred integrated regional development.

References

Akkaya, C., *Turkish Experience on Environmental Impacts of Dams*, DSI, Ankara, 2001.

Altinbilek, D., 'Water and Land Resources Development in Southeastern Turkey', *Water Resources Development*, 13(3), 1997, pp. 311–332.

Altinbilek, D., M. Bayram, and T. Hazar, 'The New Approach to Development Project-Induced Resettlement in Turkey', *Water Resources Development*, 15(3), 1999a, pp. 291–300.

—— 'The New Approach to Reservoir-Induced Resettlement and Expropriation in Turkey', in M. Turfan (ed.) *Benefits of and Concerns about Dams. Case Studies*, 67th Annual Meeting of International Commission of Large Dams (ICOLD), Antalya, 1999b.

Akpinar Construction Company, *Annual Gross Salary for Personnel Working at Akpinar Construction Company*, Turkey, 1997.

ATA Construction Company, *Number of people employed at Atatürk Dam Site, 1983–1986*, Turkey, 1996a.

——, *Monthly Gross Salaries of Unskilled Workers at Atatürk Dam, 1984–1996*, Turkey, 1996b.

Biswas, A.K. and C. Tortajada, 'Turkey´s Southeastern Anatolia Project: An Innovative Approach to Sustainable Regional Development', *Ecodecision*, Royal Society of Canada, Spring, 24, 1997, pp. 61–3.

—— *Rapid Appraisal of Social, Economic and Environmental Impacts of the Atatürk Dam*, GAP/UNDP, Ankara, 1999.

Chamber of Agricultural Engineers of Turkey, *Trends of Social Change in GAP Region*, GAP Regional Development Administration, Ankara, 1993.

Development Foundation of Turkey, *Women Status and their Integration to Development Process in GAP Region*, GAP Regional Development Administration, Ankara, 1994.

DSI, *Dams and Hydroelectric Power Plants in Turkey*, General Directorate of State Hydraulic Works, Ministry of Energy and Natural Resources, Ankara, Turkey, 1999.

—— *DSI in Brief*, General Directorate of State Hydraulic Works, Ministry of Energy and Natural Resources, Ankara, Turkey, 2000.

GAP Administration, *Southeastern Anatolia Project Social Action Plan*, GAP Regional Development Administration, Prime Ministry, Ankara, Turkey, 1995.

—— *Social Policy Objectives*, GAP Regional Development Administration, Prime Ministry, Ankara, 1998.

—— *Latest State in Southeastern Anatolia Project*, GAP Regional Development Administration, Prime Ministry, Ankara, 1999a.

—— The Atatürk Dam and Hydroelectric Power Plant (HPP), Information Series, GAP Regional Development Administration, Prime Ministry, Ankara, 1999b.

Harmancioglu, N., N. Alpaslan, and E. Boelee, *Irrigation, Health and Environment, A Review of Literature from Turkey*, Working Paper 6, Colombo: International Water Management Institute, 2001.

METU, Department of Sociology, *MOM Project Socio-economic Studies*, GAP Regional Development Administration, Ankara, 1993.

—— *Population Movements in the Southeastern Anatolia Project Region*, GAP Regional Development Administration, Ankara, 1994a.

METU, *Socio-Economic Survey on the Management, Maintenance and Administration Project*, GAP Regional Development Administration, Ankara, 1994b.

METU, Centre for Research and Assessment of the Historic Environment, 'Salvage Projects of the Archaeological Heritage of the Ilisu and Carchemish Dam Reservoirs. Activities in 1998', in Tuna, N. and J. Ozturk (ed.), Middle East Technical University, Ankara, 1999.

Safak, N., G. Berk, S. Buyukkusoglu, and Z. Oztekin, *Assessment of the Stock of Water Products and Fishing Grounds in the Atatürk Dam Lake*, DSI, Ankara, 1994.

Safak, N., N. Dirgin, and O.M. Tufek, 'Reservoir Fisheries in Turkey' in M. Turfan (ed.), *Benefits of and Concerns about Dams. Case Studies*, 67th Annual Meeting of International Commission of Large Dams (ICOLD), Antalya, 1999.

Sociology Association, *Survey on the Problems of Employment and Resettlement in the Areas which will be Affected by Damned Lakes in GAP Region*, GAP Regional Development Administration, Ankara, 1994.

Tortajada, C., 'Evaluation of Actual Impacts of the Atatürk Dam', *Water Resources Development*, 16(4), 2000, pp. 453–64.

Turan, F., *Benefits of Water-based Development Projects in Turkey, International Workshop on Water-based Development Projects, Experiences in the World*, GAP Regional Development Administration, Şanlıurfa, 8–10 November 1999.

Ünver, O., 'South-eastern Anatolia Integrated Development Project (GAP), Turkey; An Overview of Issues of Sustainability', *Water Resources Development*, 13(2), 1997, pp. 187–207.

—— 'GAP: A Pioneering Model for Water-based Regional Development', Paper presented at the International Workshop on Water-based Development Projects, Experiences in the World, GAP Regional Development Administration, Şanlıurfa, 8–10 November 1999.

—— 'The Southeastern Anatolia Project (GAP): An Overview' in *Water and Development in Southeastern Anatolia, Essays on the Ilisu Dam and GAP*, Proceedings of a Seminar held at the Turkish Embassy in London, 2000a.
—— Southeastern Anatolia Project and Financing Hydropower Projects, Second World Water Forum, The Hague, 2000b.

9

Evolution of Water Management Institutions in Select Southern African International River Basins

Anthony R. Turton

The effort required to reach any agreement increases by the cube of the number of parties involved.

—Peter Pyke

Introduction

Southern Africa has a history of political tension and civil strife which was driven in part by the anti-colonial struggle for liberation, but also fuelled, to a large extent, by the political dynamics of the cold war. As such, southern Africa became a theatre in which the main cold war rivals could conduct their proxy wars, thereby establishing a linkage between the sub-national and international political domains. In this regard southern Africa resembles a number of other regions in the world. What is significant, however, is that southern Africa has a disproportionately large number of international river basins that are shared between two or more riparian states. An analysis of institutional development in these international river basins therefore becomes instructive because it reflects the impacts of the tensions of the cold war. A central tenet to this thesis is that one cannot understand the patterns of institutional development in the water sector without contextualizing them within the broader political domain. This is important because in the post-cold-war era, there is a propensity towards regional cooperation and integration, but the impediment of these historic influences still remains. Given that southern Africa has so many shared international river basins, the proper management of water resources could become one of the major platforms on which sustained economic growth and prosperity could be developed (Chenje and Johnson, 1996: 151; Granit, 2000). In order to achieve this, however, a win-win solution is needed, which is difficult to find given the legacy of regional tension in the cold-war era. The need to unravel the mysteries of this transition from a zero-sum to a plus-sum outcome has prompted a research project that seeks to develop new knowledge on the political dynamics of institutional development in trans-boundary rivers (Turton, 2002a, 2002b). This chapter will examine the institutional dynamics in five river basins in southern Africa, using the conceptual development that has been taking place in this research programme (Turton, 2002a), in an attempt to explain, and possibly predict, strategic trends.

Regional Water Resources in Southern Africa

Southern Africa is not that easy to define, despite the seductive simplicity seemingly inherent in the name. If it were defined as Africa south of the equator, this would include Gabon, Congo, Rwanda, Burundi, Kenya, and Somalia, all of which are not commonly regarded as being in 'southern Africa'. For the purposes of this chapter, southern Africa will be defined as the area of mainland Africa covering those countries which are members of the Southern African Development Community (SADC). Even this is problematic, because it includes islands such as Mauritius, the Democratic Republic of Congo (DRC), and Tanzania, neither of which is generally regarded as being southern African. This small but significant fact betrays the complexity inherent in institutional development in the water sector at the international level.

The southern African region (which for brevity will simply be called 'southern Africa') as defined here has one unique feature. Approximately 70 per cent of the total land area is located in international river basins that are shared between two or more countries (Granit, 2000). Fifteen major international river basins occur in southern Africa. These are presented in Table 9.1, which for the purposes of this chapter will be considered to be authoritative. It should be noted, however, that this is not an exhaustive list. Heyns (Ohlsson, 1995: 51) identifies fourteen, including the Nata River basin that is shared between Zimbabwe and Botswana (which has relevance to the Okavango

Table 9.1: International river basins in southern Africa
(after Pallett, 1997: 71)

River basin	Number of states	Riparian states
Buzi	2	Zimbabwe, Mozambique
Cunene	2	Angola, Namibia
Cuvelai	2	Angola, Namibia
Incomati	3	South Africa, Swaziland, Mozambique
Limpopo	4	Botswana, South Africa, Zimbabwe, Mozambique
Maputo	3	South Africa, Swaziland, Mozambique
Nile	10	Tanzania, Burundi, Rwanda, Kenya, Uganda, Democratic Republic of Congo (DRC), Eritrea, Ethiopia, Sudan, Egypt
Okavango	4	Angola, Namibia, Zimbabwe, Botswana
Orange	4	Lesotho, South Africa, Botswana, Namibia
Pungué	2	Zimbabwe, Mozambique
Rovuma	3	Tanzania, Malawi, Mozambique
Save	2	Zimbabwe, Mozambique
Umbeluzi	2	Swaziland, Mozambique
Zaire (Congo)	9	Burundi, Rwanda, Central African Republic (CAR), Tanzania, Cameroon, Congo, DRC, Zambia, Angola
Zambezi	8	Angola, Namibia, Botswana, Zimbabwe, Zambia, Malawi, Tanzania, Mozambique

Source: Paliet, *Sharing Water in South Africa*, DRFN, Windhoek 1997.

Basin case study), whereas Chenje and Johnson (1996: 166) list eleven, including the Songwe that is shared between Malawi and Tanzania. This situation hints at the hydro-political complexity found in southern Africa.

Granit (2000) notes that the water resources in southern Africa are relatively scarce and therefore valuable. The World Bank (2000) suggests that an economic growth rate of 5 per cent is needed in southern Africa if the region is to reduce poverty and counter the trend of existing population growth. Current economic growth rates are in the order of 1–2 per cent (with some exceptions), which is far short of the needed level. This has led some commentators (Granit, 2000; Scudder et al., 1993: 262) to conclude that water could prove to be a limiting factor for the attainment of sustainable economic growth, as well as become a source of tension between states (Buzan, 1991: 132). Four of the most economically developed states in the region—South Africa, Botswana, Namibia, and Zimbabwe—are approaching the limit of their readily available water resources. Downstream riparians in particular are becoming increasingly concerned with threats to their development agenda (Heyns, 1995: 1; Vas and Pereira, 1998: 112) and ultimately their sovereignty (Granit, 2000). This means that water resource management in southern Africa's international basins is rapidly becoming a strategic issue, with the risk of securitizing the management of those resources (Turton, 2001a, 2002a). This is an undesirable condition as it has been shown that securitization tends to result in a zero-sum outcome, partly because it is ultimately unsustainable, and partly because it undermines the development of institutions (Turton, 2002a).

Patterns of Conflict in Southern Africa

In the view of the author, two quotations accurately portray the historic background that contextualizes the genesis of the patterns of conflict and cooperation found in southern Africa over the last quarter-century.

Except for Angola, the black states in the region are economically closely tied to, and in many cases heavily dependent on, South Africa. In military terms, [South Africa] is ... the regional leviathan. On the political/ideological level, South Africa is seen by the black states as the last remnant of racism and white minority rule in Africa. In addition, South Africa is regarded as something of a colonial power too, maintaining its control of Namibia. ... Being economically dominant is a feature which in itself can easily give a state the image of a 'bully'. Add to this [South Africa's] military supremacy and its political/ideological distance from its neighbours, and the scene is set for heavily strained relationships. The black states also widely believe that South Africa is bent on keeping them economically and militarily weak ... [which] they see as part of South Africa's strategy to create a regional environment conducive to the perpetuation of the status quo in [South Africa]. ... Relations between South Africa and the black states are, on both sides, characterized by suspicion, fear and even a strong dose of paranoia. Each sees its security and stability threatened by the other; each side ... perceives itself the target of destabilization by the other (Geldenhuys, 1983: 48–9).

The development of economic cooperation with South Africa, including possibly water supply from the Okavango, is likely to reinforce the respect of mutual interests which exists [between South Africa and Botswana]. A security agreement is pending. ... Water, amongst other things, is an issue between Lesotho and South Africa. Pretoria has used failure to reach a firm agreement with Lesotho on security issues, ... to delay a feasibility study on the Highlands Water Project. ... South African technicians involved were actually withdrawn from Lesotho at that time. ... [T]his is an excellent example of the two-pronged approach of South Africa to its problems— military strength, which has actually been used against targets in Lesotho, coupled with the carrot of development (Gutteridge, 1985a: 100).

In attempting to analyse the dynamics of institutional development in the water sector, the argument presented by Homer-Dixon (1996: 363) that South Africa is a 'pivotal state' will be used. To accomplish this, a brief description of three distinct phases of political dynamics will be made, in order to lay the foundation for the subsequent analysis of five river basins. These basins have been chosen because they link the four most economically developed states in the region into what can be described as a hydro-political security complex (Schulz, 1995; Turton, 2001b, 2001c, 2002c). As such they all have strategic significance for their respective riparians. These basins are the Orange, Limpopo, Okavango, Incomati, and Maputo.

The political dynamics of southern Africa can be broken down into three distinct periods of historic time.

Genesis of Conflict

This covers the period prior to 1974. The roots of this conflict date back to the Anglo-Boer War, but for brevity this will be excluded. In 1948 the National Party (NP) won the election in South Africa. This was seen as a major triumph for the Afrikaners who had strong nationalist sentiments, many of whom still had living memory of their defeat in the Anglo-Boer War and their subsequent treatment in the British concentration camps. So in 1948 the political power that was lost in the Boer War was returned to the Afrikaners, who immediately set about consolidating their position by implementing the policy of Grand Apartheid (Turton, 1999). Economic development was high on the agenda, but without water this would be impossible. For this reason early reconnaissance work was begun on the hydrology of Basutoland (Lesotho after independence) as a possible source of water for the South African goldfields (Ninham Shand, 1956) and their related industrial complex. It was against this socio-economic background that British Prime Minister Harold Macmillan made his 'Winds of Change' speech in the Cape Town Parliament (Geldenhuys, 1984: 11), which referred to the strong desire for independence that was emerging in the former colonies. This set the scene for South Africa's systematic isolation. Shortly after Macmillan's speech, events unfolded rapidly in the wake of the tragic Sharpeville massacre in which sixty-nine people were killed and 180 wounded (Spitz and Chaskalson, 2000: 7). The United Nations (UN) Security Council adopted a resolution that mandated the visit by Secretary General Hammarskjöld to South Africa. The 1961 Commonwealth Conference in London saw Prime Minister Verwoerd trying to defend South Africa's racial policies (Geldenhuys, 1984: 24), leading ultimately to its expulsion and the independence of South Africa from Britain (Geldenhuys, 1990: 205), which for many Afrikaners was a final vindication of their defeat in the Boer War. This laid the foundation for what can be described as a 'garrison state' (Frankel, 1984: 30). When Ian Smith, the Rhodesian prime minister, made his Unilateral Declaration of Independence (UDI) from Britain, South Africa offered its support. This determined the patterns of conflict that were to be unleashed from that moment onwards.

Sharpeville also reverberated across South Africa, with the banning of the African National Congress (ANC), Pan Africanist Congress (PAC), and the imprisonment of leaders like Nelson Mandela and others. This dramatic series of events saw the birth of the so-called 'armed struggle' in the face of the apparent failure of Gandhian passive resistance. The now famous 1964 'Rivonia Trial' that convicted Nelson Mandela and others (Spitz and Chaskalson, 2000: 7) was a direct outcome of this series of events. On the water and development side of the hydro-political equation, South Africa's focus of attention again turned to Lesotho, but this time as a source of water for the rapidly growing industrial complex in the Witwatersrand that was outstripping the capacity of the Vaal Basin (Carter, 1965; Young, 1961). At the same time, the Orange River Project (ORP) was launched (Conley and van Niekerk, 1998: 145). This has a profoundly political undertone to it as it was designed to restore investor confidence in South Africa in the post-Sharpeville period, and it can probably be regarded as the birth of the South African hydraulic mission. The one dam in the ORP, which was called the H.F. Verwoerd Dam as testimony to the political significance of this project, was completed in 1971 and is the largest reservoir by volume in South Africa (Conley and van Niekerk, 1998: 145). It has since been renamed the Gariep Dam. The 5.35m diameter, 8.28km long delivery tunnel was the longest in the world at that time

(Conley and van Niekerk, 1998: 145). The ORP was hailed as a triumph of Afrikaner independence and technical ingenuity.

In 1966, the guerilla war was launched in Namibia, significantly drawing South Africa's security forces into the Caprivi Strip (Frankel, 1984: 102) where the Okavango and Zambezi form a water-rich haven in the midst of surrounding aridity. Faced with this reality, which was manifest as increasing isolation for South Africa, diplomatic contact with what was known as 'Black Africa' was deemed to be vital. One of the targets of this period of détente or peaceful coexistence was Chief Leabua Jonathan, who was destined to become the prime minister of Lesotho, which gained its independence in 1966 (Geldenhuys, 1984: 19). Strong relations were forged with him, and he was even regarded as being a South African protégé, until relations began to sour (Geldenhuys, 1983: 48). In an attempt to divert growing criticism of his own domestic political style, Jonathan became one of South Africa's strongest critics, openly declaring his support to the 'liberation struggle'. This was a diplomatic slap on the face for South Africa.

Early aspects of the water, economic development, and energy connection can be found in two agreements between South Africa and Portugal during 1969. The first was on the Cabora Bassa Project on the Zambezi River in Mozambique (Treaty, 1969a), while the second was on the Cunene River (Treaty, 1969b).

In 1970, the Jonathan regime was toppled in a military coup d'état and Lesotho was plunged into political crisis. The State Security Council (SSC) was established in South Africa during 1972 (Gutteridge, 1994: 215), which was later to become an extremely important organ in the formulation of South Africa's foreign policy (Geldenhuys, 1984: 93). The end of this period was characterized by the deterioration in the threat perception and the publishing of the 1973 White Paper on Defence, which for the first time introduced the concept of a 'total strategy' (Geldenhuys, 1984: 140). The height of the détente era occurred in 1975 with the Victoria Falls Bridge meeting between the intransigent Rhodesian Government and Black Nationalists, which had been made possible by the interaction between Prime Minister Vorster of South Africa and President Kaunda of Zambia (Geldenhuys, 1984: 39).

From Détente to Total Onslaught

This covers the period 1974 to 1990. Similar to the 1960 period, when a series of events rapidly shaped a transition phase, 1974 can be called a watershed year. The start of this was signaled by the coup d'état in Portugal (Geldenhuys, 1984: 78). This event, taking place thousands of kilometres away from southern Africa, set off a domino effect that was associated with the rapid decolonization of the former Portuguese territories. 'White Africa' was getting smaller, so Macmillan's 'Winds of Change' were apparently real. The fact that in each of the former Portuguese colonies, there was an unfinished war of liberation, and the speed with which the decision was made and executed, left no time to prepare for an interim administration. The effect was startling. Overnight, the Angolan War of Liberation turned into the Angolan Civil War, which has raged on until today, a quarter-century later. South Africa was drawn into this with the launch of Operation Savannah (Geldenhuys, 1984: 79). Alarm bells were sounded as the regional balance of power changed overnight (Gutteridge, 1983: 35). The Rhodesian Bush War was already ongoing, and suddenly a second front was opened up along the border with Mozambique. The Cabora Bassa Project immediately became a target for military attack, with the long power lines proving impossible to defend. This drew in South African military support, further strengthening the garrison state mentality that had already taken root in South Africa (Frankel, 1984: 30). Suddenly the South African border with Mozambique, some 400 km from Pretoria, became a military front line. Enthused by this series of events, youths took to the streets, angry at the apparent inability of the older generation to liberate South Africa, and the now famous Soweto Riots occurred on 16 June 1976. As with the Sharpeville Massacre, retaliation by the security forces was brutal.

The 1977 Defence White Paper was largely devoted to refining the concept of a total national strategy, first mooted in 1973, as an official policy. This defined a 'total national strategy' as being 'the comprehensive plan to utilize all the means available to a state according to an integrated pattern in order to achieve the national aims within the framework of specific policies' (Geldenhuys, 1984: 140). This total strategy had its roots in the counter-revolutionary experiences of the Americans in Vietnam, the British in Malaya, and the French in Algeria and Indo-China. The term 'total strategy' is derived directly from André Beaufre's work *An Introduction to Strategy* (Frankel, 1984: 46). As such it resonated well with the security elites in the emerging south African garrison state, with its peculiar threat perception that interpreted the Cuban and East German support of the African liberation movements in southern Africa, as being evidence of a 'total onslaught', driven by

Soviet imperialism (Frankel, 1984: 55). This saw the development of a two-pronged approach to security-related issues, and heralded the start of the gradual securitization of water resource management. One element of this approach was based on a strong military response to any threat, supported by destabilization via economic means (Gutteridge, 1983: 38). The olive branch of economic 'development'—as the second element of this approach—thus became securitized, with far-reaching ramifications that are being felt today. This was given greater structure when P.W. Botha produced a twelve-point plan for survival at the NP Congress in 1979 (Gutteridge, 1985a: 93).

Central to this 'total strategy' was economic development and the resultant dependencies that would emerge from this. The foundation for this thinking can be traced to the speech made by Prime Minister Vorster in 1974, in which he spoke of a 'power block' of states (Geldenhuys, 1984: 39). This was subsequently refined when he spoke of a constellation of politically independent states maintaining close economic ties. When P.W. Botha came to power, he used what he called a Constellation of Southern African States (CONSAS) as the basis of his policy (Geldenhuys, 1984: 41). Foreign Minister Pik Botha subsequently announced in 1979 that this vision embraced some forty million people south of the Cunene and Zambezi Rivers all joining forces to design a common approach to the security, economic, and political field. During the same year, a scheme to divert up to $3,000 \times 10^6 m^3$ of water from the Zambezi, via the Thamalakane and Boteti rivers in the lower Okavango Basin downstream of the Delta was found to be economically competitive with the Tugela-Vaal scheme (Midgley, 1987: 15; Scudder et al., 1993: 263). The nexus between water, economic development, and state security was becoming stronger.

At the Lancaster House Conference in late 1979, the foundation was laid for the cessation of hostilities in Rhodesia. Bishop Abel Muzorewa was widely tipped to win the elections. It therefore came as a great shock to the security elites in Pretoria when Robert Mugabe swept to victory in 1980. Mugabe immediately announced that he had no intention of joining the proposed CONSAS. Instead, Zimbabwe, along with Botswana, Lesotho, Swaziland, Mozambique, Angola, Zambia, Malawi, and Tanzania, joined forces in the Southern African Development Coordination Conference (SADCC), which was formally launched in Lusaka in 1980 (Pallett, 1997: 70). This new grouping was specifically designed to reduce their combined dependence on South Africa (Baynham, 1989: 88; Conley and van Niekerk, 1998: 145), and was quickly dubbed the 'counter-constellation' (Geldenhuys, 1984: 41). The establishment of the SADCC was thus a direct response to South Africa's policy of destabilization. The linkage between water and development became manifest at the fourth SADCC consultative conference which was held in Lusaka during 1984. Opening the conference, President Kaunda of Zambia said that the effects of water scarcity and drought had resulted in food deficits and poor prospects for agricultural development in southern Africa (Africa, 1984).

The emergence of this total strategy approach saw South Africa's foreign policy becoming captive to the SSC (Frankel, 1984: 149), which had an all-consuming security focus to it. Seen in this light, every aspect of foreign relations became securitized, including cooperation over water resources. An example of the impact of the so-called total strategy in the water sector can be found in a paper that was written by the chief engineer of the Rand Water Board, who used the concept to contextualize the need for the South African economic heartland to gain access to secure supplies of water (James, 1980). Listed in this document are inter-basin transfer (IBT) schemes such as the Lesotho Highlands Water Project (LHWP), the Tugela-Vaal link, and the mooted Okavango development. Significantly, gaining access to the Okavango is referred to in the context of CONSAS (Blanchon, 2001: 123), thereby establishing water as a strategic element of this overall policy. This is the basis of the so-called 'pipelines of power thesis' (Turton, 2000).

In 1980, the armed struggle intensified after an announcement to this effect by the ANC during festivities to mark the occasion of its seventieth anniversary (Gutteridge, 1990: 167). A hostage incident at a bank in Silverton involving guerillas, along with a rocket attack on the SASOL oil-from-coal refinery, and the subsequent derailment of a train near Richards Bay, all came in quick succession (Gutteridge, 1981: 5). This was punctuated by the political energy that the newly independent Zimbabwe had given to the creation of SADCC (Geldenhuys, 1984: 41), which was 'devoted to mutual cooperation for development, and the reduction of members' collective dependence on South Africa' (Simon, 1991: 205) and therefore seen as further evidence of the 'total onslaught'.

In 1981, the first military retaliation was launched (Geldenhuys, 1984: 140), with a South African Defence Force (SADF) Special Forces raid on ANC bases at Matola near Maputo in Mozambique (Gutteridge, 1981: 14). This was followed in 1982 with retaliatory attacks against ANC targets in Maseru, Lesotho (Gutteridge, 1983: 35). These signalled that South Africa was not prepared to countenance what they perceived as terrorist or guerilla bases in neighbouring states (Geldenhuys, 1983: 47) using rhetoric that resembles the contemporary American-led 'War on Terror'. This was manifest in subsequent attacks on targets in Angola, Mozambique, Lesotho, Botswana, and Zimbabwe. A southern African defence zone was conceived embracing Namibia, Botswana, Swaziland, and Zimbabwe to counter the presence of East German troops in Angola and Mozambique (Gutteridge, 1981: 19). This marked an escalation in South African destabilization tactics, involving military action and economic pressure (Geldenhuys, 1983: 43). Swaziland was seemingly exempt from this practice (Geldenhuys, 1983: 46). This approach simply strengthened the SADCC resolve to liberate their members from the stranglehold of South African economic power (Geldenhuys, 1983: 47).

In 1983, a car bomb was detonated in front of the military intelligence headquarters in Pretoria with significant casualties. This marked the escalation of the conflict into previously neutral areas, as evidenced by the subsequent bombing of the ANC offices in London, assassination attempts on ANC figures in Brussels, and the assassination of Dulcie September in Paris. Elements of this were revealed during testimony to the Truth and Reconciliation Commission (TRC). This is mentioned to illustrate the gravity of the situation, as it was being interpreted by decision-making elites in South Africa at that time (Turton and Bernhardt, 1998). This series of events had unforeseen circumstances, and South Africa increasingly became isolated (Geldenhuys, 1990: 206) as a pariah state, associating closely with the experiences of Taiwan and Israel (Frankel, 1984: 65). Central to this was the notion that these three states were strategic pillars against a global Marxist onslaught that the Free World would not be able to ignore (Geldenhuys, 1984: 116).

This round of tit-for-tat exchanges ushered in a new era when in 1984 the South African Constitution was changed and P.W. Botha was elevated to the status of executive president. During the build-up to this constitutional watershed event, a security agreement between South Africa and Swaziland was reached (Treaty, 1982), supported by an economic cooperation agreement (Treaty, 1983a). This was followed shortly afterwards by the Nkomati Peace Accords (Treaty, 1984a) signed by President Samora Machel of Mozambique and Prime Minister P.W. Botha of South Africa, in March 1984 (Gutteridge, 1985a: 94). Water was intimately linked to the Nkomati Accord process when an agreement was signed during May in Cape Town between Mozambique, Portugal, and South Africa on the revival of the Cabora Bassa Project (Treaty, 1984b). Similar security agreements were mooted between South Africa and Botswana, where economic cooperation and possible access to the Okavango River was discussed; and between South Africa and Lesotho, where access to water was also a feature (Gutteridge, 1985a: 100). Water and energy were increasingly being seen as elements of this 'total strategy'. The need for such a security agreement was underscored by unrest within South Africa that was escalating uncontrollably, with the SADF being committed to internal riot control. This started to blur the lines between police and army responsibilities. The ANC held a high-level meeting in Kabwe, Zambia, during June 1985 at which time a decision was taken to allow attacks on soft targets (Gutteridge, 1985b: 129). Special forces reprisal was launched in Kabwe a few days later. A state of emergency was announced in 1985, giving security forces wider powers (Gutteridge, 1985b: 124). The ANC leadership started to regard the internal situation as a 'peoples' war' from this moment onwards (Gutteridge, 1985b: 130). The situation deteriorated rapidly with a flight of foreign capital threatening a total collapse of the economy. All foreign currency trading was suspended in South Africa on 27 August 1985 (Gutteridge, 1985b: 144).

During 1986, violence erupted in Natal between comrades from the ANC and Zulu impis from Chief Buthelezi's Inkatha Freedom Party (IFP) (Gutteridge, 1990: 168). This degenerated into a localized low-intensity civil war, which endured until after the election of Nelson Mandela as the first democratic president of South Africa (Percival and Homer-Dixon, 1995: 3). In neighbouring Lesotho, Major General Justin Lekhanya overthrew Leabua Jonathan during a military coup d'état on 20 January 1986 (Esterhuysen, 1992: 46; Lawrence, 1986). Shortly after this the treaty on the Lesotho Highlands Water Project was signed on 24 October 1986 between 'Pik' Botha of South Africa and Col. Thaabe Letsie of Lesotho (Treaty, 1986a), fuelling speculation about possible South African involvement (Homer-Dixon, 1994: 19). Commentary on this project from that time reflects the socio-economic benefit angle that was central to the 'total strategy' approach (Vorster, 1988: 95). During 1987, another study on the feasibility of transferring water from the Zambezi via Botswana found that the cost of water delivered to Pretoria was competitive with existing water supply schemes (Midgley, 1987: 15; Scudder et al., 1993: 263). This plan had been developed from earlier studies (Borchert and Kempe, 1985), with the most refined version consisting of a 1,116-km concrete structure feeding 2,500 × 10^6m^3 of water from the Zambezi/Chobe confluence via Botswana to a dam in South Africa, from where it would be reticulated to the Vaal River Supply Area (Borchert, 1987; Scudder et al., 1993: 268). This water was needed to meet the estimated demand by 2015 even with the LHWP functioning (Scudder et al., 1993: 268; Williams, 1986). Botswana would have been supplied $60 × 10^6\text{m}^3$ from this aqueduct (Borchert and Kemp, 1985; Scudder et al., 1993: 268). The existing trans-boundary water supply from the Molatedi Dam in South Africa to Gaborone should be seen in light of this total strategy approach.

Significantly in the context of this chapter on water institutions, this era drew to an end in the upper reaches of the Okavango River basin, where the battle of Cuito Cuanavale took place in 1988. This battle saw the first significant defeat of the SADF since its initial incursion into Angola during Operation Savannah in 1975–6, thereby shattering the myth of South African invincibility. Although officially denied at first, General Magnus Malan subsequently admitted that this event turned the balance in favour of ditching 'the millstone which Namibia had become' (Simon, 1991: 187).

Post-Cold War Normalization of Political Dynamics

This covers the period from 1990 onwards, and was ushered in by the demise of P.W. Botha and the assumption of power by F.W. de Klerk. On 2 February 1990, de Klerk made a watershed speech in which he appealed for a united South Africa as a way to overcome the divisions of violently conflicting nationalisms (Gutteridge, 1994: 214). This effectively marked the end of the SSC (Spitz and Chaskalson, 2000: 15) and the total onslaught mentality that they had established in South Africa. Almost immediately, Nelson Mandela was released from prison. The security forces had become deeply divided during the latter parts of the 1980s, with hawkish elements of the police and army combining to form the now deeply discredited paramilitary Vlakplaas Unit and Citizens Cooperation Bureau (CCB), whose antics included the bombing of the London ANC offices and the assassination of a senior South-west African People's Organization (SWAPO) activist in Namibia. Dovish elements clustered under the leadership of Dr Neil Barnard, Director General of the National Intelligence Service (NIS), with various special operations units being tasked with the sensitive role of determining the strategic implications, strategies, and pitfalls of a negotiated settlement. This dovish element played a major but low-profile role in establishing the enabling environment in which a number of strategic actions could occur. These included the Cuban troop withdrawal from Angola and the implementation of UN Resolution 435 in Namibia; the Convention for a Democratic South Africa that negotiated the necessary transitional arrangements needed to ensure that the process of democratization could proceed with relative peace and stability; and the cessation of hostilities in Mozambique, in particular bringing RENAMO (Resistencia Nacional Mocambicana [Mozambican National Resistance Movement])into the elections.

Namibian independence followed shortly after the release of Mandela, heralding the end of a liberation struggle that was second in duration only to that of South Africa itself (Simon, 1991: 185). This series of events threatened to outpace SADCC, whose raison d'être was now questioned. A decision was therefore made to transform SADCC into the Southern African Development Community (SADC), which was concluded formally in Windhoek, Namibia, in 1992 (Granit, 2000; Pallett, 1997: 70). A small blemish on South African–Namibian relations emerged in the form of a dispute over the border between the two countries along the shared portion of the Orange River (Maletsky, 1999; Meissner, 2001). The first democratic elections took place in South Africa during 1994, marking the end of isolation and the policy of destabilization. One of the first tasks of the newly elected ANC government was to resume full state control over water, most of which was linked to the land rights of 60,000 white commercial farmers, on behalf of the majority of South Africans (Conley, 1997: 23). Significantly, the first protocol that was agreed on within the context of SADC after the admission of South Africa as a full member was the SADC Protocol on Shared Watercourse Systems (Treaty, 1995), signed in Johannesburg in 1995. This was amended in 1997 in order to incorporate the principles found in the United Nation Convention on the Non-Navigational Uses of International Watercourses (Granit, 2000; Ramoeli, 2002). While this has laid the foundation for greater cooperation in the water sector, economic development is threatened by the current political turmoil in Zimbabwe, and the ongoing civil wars in Angola and the DRC (Granit, 2000).

In 1998, political instability again erupted in Lesotho. This became a major test for the SADC in general, and South Africa in particular given its past history. The SADC decided to send in a peacekeeping force, comprising soldiers from South Africa and Botswana. This became known as Operation Boleas, which moved across the border and immediately came under heavy and unanticipated fire. Boleas forces split into two with one element concentrating on Maseru while the other moved in to secure the infrastructure related to the LHWP. A number of casualties were sustained on all sides. This has unfortunately caused strained relations between South Africa and Lesotho (Laurence, 1998; Mills, 1998; Mopheme, 1998a, 1998b; Turton, 2002c). For the purposes of our analysis, this is where we will end.

Southern African Regional Politics and the Cold War

If the events noted here contextualize the patterns of regional conflict and cooperation, it is necessary to locate these within the broader theatre of the cold war. The interaction between the sub-regional and international political milieu relates to what Buzan (1991: 216–20) and Buzan et al. (1998: 13–4) have called 'overlay'. The link between these regional dynamics and global political interaction can be traced back to 1957. At that time, a Soviet strategist and expert on economic warfare by the name of Major General A.N. Lagovsky, formulated what became known as the 'weak link principle' (Gutteridge, 1984: 60–1). In terms of this thinking, the Western powers such as NATO and its allies, were almost entirely dependent on a wide range of strategic minerals that were imported from countries in the developing world. By contrast, the Warsaw Pact countries were more or less self-sufficient in those strategic resources. This prompted General Alexander Haig to conclude, in a presentation to the House of Representatives in 1980, that the era of the 'resource war' had arrived (Gutteridge, 1984: 61).

This made southern Africa a theatre in which proxy wars were played out. Central to this was the theory of limited war that had been developed by Henry Kissinger (Dougherty and Pfaltzgraff, 1981: 111–6), who postulated that in the thermonuclear age, the risk of total war was so high that it was to be avoided at all costs. Instead of total war between nuclear powers, a series of local proxy wars could be fought, each allowing global political tensions to be dissipated in a controlled way, and each allowing military technology to be developed and tested through the surrogate forces. The linkage was established after the 1974 Portuguese coup d'état, when through a series of rapid political events, the Angolan war of liberation became the Angolan civil war. South Africa was deeply alarmed at this turn of events. Encouraged by Kissinger's statement that the Soviet and Cuban support of the Popular Movement for the Liberation of Angola (MPLA) in the Angolan civil war was a 'serious matter', and that Moscow's 'hegemonial aspirations' would not be tolerated (Geldenhuys, 1984: 79), South Africa decided to become the American proxy force in the region. This notion of limited warfare was played out in classic fashion when American support for the SADF proxy force, which at that time was literally in sight of Luanda (Turton and Bernhardt, 1998), was suddenly withdrawn. This was seen as a humiliation by South Africa (Gutteridge, 1985a: 97) whose soldiers were left stranded and therefore forced to withdraw without capturing the capital city.

The final link to the cold war is related to the cessation of hostilities associated with the collapse of the former Soviet Union. On the same day in January 1990 on which President Gorbachev was in Vilnius remonstrating with Lithuanian secessionists, President de Klerk was in Umtata trying to persuade General Holomisa to reintegrate the so-called 'independent homeland' of Transkei back into South Africa (Gutteridge, 1990: 176), which was now firmly on the road to negotiations with the previously banned ANC. The demise of apartheid is intimately linked with the collapse of the former Soviet Union and consequently the ending of the cold war. As such, this removed the influence of overlay (Buzan, 1991: 217) and consequently unleashed a set of political dynamics that would start to shape a new pattern of conflict and cooperation (Buzan, 1991: 216–20; Buzan et al., 1998: 13–4), and consequently impact on institutional dynamics within the southern African region.

Regional Water Management Institutions in International River Basins

Given the importance of shared river basins noted earlier, it becomes instructive to assess the relative state of institutional development within some of the more important shared basins listed in Table 9.1. No systematic study has ever been done of the institutional development in southern African trans-boundary river basins. The best approximation of a systematic study known to the author is the work by Heyns (Ohlsson, 1995: 59–60), who listed the more important regimes, and then classified them using an arbitrary set of criteria. There are, however, known factual errors in this study. Significantly, this material immediately found its way into other publications (Chenje and Johnson, 1996: 164–5; Pallett, 1997: 69–70) where it was mostly repeated without any further attempts at analysis or explanation. None of these efforts have shed any light on institutional dynamics.

The concept of an institution can be understood within two contexts. Firstly, institutions can be defined in terms of economic theory. Thus we have Schneller (cited by Furubotn and Richter, 2000: 6) defining an institution as 'sets of formal and informal rules, including their enforcement arrangements' (Perret, forthcoming). Second, a regime has been defined as a set of implicit or explicit principles, norms, rules, and decision-making procedures around which actors' expectations converge in a given area of international relations (Krasner, 1982: 186; 1983: 2). In the water sector, regimes refer to water-sharing agreements and their associated rules, procedures, and institutional arrangements (Turton and Kgathi, 2002). More specifically, a water regime exists when the affected states observe a set of rules designed to reduce the conflict potential caused by the use, pollution, or division of a given water resource; or the reduction of the standing costs; and the observance over time of these rules (Haftendorn, 2000: 65). Seen in this light, a water-sharing regime is a specific species of regime, which in turn is an institution involving sovereign states (Turton, 2002a: 29). Using this definition, we can now examine the evolution of water regimes in southern Africa, by focusing on five key basins that have been selected using the criteria noted here. The evolution of these institutions can then be understood within the context of the broader political scenario presented earlier.

Orange River Basin

The Orange River basin is the most developed of all the rivers in southern Africa (Heyns, 1995: 10). It has a total basin area of 964,000 km^2 with a mean annual run-off (MAR) of $11,200 \times 10^6 m^3$ (Basson, 1999). There are four riparians, with 4 per cent of the basin area lying in Lesotho (upstream riparian), 62 per cent in South Africa, 9 per cent in Botswana, and 25 per cent in Namibia (downstream riparian), (Basson, 1999). The contribution of each riparian to the MAR is unequally distributed, with 55 per cent coming from South Africa, nothing from Botswana, 41 per cent coming from Lesotho and 4 per cent coming from Namibia (Basson, 1999). There are slight variations in this data between the riparians (Savenije and van der Zaag, 1998: 30) but this is not contested in any way, so this variation is irrelevant. The Orange River carries approximately 20 per cent of the total river flow in South Africa, with the Vaal being an important tributary (Basson et al., 1997: 40). The Vaal River is regarded as being a river basin in its own right and provides Gauteng with all of its water. Gauteng (formerly Witwatersrand) in turn houses 40 per cent of the South African population, creates 50 per cent of the country's wealth, and generates 85 per cent of the electricity in the entire country (Conley and van Niekerk, 1998: 146). In order to support this economic activity, the Vaal has links to eight other river basins in a complex arrangement of IBTs that range from the Limpopo in the north, to the Sundays in the south (Heyns, 1995: 18; Turton, 2001b, 2002c). A staggering 100 per cent of the economic activity in Gauteng is reliant on IBTs (Basson et al., 1997: 55). This makes the Orange River of great strategic importance to South Africa, hence the significance of the LHWP (Blanchon, 2001; James, 1980; Davies and Day, 1998: 299–304; Davies et al., 1993: 169). The Orange River is in deficit (Conley, 1996: 17) and further opportunities for development are limited (Conley, 1995: 7). Namibia has expressed an interest in obtaining more water from the Orange, but for transfers to occur the large losses that are experienced in the Lower Orange would have to be taken into account (Conley, 1995: 11). The Orange River forms the border between South Africa and Namibia. There has been confusion over the actual location of the border, with a demarcation in 1890 being the high-water level on the northern bank (Hangula, 1993: 105), effectively depriving Namibia of access (Heyns, 1995: 11). There is a border dispute between South Africa and Namibia as a result of promises that were allegedly made during the run-up to Namibian independence that the border would be moved to the middle of the river (Ashton, 2000: 86–9; Maletsky, 1999; Meissner, 2001: 35). This never took place and allegations are being made that South Africa has reneged on its agreement. This has the potential to further tarnish South Africa's image which was damaged during Operation Boleas in Lesotho.

Institutional development within the basin has been fragmented but intense where it has occurred, reaching degrees of sophistication not evident in any of the other basins under review. It began with the establishment of the Southern African Regional Commission for the Conservation and Utilization of the Soil (SARCCUS) in 1948. This has ten standing committees, one of which deals with water (Ohlsson, 1995: 60). For a long period of time there was no institutional development at all, until 1978 when a Joint Technical Committee (JTC) was established between South Africa and Lesotho to investigate the feasibility of the proposed LHWP (Heyns, 1995: 11). During 1979, the JTC tabled its preliminary feasibility investigation, and a decision was taken to proceed to the next stage of the work (Heyns, 1995: 11). In 1986, the LHWP Treaty was signed (Treaty, 1986a), but this did not constitute a determination of the apportionment of water according to Conley and van Niekerk (1997: 11). This treaty has four protocols covering in detail aspects of design, construction, operation, and maintenance, and the institutional arrangements needed to manage such a complex project. The treaty is the most comprehensive in existence in the southern African water sector, with the main document being eighty-five pages long (compare this to the one-and-a-half-page document establishing the agreement between Mozambique and Swaziland on the Umbeluzi River [Vas and Pereira, 1998: 119]) excluding annexures. From an institutional perspective, this 1986 treaty established two autonomous statutory parastatal bodies (Heyns, 1995: 11). The Lesotho Highlands Development Authority (LHDA) is responsible for the management of the dam construction and related issues within Lesotho itself (Treaty, 1986a: 23–32), whereas the Trans-Caledon Tunnel Authority (TCTA) is responsible for the management of the complex set of delivery tunnels into South Africa (Treaty, 1986a: 33–9). In addition to these, a Joint Permanent Technical Commission (JPTC) was established, consisting of delegates from both riparian states, with the responsibility to coordinate the two parastatals, as well as to report back to their respective governments (Treaty, 1986a: 41–6). Article 10 stipulates that South Africa is responsible for the costs of the project except for the Muela hydroelectric power station, which Lesotho has to pay for (Treaty, 1986a: 47–50). Paragraph 6 of Article 11 stipulates that South Africa will guarantee the loans (Treaty, 1986a: 52). Article 5 stipulates the calculation of royalty payments (Treaty, 1986a: 55–66), which has been determined as half of the difference in cost for supplying $70m^3s^{-1}$ from the Lesotho Highlands Project, and the least cost of the alternative Orange Vaal Transfer Scheme (Treaty, 1986a: 55). Annexure II stipulates minimum quantities of water to be delivered by the LHWP over time, starting with $57 \times 10^6m^3$ in 1995, and ending with $2,208 \times 10^6m^3$ after 2020 (Treaty, 1986a: Annexure II). A related treaty deals with issues of immunity for the JPTC members (Treaty, 1986b).

This was further strengthened in 1999 with the agreement on what became known as Protocol VI, which upgraded the JPTC into the Lesotho Highlands Water Commission (LHWC), (Treaty, 1999a). This in turn resulted in the implementation of a new governance model (Treaty,

1999b) that retained the two parastatal bodies (TCTA and LHDA). This change was the result of a study that highlighted problems with respect to reporting relationships and lines of authority between the LHDA and the JPTC. The final proposals regarding those changes were accepted by the two governments on 22 November 1995, and implemented in the 1999 Agreement (Treaty, 1999b). In essence, the implementation of the new governance model marked the end of the initial construction phase (Phase 1a) and the commencement of water delivery.

In 1992, an agreement was signed between South Africa and Namibia (Treaty, 1992) on the establishment of a Permanent Water Commission (PWC) (Chenje and Johnson, 1996: 165; Pallett, 1997: 70). At the same time an agreement was signed on the establishment of a Joint Irrigation Authority (JIA) to implement the Treaty on the Vioolsdrift and Noordoewer Joint Irrigation Schemes (VNJIS) (Chenje and Johnson, 1996: 165; Pallett, 1997: 70). Subsequent to this, negotiations were started between all of the riparian states, motivated largely by Namibia, on the establishment of a basin-wide regime. This came to fruition when the Orange–Senqu River Commission (ORASECOM) was formally established on 3 November 2000 (Treaty, 2000). This agreement is the fourth basin-wide regime to be established in southern Africa and the first under the SADC Protocol on Shared Watercourse Systems (Treaty, 1995). A significant aspect of ORASECOM is the fact that Botswana is a recognized riparian state, even though it contributes no stream-flow and makes no use of the water from the Orange River (Turton, 2001b, 2002c). This gives Botswana a wider range of diplomatic options by allowing concessions to be granted to other riparian states in return for political support in river basin commissions where they have a greater strategic interest such as in the Limpopo and Okavango Basins. This makes Botswana the balancer of political power in ORASECOM, suggesting the emergence of a hydro-political security complex in southern Africa based on the Okavango, Orange, and Limpopo Rivers shared between the most economically developed states—South Africa, Botswana, Namibia, and Zimbabwe—and their less developed co-riparians Angola, Lesotho, and Mozambique (Turton, 2001b, 2001c).

The ORASECOM (Treaty, 2000) recognizes the Helsinki Rules, the UN Convention on the Non-Navigational Uses of Water and the SADC Water Protocol. It also refers to the amended SADC Protocol (Ramoeli, 2002) with respect to definitions of the key concepts 'equitable and reasonable' and 'significant harm'. Dispute resolution is formally vested in the SADC tribunal, which is a first for institutional development in the water sector. It recognizes the right of the parties to form bilateral arrangements (such as the LHWC and PWC, although these are not mentioned by name), and it says that any new commission will be subordinate to ORASECOM, while existing commissions must liaise with ORASECOM (Treaty, 2000: Article 1, para. 1.4). This means that the LHWC will essentially continue to function as a bilateral arrangement, but that downstream riparians will be kept informed of upstream developments. As such South Africa will still have direct control over its strategic interest in the basin.

Limpopo River Basin

The Limpopo River is also highly developed, with some forty-three dams having a storage capacity of greater than $12 \times 10^6 m^3$ each (Heyns, 1995: 7). Thirteen of these dams have a storage capacity in excess of $100 \times 10^6 m^3$ (Heyns, 1995: 7). It has a total basin area of 183,000 km^2 with an annual MAR of $5,750 \times 10^6 m^3$ (Basson, 1999). There are four riparians, with 20 per cent of the basin area lying in Botswana (upstream riparian), 45 per cent lying in South Africa, 15 per cent in Zimbabwe, and 20 per cent in Mozambique (downstream riparian) (Basson, 1999). Contribution to MAR by each riparian is disputed, with 66–88 per cent coming from South Africa, 3–6 per cent coming from Botswana, 7–16 per cent coming from Zimbabwe, and 9–12 per cent coming from Mozambique, depending on whose data is being used (Basson, 1999; Savenije and van der Zaag, 1998: 30). The water for Gaborone, the industrial hub of Botswana, was initially supplied from South Africa via the Molatedi Dam and associated pipeline, at a rate of $7.3 \times 10^6 m^3 yr^{-1}$, although the design parameters will allow for the delivery of $9 \times 10^6 m^3 yr^{-1}$ (Conley, 1995: 13). A second source of supply has subsequently been developed via the North–South Carrier and the Letsibogo Dam on the Moutloutse River, which is a tributary of the Limpopo.

Institutional development dates back to a general agreement in 1926 between South Africa and Portugal (Treaty, 1926). In 1948, the SARCCUS was established, with relevance to the Limpopo, Incomati, Maputo, and Okavango riparian states. A Second Water Use Agreement was signed in 1964 by South Africa and Portugal, which spoke of rivers of 'mutual interest' including the Cuvelai, Okavango, Limpopo, Maputo, and Incomati (Heyns, 1995: 5; 1996: 263), but which focused in detail on the Cunene (Treaty, 1964). Another formal agreement was reached between

South Africa and Portugal in 1971 for the purposes of constructing the Massingir Dam thirty kilometres downstream of the South African border (Treaty, 1971), but it placed no restrictions on South Africa, recognizing that the inflow would decrease as South Africa developed more dams in the future (Conley, 1995: 13). In February 1983, the Tripartite Permanent Technical Committee (TPTC) was established between Mozambique, South Africa, and Swaziland, with the purpose of making recommendation on the management of the water shortages being experienced in the Limpopo, Incomati, and Maputo Rivers (Chenje and Johnson, 1996: 164; Ohlsson, 1995: 60; Pallett, 1997: 70; Treaty, 1983b). The TPTC did not function from its inception (Heyns, 1995: 7; Ohlsson, 1995: 60), primarily because of the deteriorating political situation at that time (Vas and Pereira, 1998: 119–20), so a bilateral agreement was reached in November 1983 between South Africa, and Botswana establishing the Joint Permanent Technical Committee (JPTC) to deal with matters of mutual interest (Chenje and Johnson, 1996: 164; Pallett, 1997: 70). In 1986, the Limpopo Basin Permanent Technical Committee (LBPTC) was established with Botswana, Mozambique, South Africa, and Zimbabwe as parties (Chenje and Johnson, 1996: 164; Pallett, 1997: 69; Vas and Pereira, 1998: 120), but this did not function well, much like its predecessor the TPTC (Ohlsson 1995: 59). Vas and Pereira (1998: 120) attribute this situation to 'the passive attitude of DNA' (the Mozambican National Department of Water Affairs). This became the second basin-wide agreement to be reached in southern Africa (refer to the TPTC in the Incomati and Maputo case study), but it also failed. Consequently, this was replaced by the upgrading of the 1983 bilateral agreement between South Africa and Botswana to the JPTC on the Limpopo River basin as far as it constitutes the border between the two countries in June 1989 (Chenje and Johnson, 1996: 164; Pallett, 1997: 70). The JPTC is functioning well and has been responsible for the Joint Upper Limpopo Basin Study, which is investigating a range of issues including three possible new dams at Cumberland, Martins Drift, and Pont Drift (Heyns, 1995: 7). Mozambique has since expressed concern over reduced run-off, and renewed attempts are being made to revive the defunct LBPTC, with meetings between Mozambique and Zimbabwe on the issue (Heyns, 1995: 7). Commentators have noted that this is 'clearly an opportunity that DNA must use to have … serious involvement and … active participation' (Vas and Pereira, 1998: 120). This may herald in a new phase of cooperation and institutional development in the basin.

Incomati River Basin

The Incomati River has a total basin area of 50,000 km^2 with an annual MAR of 3,600 × 10^{6}m^3 (Basson, 1999). There are three riparians, with 62 per cent of the basin area lying in South Africa (upstream riparian), 5 per cent lying in Swaziland, and 33 per cent in Mozambique (downstream riparian) (Basson, 1999). Contribution to the MAR by each riparian is disputed, with 64–81 per cent coming from South Africa, 13–20 per cent coming from Swaziland, and 6–16 per cent coming from Mozambique, depending on whose data is being used (Basson, 1999; Savenije and van der Zaag, 1998: 30). This is partly because Mozambique did not get involved in earlier joint studies due to institutional problems and political tensions (Vas and Pereira, 1998: 119). The basin is of great strategic importance to South Africa, because it supports a large amount of economic activity in that country. One of the key elements of this basin is that an IBT is used to sustain the generation of electricity in the adjacent Olifants catchment (a tributary of the Limpopo River), (Ohlsson, 1995: 51) on which a significant portion of the South African economy is dependent. There are consequently a number of dams in this basin, with ten in excess of 12 × 10^{6}m^3 (Heyns, 1995: 6). The combined storage capacity of twenty-two dams in the basin is 400 × 10^{6}m^3/yr^{-1}, with two new dams under development, or having just been completed (Conley, 1995: 22). The Sterkspruit Dam in South Africa has a storage capacity of 167 × 10^{6}m^3 (Heyns, 1995: 6). The stream-flow in this basin is highly variable, ranging in recorded time from 4,926 × 10^{6}m^3 to 28 × 10^{6}m^3 (Conley, 1995: 22). One of the tributaries is the Sabie River, which sustains the Kruger National Park and is probably the most biologically diverse river in South Africa (Davies et al., 1993: 179). In Swaziland, water is diverted into the Umbeluzi River in order to irrigate sugar cane (Heyns, 1995: 7). An unusual aspect of this basin is that South Africa is both an upstream and downstream riparian relative to Swaziland, so dams built in that country increase the yield for subsequent release downstream, and are therefore to South Africa's advantage.

In 1967, Swaziland acceded to the Second Water Use Agreement signed by Portugal and South Africa in 1964 (Heyns, 1995: 5). In 1983, the TPTC became the first basin-wide water regime in southern Africa (refer to the Limpopo case study) applying to the Incomati River, but this did not function (Ohlsson, 1995: 60) because of institutional incapacity and political tension in Mozambique (Vas and Pereira, 1998:

119). As a result of the collapse of the TPTC, a bilateral agreement establishing the Joint Permanent Technical Water Commission between Mozambique and Swaziland was reached in 1991 (Chenje and Johnson, 1996: 164; Pallett, 1997: 70), but it also failed (Ohlsson, 1995: 60). In response to the growing need to develop the water resources, and in recognition of the failure of the TPTC, two bilateral agreements were reached between South Africa and Swaziland in 1992. A treaty was signed between South Africa and Swaziland establishing the Joint Water Commission on 13 March 1992 (Chenje and Johnson, 1996: 165) acting as an advisory body for matters of common interest (Pallett, 1997: 70). The Komati Basin Water Authority was established at the same time, but as a separate institution, and is responsible for the implementation of the Komati River Basin Development Project (Chenje and Johnson, 1996: 165; Pallett, 1997: 70). This bilateral arrangement is functioning well and is responsible for the construction of the Maguga Dam in Swaziland, which is part of a more complex water management scheme involving the recently constructed Driekoppies Dam. Any renewed cooperation between all riparian states will open up the thorny issue of water allocation between the various countries. At present the agreement between South Africa and Mozambique allows a mere $2m^3s^{-1}$ flow at the border (Vas and Pereira, 1998: 120), which is insufficient for Mozambique to develop. Salinization of the estuary in Mozambique is a serious problem during periods of low flow, impacting negatively on ecological and socio-economic aspects.

Maputo River Basin

The Maputo River has a total basin area of 35,000 km^2 with an annual MAR of $3,900 \times 10^6 m^3$ (Basson, 1999). There are three riparians, with 56 per cent of the basin area lying in South Africa (upstream riparian), 34 per cent lying in Swaziland, and 10 per cent in Mozambique (downstream riparian) (Basson, 1999). The contribution to the MAR by each riparian is not disputed, with 56 per cent coming from South Africa, 38 per cent coming from Swaziland, and 6 per cent coming from Mozambique (Basson, 1999). There are six dams with a storage capacity in excess of $12 \times 10^6 m^3$, with the largest being the Pongolapoort Dam in South Africa that inundates part of Swaziland (Heyns, 1995: 8). Ironically, the water that this dam stores has never been used, but it serves to stake a claim over the resource for future development as a bizarre monument to the hydro-politics of the 'total strategy' era.

Currently under consideration are plans to divert this water to other stressed basins inland, but no final decision has yet been taken. There is a significant IBT from the Usuthu catchment for industrial use and the cooling of power stations in the Limpopo and Orange River basins (Heyns, 1995: 8). The Maputo River can therefore be regarded as having been captured by South Africa for transfer elsewhere as the strategic need dictates.

The institutional development of the Maputo River basin is similar to that of the Incomati basin because it involves the same riparian states.

Okavango River Basin

The Okavango is an endoreic river, making it a complicated basin that is hard to understand. There is a paucity of accurate data because of the ongoing Angolan civil war. This is further exacerbated by the fact that significant portions of the basin do not contribute any stream-flow. An additional complication is the lack of agreement as to whether the Nata River and Makgadikgadi salt pans are in fact part of the basin, or whether the river ends at the Okavango delta. Noting the fact that the downstream riparians are both extremely arid (Namibia and Botswana), with the highest resource need, they also contribute the least volume of MAR. Numbers are thus extremely important and therefore highly contestable. It has a total basin area of 333,110 km^2 (CSIR, 1997: 13–31) with an annual MAR of $11,650 \times 10^6 m^3$ (Pitman, in Conley, 1995: 18). Data sequences covering 1933 to 1946 for upstream of the panhandle exist (Scudder et al., 1993: 343) but these are too short to be reliable. The Angolan civil war has prevented the collection of more reliable data. Meissner (2000: 118) has suggested that a deliberate strategy to contain the fighting in the Okavango basin, and therefore away from the more strategically important Cunene hydraulic infrastructure, may exist, but this cannot be confirmed. The MAR data is known to be inaccurate. The most comprehensive available data (CSIR, 1997: 13–44) shows that 142,720 km^2 (42.84 per cent) is considered to be non-functional (contributing no stream-flow). This complicates the relevance of distributional data. There are three riparian states (which is immediately at odds with Table 9.1) if one excludes the Nata River. Angola is the upstream riparian and has a total basin area of 45.39 per cent. Namibia has a total basin area of 37.1 per cent while Botswana (downstream riparian) has a total basin area of 17.51 per cent

(CSIR, 1997: 13–44). If one considers functional areas only (those generating run-off), this situation changes. Angola has 45.21 per cent, Namibia 5.38 per cent, and Botswana 6.57 per cent of the functional area (CSIR, 1997: 13–44). The estimated run-off contribution for each country is 95 per cent for Angola, 3 per cent for Namibia, and 2 per cent for Botswana (CSIR, 1997: 13–46). A staggering 84 per cent of the water that flows into the Okavango delta is lost to evapotranspiration (CSIR, 1997: 13–55), a fact that stimulated an earlier project to 'enhance the yield' by dredging the Boro distributary. This was known as the Southern Okavango Integrated Water Development Project (SOIWDP), (Scudder et al., 1993), which was vigorously opposed by various actors (Conley, 1995: 19), ultimately leading to its cancellation in May 1992 (Heyns, 1995: 10; Scudder et al., 1993: 265). Namibia is somewhat of a unique country because it has no permanently flowing water within its own border, with the exception of the Okavango River that flows for a short distance across the Caprivi strip into Botswana, which is in a similar position. Both countries have plans to develop these resources, which has raised tensions (Ashton, 1999; 2000: 80–2; Ramberg, 1997; *Weekly Mail* and *Guardian* 1996a, 1996b).

In 1969, a Third Water Use Agreement was signed, ostensibly to allow for development of the Cunene River (Heyns, 1995: 5), which also established the Permanent Joint Technical Commission (PJTC), (Shivuti, 2000). This institutionalized contact between Angola and Namibia and therefore facilitated negotiations over the Okavango basin. Very little institutional development occurred from 1969 until 1990 as a result of the guerilla wars in Namibia and Angola. In 1969, construction began on the Eastern National Water Carrier in Namibia, which derives its water from the Cunene River, but with the intention of eventually linking into the Okavango River (Davies et al., 1993: 167; Davies and Day, 1998: 296–9; Heyns, 1995: 10). This became the target of military operations, however, so construction on the Calueque Dam was halted in 1975 (Conley, 1995: 14). During 1990, Namibia and Botswana signed a treaty establishing the Joint Permanent Water Commission (JPWC), (Chenje and Johnson, 1996: 164; Heyns, 1995: 10; Ohlsson, 1995: 59; Scudder et al., 1993: 266; Treaty, 1990). The JPWC focused on the Okavango, the Chobe–Linyanti system in the Zambezi, and on groundwater development (Heyns, 1995: 10). It was bilateral in nature. A parallel bilateral agreement was signed between Namibia and Angola, which endorsed the Third Water Use Agreement that was reached between the colonial powers in 1969 (Heyns, 1995: 6). This reinstated the PJTC and the Joint Operating Authority (JCA) for the management of the Cunene basin (Chenje and Johnson, 1996: 164; Heyns, 1995: 10; Ohlsson, 1995: 59; Pallet, 1997: 69). Linked with this was the Angolan-Namibian Joint Commission of Cooperation, which was an umbrella institution that included water (Ohlsson, 1995: 59).

The strictly bilateral nature of the agreements started to change when Namibia took the initiative to involve all riparian states, with initial discussions starting in 1992 (Heyns, 1995: 10). This occurred shortly after Namibian independence and coincided with the abandonment of the SOWIDP in Botswana where the need for a more integrated management approach had been highlighted. This came to fruition shortly thereafter with the establishment in 1994 of the Permanent Okavango River Basin Water Commission (OKACOM), (CSIR, 1997: 13–6; Treaty, 1994). This became the third water regime to be established in southern Africa that involved all of the riparian states (refer to the Limpopo case), assuming that the basin terminates at the Okavango delta and not in the Makgadikgadi pans. Significantly, the document is short and contains no water-sharing agreement. It precedes the SADC Protocol on Shared Watercourse Systems, but it does recognize, the concept of 'equitable utilization of shared watercourse systems as reflected in the relevant provisions of Agenda 21 of the United Nations Conference on the Environment and Development held at Rio de Janeiro in June 1992' (Treaty, 1994). It also recognizes the Helsinki Rules. The OKACOM is therefore not a sophisticated water management institution because it has no permanent structures and has no institutionalized database of basin-wide hydrological information, but it is an important step towards the development of an integrated management approach. The institutional integrity was tested as a result of the emotional response to what occurred when Namibia announced its plans to develop the Okavango River—the Grootfontein Pipeline link into the Eastern National Water Carrier (Ashton, 2000: 82; Ramberg, 1997; *Weekly Mail and Guardian*, 1996a, 1996b), but these initial tensions seem to have dissipated. In the opinion of the author, despite rhetoric in the public media to the contrary, the OKACOM is functioning well given its historic setting and short existence. It shows how a downstream riparian with a pressing resource need can take the lead in establishing a regime. It remains to be seen whether the upstream riparian (Angola) cooperates, as this may not be in their strategic long-term interest. An ironic outcome of possible peace in Angola will be the clearing of landmines from the upper Okavango basin, and the subsequent development of agriculture and industry, all of which will impact negatively downstream.

Key Institutional Dynamics and Processes: The Concept of Second-Order Resources

It is widely recognized that governance is a key constraint to sustainable water resource management (Granit, 2000; GWP, 2000; Turton, 2001d: 157), primarily because of the weak institutional and legal frameworks in existence in southern Africa (Hirji and Grey, 1998: 89). Despite the SADC Protocol, the majority of rivers in southern Africa still have no functional agreements (Arnestrand and Hanson, 1993: 25). In order to overcome these impediments, there are a number of regional initiatives to promote institutional development. These include the SADC/UNDP Round Table Conference in 1998, the development of the SADC Regional Strategic Action Plan (RSAP) in the same year (Granit, 2000), and the SADC/European Union Conference on the Management of Shared River Basins (Savenije and van der Zaag, 1998). In the opinion of the author, while these initiatives are praiseworthy, they are probably insufficient because they do not fully recognize the importance of second-order resource scarcity.

Second-order resources can be loosely defined as the ability of a given society to develop and adapt institutions with which to effectively manage scarce natural resources like water (Allan, 2000: 322–23; Ohlsson, 1999: 161; Ohlsson and Turton, 1999; xvi; Turton, 2002a: 23; 2002b; Turton and Kgathi, 2002; Turton and Warner, 2001). For example, the absence of hydro-meteorological and other data in Mozambique has resulted in the failure to reach agreement with Zimbabwe on the development of the Pungué River (Granit, 2000), which can be interpreted as being the manifestation of a form of second-order scarcity, or what Homer-Dixon (1994: 16) calls 'technical ingenuity'. Significantly, other institutions involving Mozambique are also problematic (namely, the JPTWC), for the same reason (Vas and Pereira, 1998: 119). A similar situation exists in the Okavango basin, where no accurate and uncontested basin-wide data occurs because of capacity problems in Angola. In all of these cases one of the riparian states has been ravaged by civil war, which has undermined its capacity to develop and sustain institutions (Vas and Pereira, 1998: 119). The case studies in this chapter have shown that four basin-wide regimes have been established in southern Africa, two of which failed (TPTC and LBPTC). The other two (OKACOM and ORASECOM) are so recent that their performance cannot really be evaluated yet. The author is of the opinion that this failure has been the result of a specific form of second-order scarcity, namely the inability to negotiate an agreement and then sustain that agreement in the face of opposition. This form of second-order scarcity coincides with what Homer-Dixon (1994: 16–17) calls 'social ingenuity'.

Current research (Turton, 2002a, 2002b; Turton and Kgathi, 2002; Turton and Warner, 2001) suggests that there are two key elements that are a necessary condition for the development of an effective water management institution at the international level. The first element is the ability to generate accurate data on which decision making can be based. While this is a necessary condition, it is not a sufficient condition for successful institutional development. For this to occur, a second element is needed. This can be defined as the ability to negotiate with other actors, representing sovereign states with potentially antagonistic national interests, and reach agreement on the data and methodology used to process that data. Stated differently, where the basin-wide data is uncontested, it is also perceived as being legitimate. The process of legitimizing that data causes a redefinition of the core problem needing management, and consequently results in the synergistic process of institutional learning. It is the opinion of the author that unless this process of legitimization is present, institutional development will be unsustainable and the conflict potential will increase in river basins facing deficit. This hypothesis is being tested on the OKACOM and ORASECOM cases, and results will be published as they become available.

Discussion

The raison d'être of the SADCC was to develop economic independence away from South Africa. This has not changed with its transformation into the SADC, where fears of potential South African economic hegemony still run deep. This impacts on the possible effectiveness of the SADC Protocol on Shared Watercourses as amended. One is confronted by the paradoxical situation where the major instrument that is supposed to foster institutional development in the water sector, is the product of a political structure that has a deep-seated history of resisting South African economic and military hegemony, a situation that has not changed in post-apartheid South Africa. The ill-conceived Operation Boleas into Lesotho, and the ongoing South Africa–Namibia border dispute has merely reinforced this belief. Unfortunately, one cannot undo history—a theme that has run throughout this whole chapter—so we sit with a fait accompli.

Lowi (1993) found that a variant of hegemonic stability theory can be used to explain riparian behaviour in some international river basins. She concluded that in the four cases she studied—the Nile, Indus, Jordan, and Euphrates—the institutional outcome reflects the distribution of power in the basin. 'Cooperation is not achieved unless the dominant power in the basin accepts it, or has been induced to do so by an external power. Moreover, the hegemon will take the lead in establishing a regime or accept regime change, and will enforce compliance to the regime, only if it serves to gain as a result' (Lowi, 1993: 203). This seems to be true in the river basins under review, although there are some subtle differences. For example, Lowi (1993: 203) goes on to state that 'in the absence of coercion from outside, this occurs in rivers *only* if: (1) the dominant power's relationship to the water resources in question is one of critical need, linked to its national security concerns, and (2) it is not the upstream riparian. Cooperation in international river basins is brought about by hegemonic powers' (emphasis in original text).

Namibia is the driving force in both the OKACOM and the ORASECOM. This can be explained because of its critical need, the enthusiasm of a newly independent state, and the extraordinary negotiating skills that Namibia seems to be able to muster. It can also be explained by the fact that South Africa desperately wants to be seen as the cooperative neighbour, given its history of regional destabilization, and as such the development of these regimes suits its national interest. This may be short-lived, however, so the final validity of Lowi's conclusion cannot be fully assessed with any degree of certainty. For example, the LHWC is strategically more important than the ORASECOM from a South African perspective, so its subordination to the basin-wide institution is unlikely, and is certainly not part of the agreement. The collapse of the TPTC and LBPTC is not easily explained by Lowi's (1993: 203) conclusion, although the reincarnation of these into well-functioning bilateral regimes is entirely compatible with her interpretation of the hegemonic stability theory. What is significant about Lowi's conclusion is the role that third-party actors play in inducing potentially non-compliant riparians into cooperation. This will be particularly useful in OKACOM, to ensure Angolan support and compliance in a post-conflict scenario. In the opinion of the author, this is a valid finding, explaining the role of third parties that already exist in the region. It also strengthens the case for a hydro-political security complex because it opens the door to diplomatic engagement outside of the previously narrowly defined water sector.

Conclusion

It must be remembered that the original SADCC was founded on two core principles—the commitment to ending apartheid and the avoidance of South African economic hegemony. While the former has disappeared, the latter is still very evident. Even though the SADCC has reinvented itself as a development community—the SADC—along EU lines, with South Africa as a member state, this does not mean that the ANC Government is viewed with less distrust that the former NP government was. Operation Boleas and the South Africa–Namibia border issue has merely reinforced that old attitude. This is impacting on institutional development in the water sector. However, if one considers a variant of the hegemonic stability theory to be valid, it actually comes as no surprise that regime creation is usually to the benefit of the stronger riparian state, and that where they exist, regimes are more likely to be bilateral rather than involving every riparian in the entire river basin. What is evident is the emergence of a hydro-political security complex centered on the co-dependence of the four most economically developed states in southern Africa on river basins that are either highly developed, or difficult to develop further. Two key issues emerge from this chapter. First, South Africa is a pivotal state, with the capacity to either assist or resist as it sees fit. Second, third-party actors are likely to become more important as the playing field is levelled. International relations are, in the final analysis, about dynamic patterns of conflict and cooperation.

References

Africa, 'SADCC Moves to Tackle Drought', *Africa*, 153, 1984, pp. 74–5.
Allan, J.A., *The Middle East Water Question: Hydropolitics and the Global Economy*, IB Tauris, London, 2000.

Arnestrand, M. and Hanson, G., *Management of Scarce Water Resources in Southern Africa. Main Report. Sustainable Use of Water Resources (SUWAR) Report No. D3, 1993*, Stockholm: Royal Institute of Technology, 1993.

Ashton, P., 'Potential Environmental Impacts Associated with the Abstraction of Water from the Okavango River in Namibia', in the *Proceedings of the Annual Conference of the Southern African Association of Aquatic Scientists, Swakopmund, Namibia, 23–26 June 1999*, pp. 12.

Ashton, P., 'Southern African Water Conflicts: Are They Inevitable or Preventable?', in H. Solomon, and A.R. Turton (ed.), *Water Wars: Enduring Myth or Impending Reality? African Dialogue Monograph Series No. 2*, pp. 65–102, Durban: ACCORD, 2000.

Basson, M.S., P.H. van Niekerk, and J.A. van Rooyen, *Overview of Water Resources Availability and Utilization in South Africa*, Pretoria: Department of Water Affairs and Forestry, 1997.

Basson, M.S., 'South Africa Country Paper on Shared Watercourse Systems', Paper presented at the SADC Water Week Workshop held in Pretoria, South Africa, 1999.

Baynham, S., 'SADCC Security Issues', *Africa Insight*, 19(2), 1989, pp. 88–95.

Blanchon, D., 'Les Nouveaux Enjeux Géopolitiques de l'eau en Afrique Australe', *Hérodote Revue de Géographie et de Géopolitique*, Troiseme Trimestre, 102, 2001, pp. 113–7.

Borchert, G. and S. Kemp, 'A Zambezi Aqueduct', *SCOPE/UNEP Sonderband Heft*, 58, 1985, pp. 443–57.

Borchert, G., *Zambezi-Aqueduct*, Hamburg: Institute of Geography and Economic Geography, University of Hamburg, 1997.

Buzan, B., *People, States and Fear. An Agenda for International Security Studies in the Post-Cold War Era*, London: Harvester Wheatsheaf, 1991.

Buzan, B., O. Wæver, and J. de Wilde, *Security: A New Framework for Analysis*, London: Lynne Rienner, 1998.

Carter, C.A., 'Basutoland as a Source of Water for the Vaal Basin', *The Civil Engineer in South Africa*, 7(10), 1965, pp. 217–28.

Chenje, M. and P. Johnson (eds), *Water in Southern Africa*, Harare: Southern African Research and Documentation Centre (SARDC), 1996.

Conley, A.H., 'A Synoptic View of Water Resources in Southern Africa', Paper presented at the Conference of Southern Africa Foundation for Economic Research on Integrated Development of Regional Water Resources, November 1995, Nyanga, Zimbabwe.

—— 'The Need to Develop the Water Resources of Southern Africa', Paper presented at the Conference of Southern African Society of Aquatic Scientists, Zimbabwe, 16 July 1996, Pretoria: Department of Water Affairs and Forestry.

—— 'To be or not to be? South African Irrigation at the Crossroads', in Kay, M., T. Franks and L. Smith (ed.), *Water: Economics, Management and Demand*, London: E & FN Spon, 1997, pp. 21–8.

Conley, A.H. and P.H. van Niekerk, 'Sustainable Management of International Waters: The Orange River Case', Department of Water Affairs and Forestry (DWAF), Pretoria, 1997.

Conley, A. and P. van Niekerk, 'Sustainable Management of International Waters: The Orange River Case', in H. Savenije and P. van der Zaag (ed.), *The Management of Shared River Basins: Experiences from SADC and EU*, pp. 142–59, Netherlands Ministry of Foreign Affairs, The Hague, 1998.

CSIR, *An Assessment of the Potential Downstream Impacts in Namibia and Botswana of the Okavango River—Grootfontein Pipeline Link to the Eastern National Water Carrier in Namibia: Initial Environmental Evaluation Report*, Contract Report to Water Transfer Consultants, Windhoek, Namibia by Division of Water, Environment and Forestry Technology, CSIR, Report No. ENV/P/C 97120, 1997.

Davies, B.R., J.H. O'Keefe, and C.D. Snaddon, *A Synthesis of the Ecological Functioning, Conservation and Management of South African River Ecosystems*, Water Research Commission Report No. TT 62/93, 1993.

Davies, B.R. and J. Day, *Vanishing Waters*, University of Cape Town Press, Cape Town, 1998.

Dougherty, J.E. and R.L. Pfaltzgraff, *Contending Theories of International Relations: A Comprehensive Survey*, Harper & Row, New York, 1981.

Esterhuysen, P. (ed.), *Africa at a Glance*, Africa Institute, Pretoria, 1992.

Frankel, P.H., *Pretoria's Praetorians: Civil-Military Relations in South Africa*, Cambridge University Press, London, 1984.

Furubotn, E.G. and R. Richter, *Institutions and Economic Theory: The Contribution of the New Institutional Economics*, University of Michigan Press, Lansing, 2000.

Geldenhuys, D., 'The Destabilization Controversy: An Analysis of a High-Risk Foreign Policy Option for South Africa', *Politikon*, 9(2), 1982, pp. 16–31, Reprinted as Geldenhuys (1983).

—— 'The Destabilization Controversy: An Analysis of a High-Risk Foreign Policy Option for South Africa', *Conflict Studies*, 148, 1983, pp. 11–26.

—— *The Diplomacy of Isolation: South African Foreign Policy Making*, Macmillan South Africa, Johannesburg, 1984.

—— *Isolated States: A Comparative Analysis*, Jonathan Ball Publishers, Johannesburg, 1990.

Granit, J., 'Swedish Experiences from Transboundary Water Resources Management in Southern Africa', *SIDA Publications on Water Resources*, 17, Swedish International Development Cooperation Agency (SIDA), Stockholm, 2000.

Gutteridge, W., 'South Africa: Strategy for Survival?', *Conflict Studies*, 131, 1981, pp. 1–33, Reprinted in Gutteridge (1995).

—— 'South Africa's National Strategy: Implications for Regional Security', *Conflict Studies*, 148, 1983, pp. 3–9, Reprinted in Gutteridge.

—— 'Mineral Resources and National Security', *Conflict Studies*, 162, 1984, pp. 1–25, Reprinted in Gutteridge (1995).

—— 'South Africa: Evolution or Revolution?', *Conflict Studies*, 171, 1985 (a), pp. 3–39, Reprinted in Gutteridge (1995).

—— 'The South Africa Crisis: Time for International Action', *Conflict Studies*, 179, 1985(b), pp. 1–23, Reprinted in Gutteridge (1995).

—— 'South Africa: Apartheid's Endgame', *Conflict Studies*, 228, 1990, pp. 1–37, Reprinted in Gutteridge (1995).

—— 'The Military in South African Politics: Champions of National Unity?', *Conflict Studies*, 271, 1994, pp. 1–29, Reprinted in Gutteridge (1995).

—— (ed.), *South Africa: From Apartheid to National Unity, 1981–1994*, Aldershot, Hants & Brookfield, VT, Dartmouth Publishing, 1995, pp. 42–57.

Gutteridge, W. (ed.), *South Africa: From Apartheid to National Unity*, 1981–1994, Aldershot, Hants & Brookfield, V.T.: Dartmouth Publishing, 1995.

GWP, *Towards Water Security: A Framework for Action—Executive Summary*, The Hague: Global Water Partnership, 2000.

Haftendorn, H., 'Water and International Conflict', *Third World Quarterly*, 21(1), 2000, pp. 51–68.

Hangula, L., *The International Boundary of Namibia*, Windhoek: Gamsberg Macmillan (Pty) Ltd, 1993.

Heyns, P.S., 'Existing and Planned Development Projects on International Rivers within the SADC Region', in *Proceedings of the Conference of SADC Ministers Responsible for Water Resources Management*, 23–24 November 1995, Pretoria.

—— 'Managing Water Resource Development in the Cunene River Basin', in P. Howsam and R.C. Carter (ed.), *Water Policy: Allocation and Management in Practice*, pp. 259–66, Proceedings of International Conference on Water Policy, Cranfield University, 23–24 September 1996, London: E & FN Spon, 1996.

Hirji, R. and D. Grey, 'Managing International Waters in Africa: Process and Progress', in S.M.A. Salman and L.B. de Chazournes (ed.), *International Watercourses: Enhancing Cooperation and Managing Conflict*, World Bank Technical Paper No. 414, The World Bank, Washington, D.C., 1998.

Homer-Dixon, T.F., 'Environmental Scarcities and Violent Conflict: Evidence from Cases', *International Security*, 19(1), 1994, pp. 5–40.

—— 'Environmental Scarcity, Mass Violence and the Limits to Ingenuity, in *Current History*, 95, 1996, pp. 359–65.

James, L.H. 'Total Water Strategy Needed for the Vaal Triangle: Meeting the Challenge of the Eighties', *Construction in Southern Africa*, May 1980, pp. 103–11.

Krasner, S.D., 'Structural Causes and Regime Consequences: Regimes as Intervening Variables', *International Organization*, 36(2), Spring, 1982, pp. 185–205.

—— 'Structural Causes and Regime Consequences: Regimes as Intervening Variables', in S.D. Krasner (ed.), *International Regimes*, Ithaca: Cornell University Press, 1983.

Lawrence, P., 'Pretoria has its way in Lesotho', *Africa Report*, March–April 1986, pp. 50–1.

—— 'The Cordite of Cooperation', *Financial Mail*, 30 September 1998.

Lowi, M.R., *Water and Power: The Politics of a Scarce Resource in the Jordan River Basin*, Cambridge: Cambridge University Press, 1993.

Maletsky, C., 'Orange River Dispute Rambles On', *The Namibian*, 9 July 1999.

Meissner, R., 'Hydropolitical Hotspots in Southern Africa: Will there be a Water War? The Case of the Cunene River', in H. Solomon and A.R. Turton (ed.), *Water Wars: An Enduring Myth or Impending Reality? African Dialogue Monograph Series No. 2*, pp. 103–31, Durban: ACCORD Publishers, 2000.

—— 'Drawing the line: A Look at the Water-related Border Disagreement between South Africa and Namibia', *Conflict Trends*, 34–7, Durban: ACCORD Publishers, 2001.

Midgley, D.C., *Inter-State Water Links for the Future*, The South African Academy of Science and Arts Symposium 'Water for Survival', August, 1987.

Mills, G., 'Is Lesotho Foray a Lesson Learned?', *Business Day*, 28 October 1998.

Mopheme, 'Lesotho: Can the Scars turn to Stars?', *The Survivor*, 20 October 1998, 1988 (a).

—— 'The Miseducation of Ronnie Kasrils—RSA and the Lesotho Crisis', *The Survivor*, 20 October 1998 (b).

Ninham Shand, *Report on the Regional Development of the Water Resources of Basutoland*, Ninham Shand, Cape Town, 1956.

Ohlsson, L., *Water and Security in Southern Africa*, Publications on Water Resources, No. 1, SIDA, Department for Natural Resources and the Environment, 1995.

—— *Environment, Scarcity and Conflict: A Study of Malthusian Concerns*, University of Göteborg, Department of Peace and Development Research, 1999.

Ohlsson, L. and A.R. Turton, 'The Turning of a Screw', Paper presented in the Plenary Session of the 9th Stockholm Water Symposium Urban Stability through Integrated Water-Related Management, 9–12 August, Sweden: Stockholm International Water Institute (SIWI), 1999. Also available as *MEWREW Occasional Paper No. 19* from Website *http://www.soas.ac.uk/Geography/WaterIssues/OccasionalPapers/home.html*.

Pallett, J. (ed.), *Sharing Water in Southern Africa*, Windhoek: Desert Research Foundation of Namibia (DRFN), 1947.

Percival, V. and T. Homer-Dixon, *Environmental Scarcity and Violent Conflict: The Case of South Africa*, Washington: American Association for the Advancement of Science, 1995.

Perret, S.R. (forthcoming), *Water Policies and Smallholding Irrigation Schemes in South Africa: A History and New Institutional Challenges*, Pretoria University.

Ramberg, L., 'A Pipeline from the Okavango River?', *Ambio*, 26(2), 1997, p. 129.

Ramoeli, P., 'SADC Protocol on Shared Watercourses: Its History and Current Status', in A.R. Turton and R. Henwood (ed.), *Hydropolitics in the Developing World: A Southern African Perspective*, African Water Issues Research Unit (AWIRU), Pretoria, 2002.

Savenije, H.G. and van der Zaag, P. (ed.), *The Management of Shared River Basins. Experiences from SADC and EU*, Ministry of Foreign Affairs, The Hague, 1998.

Schulz, M., 'Turkey, Syria and Iraq: A Hydropolitical Security Complex', in L. Ohlsson (ed.), *Hydropolitics: Conflicts over Water as a Development Constraint*, Zed Books, London, 1995.

Scudder, T., R.E. Manley, R.W. Coley, R.K. Davis, J. Green, G.W. Howard, S.W. Lawry, P.P. Martz, P.P. Rogers, A.R.D. Taylor, S.D. Turner, G.F. White and E.P. Wright, *The IUCN Review of the Southern Okavango Integrated Water Development Project*, IUCN Communications Division, Gland, 1993.

Shivuti, Statement by the Namibian Representative to the Sovereignty Panel at the Second World Water Forum, March 2000, The Hague.

Simon, D., 'Independent Namibia One Year On', *Conflict Studies*, 239, 1991, pp. 1–27, Reprinted in Gutteridge (1995).

Spitz, R. and M. Chaskalson, *The Politics of Transition: A Hidden History of South Africa's Negotiated Settlement*, Witwatersrand University Press, Johannesburg, 2000.

Treaty, 'Agreement Between South Africa and Portugal Regulating the Use of the Kunene [*sic*] River for the Purposes of Generating Hydraulic Power and of Inundation and Irrigation of the Mandated Territory of South-West Africa', *League of Nations Treaty Series*, LXX (1643), 1926, p. 315.

Treaty, 'Agreement Between the Government of the Republic of South Africa and the Government of Portugal in Regard to Rivers of Mutual Interest and the Cunene River Scheme', *Republic of South Africa Treaty Series*, Government Printer, Pretoria, 1964.

Treaty, 'Agreement Between the Governments of the Republic of South Africa and Portugal Relative to the Cabora Bassa Project', *South Africa Treaty Series*, 7, 1969a.

Treaty, 'Agreement Between the Republic of South Africa and the Government of Portugal in Regard to the First Phase Development of the Water Resources of the Kunene [*sic*] River Basin', *South Africa Treaty Series*, 1, 1969b.

Treaty, 'Agreement Between the Government of the Republic of South Africa and the Government of the Republic of Portugal in Regard to Rivers of Mutual Interest, 1964 Massingir Dam', *South Africa Treaty Series*, 5, 1971.

Treaty, 'Agreement Between South Africa and Swaziland', 1982.

Treaty, 'Agreement Between the Government of the Republic of South Africa and the Government of the Kingdom of Swaziland with Regard to Financial and Technical Assistance for the Construction of a Railway Link in the Kingdom of Swaziland', 1983a.

Treaty, 'Agreement Between the Government of the Republic of South Africa the Government of the Kingdom of Swaziland and the Government of the People's Republic of Mozambique Relative to the Establishment of a Tripartite Permanent Technical Committee', 1983b.

Treaty (1984a), 'Agreement of non-Aggression and Good Neighbourliness Between the Government of the Republic of South Africa and the Government of the People's Republic of Mozambique', *South Africa Treaty Series*, 14, 1986.

Treaty (1984b), 'Agreement Between the Governments of the Republic of South Africa, the People's Republic of Mozambique and the Republic of Portugal Relating to the Cahora [*sic*] Bassa Project', *South Africa Treaty Series*, 15, 1986.

Treaty, 'Treaty on the Lesotho Highlands Water Project between the Government of the Republic of South Africa and the Government of the Kingdom of Lesotho', 1986a, p. 85.

Treaty, 'Exchange of Notes Regarding the Privileges and Immunities Accorded to the Members of the Joint Permanent Technical Commission', 1986b.

Treaty, 'Agreement Between the Government of the Republic of Botswana and the Government of the Republic of Namibia on the Establishment of a Joint Permanent Water Commission', 1990, pp. 6.

Treaty, 'Agreement Between the Government of the Republic of South Africa and the Government of the Republic of Namibia on the Establishment of a Permanent Water Commission', 1992, pp. 10.

Treaty, 'Agreement Between the Governments of the Republic of Angola, the Republic of Botswana and the Republic of Namibia on the Establishment of a Permanent Okavango River Water Commission (OKACOM)', Signatory Document, signed by Representatives of the Three Governments, 15 September 1994, pp. 7, Windhoek.

Treaty, 'Protocol on Shared Watercourse Systems in the Southern African Development Community (SADC) Region', 1995, p. 19.

Treaty, 'Protocol VI to the Treaty on the Lesotho Highlands Water Project. Supplementary Arrangements Regarding the System of Governance for the Project', 1999a, p. 25.

Treaty, 'Implementation of New Governance Model. Joint Permanent Technical Commission of the Lesotho Highlands Water Project', 1999b, p. 34.

Treaty, 'Agreement Between the Governments of the Republic of Botswana, the Kingdom of Lesotho, the Republic of Namibia, and the Republic of South Africa on the Establishment of the Orange-Senqu River Commission', 2000, p. 13.

Turton, A.R., 'Statutory Instruments for the Maintenance of Ethnic Minority Interests in a Multi-Cultural Community: The Case of the Afrikaners in

South Africa', Translated into Russian as 'Pravovye mery zashchity interesov etnicheskih menshinstv v mnogonatsionalnom obshchestve: afrikanery Yuzhnoy Afriki (Legal measures of defending interests of ethnic minorities in a multinational society: The Afrikaners of South Africa)', in N.I. Novikova and V. Tishkov (ed.), *Folk Law and Legal Pluralism*, pp. 38–47 (Proceedings of the 11th International Congress on Folk Law and Legal Pluralism, August 1997, Moscow), Moscow: Institute of Ethnology and Anthropology, 1999.

—— 'Precipitation, People, Pipelines and Power: Towards a Political Ecology Discourse of Water in Southern Africa', in P. Stott and S. Sullivan (ed.), *Political Ecology: Science, Myth and Power*, London: Edward Arnold, 2000.

—— 'Towards Hydrosolidarity: Moving from Resource Capture to Cooperation and Alliances', *Stockholm International Water Institute (SIWI) Report No. 13, Proceedings of the SIWI Seminar on Water Security for Cities, Food and Environment—Towards Catchment Hydrosolidarity*, 18 August 2001, pp. 19–26, Stockholm, SIWI, 2001a.

—— 'Hydropolitics and Security Complex Theory: An African Perspective', Paper presented at the 4th Pan-European International Relations Conference, Canterbury (UK): University of Kent, 8–10 September 2001, 2001b.

—— 'A Hydropolitical Security Complex and its Relevance to SADC', *Conflict Trends*, 1/2001, 2001c, pp. 21–23.

—— 'Water for Peace in Southern Africa: Conference Report', *Water Policy*, 3, 2001d, pp. 155–7.

—— 'The Political Aspects of Institutional Developments in the Water Sector: South Africa and its International River Basins', Unpublished draft of a D.Phil. Thesis, Pretoria: University of Pretoria, Department of Political Science, 2002a.

—— 'The Political Dynamics of Institutional Development in the Water Sector: What do we Know so Far?', Proceedings of the WISA Biennial Conference, 19–23 May 2002, Durban, 2002b.

Turton, A.R., 'An Introduction to the Hydropolitical Dynamics of the Orange River Basin', in Nakayama, M. (ed.), *International Waters in Southern Africa*, Tokyo: United Nations University Press, 2002c.

Turton, A.R. and W. Bernhardt, 'Policy-Making Within an Oligarchy: The Case of South Africa under Apartheid Rule', Paper presented at the Congress of the International Union of Anthropological and Ethnological Sciences (IUAES), Symposium on Policy, Power and Governance, 27 July 1998, Williamsburg, Virginia, USA, 1998.

Turton, A.R. and J.F. Warner, 'Exploring the Population/Water Resources Nexus in the Developing World', *Environmental Change and Security Project Report*, The Woodrow Wilson Centre, Issue 7 (Summer 2001), pp. 129–132.

Turton, A.R. and D.L. Kgathi, *The Problematique of WDM as a Concept and a Policy: Towards the Development of a Set of Guidelines for Southern Africa*, Pretoria: IUCN, 2002.

Vas, A.C. and A.L. Pereira, 'The Incomati and Limpopo International River Basins: A View from Downstream,' H.G. Savenije and P. van der Zaag (ed.), *The Management of Shared River Basins. Experiences from SADC and EU*, Ministry of Foreign Affairs, The Hague, 1998, pp. 112–24.

Vorster, M.P., 'The Lesotho Highlands Water Project', *South African Yearbook of International Law*, 13, 1988, pp. 95–118.

Weekly Mail and Guardian (1996a) 'Plan could turn Okavango to dust', 29 November 1996.

—— (1996b), 'Namibia almost certain to drain Okavango', 6 December 1996.

Williams, G.J., 'Zambezi Water for South Africa?', *Zambia Geographical Journal*, 36, 1986, pp. 57–60.

World Bank, *Can Africa Claim the 21st Century?*, The World Bank, Washington, DC, 2000.

Young, B.S., 'Projected Hydro-Electric Schemes in Basutoland', *Journal of Geography*, 60(5), 1961, pp. 225–230.

10

Water Resources Development in California

Naser James Bateni

Introduction

With thirty-four million people, California is the most populous state in the United States (US), and ⸱ is the top-ranked state in dollar value of agricultural production (California Department of Water Resources (CDWR, 1998). Even with its large population, there are still vast areas of open space and land set aside for public use and enjoyment. Population density ranges from over 16,000 people per square mile in the city and county of San Francisco to less than two people per square mile in rural Alpine County in the Sierra Nevada mountains. To put California's population into perspective, about one of every eight US residents now lives in California. Over the next two decades, California's population is forecasted to increase by more than fifteen million people (CDWR, 1998). Today, four of the nation's fifteen largest cities (Los Angeles, San Diego, San Jose, and San Francisco) are located in the state.

Figure 10.1 is a relief map of California. In roughly north to south order, the major geomorphic features are: the Klamath Mountains, Modoc Plateau, Cascade Range, Central Valley, Sierra Nevada, Coast Range, Great Basin, Transverse Ranges, Mojave Desert, Peninsular Ranges, and Colorado River Desert. The Central Valley, an alluvial basin which makes up about 38 per cent of the state's land area, is bounded by the Coast Range on the west and the Sierra Nevada on the east (CDWR, 1998). Within the Central Valley, the Sacramento River (the state's largest river) flows through the northern portion which is called the Sacramento Valley, and the San Joaquin River flows through the southern portion called the San Joaquin Valley. The two rivers

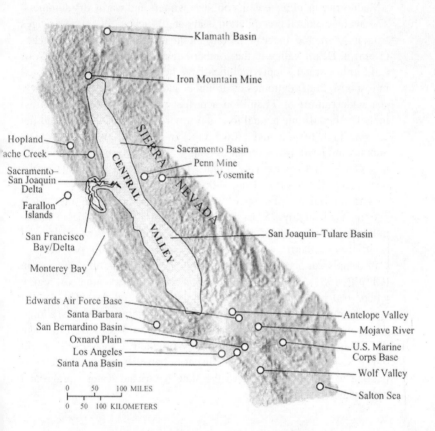

Figure 10.1: Relief Map of California

Source: US Geological Survey, Programs in California, Fact Sheet FS-005-96

converge to form the Sacramento-San Joaquin River Delta estuary, the hub of California's surface water delivery system.

California is a state of diverse climates. Much of California enjoys a Mediterranean climate with cool, wet winters and warm, dry summers. The average annual precipitation is about 23 inches (58 cm) but ranges from more than 90 inches (229 cm) on the north coast to about 2 inches (5 cm) in Death Valley in the southern desert (CDWR, 1994). Most of California's water supply facilities have been planned and constructed in response to the extremes of droughts and floods. The average yearly statewide run-off of 71 million acre-feet maf (87.6 cubic kilometres) includes the all-time annual low of 15 maf (18.5 km^3) in 1977 and the all-time high of 135 maf (166.5 km^3) in 1983 (CDWR, 1994). The Sacramento River flows for these years were 5.1 maf (6.3 km^3) and 37.6 maf (46.4 km^3) respectively. Figure 10.2 illustrates variations in the Sacramento River's flow from 1906 to 1996.

The uneven distribution of water resources is part of the state's geography. Roughly 75 per cent of the natural run-off occurs in the northern one-third of the state; about 75 per cent of the water demand is in the southern two-thirds of the state. About 40 per cent of California's surface water run-off originates in the north coast region (CDWR, 1951). The largest urban water use is in the south coast region where roughly half of California's population resides. The largest agricultural water use is in the Central Valley where fertile soils, a long, dry growing season, and availability of water have combined to make this area one of the most agriculturally productive areas in the world (CDWR, 1998). This geographic disparity between supply and demand has led to the development of the state's intricate water storage and conveyance systems.

Story of Water Development in California

Water development in California started under the authority of Spain with the founding of the San Diego Mission in 1769 when water was diverted from the San Diego River to irrigate fields surrounding the mission (Hundley, 1992). After 1850, irrigation expanded significantly as the amount of irrigated agricultural land increased dramatically. This increase was encouraged by the mining boom, which provided a nearby market for agricultural products. During the 1920s, large reservoirs were built in northern California; in many cases, they were partially

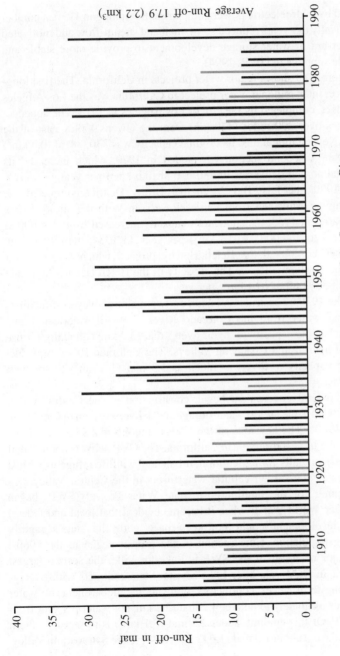

Figure 10.2: Annual variability of the flow of the Sacramento River

Source: California Department of Water Resources.

293

funded by hydroelectric power companies. Beginning in 1930, a number of critically dry years reduced snow-melt and stream-flow and motivated another era of water storage development to provide more stable and reliable supplies (CDWR, 1994).

Figure 10.3 shows major water projects in California. The first long-distance, inter-regional water project in California was the Los Angeles Aqueduct, completed by the city of Los Angeles in 1913. The aqueduct stretches over 290 miles (470 km) from the Owens Valley east of the Sierra Nevada mountains; its original capacity was 330,000 af (0.4 km^3) per year. A second section was added in 1970, which increased its potential annual deliveries to 480,000 af (0.6 km^3) per year.

In 1929, the East Bay Municipal Utility District completed the Mokelumne Aqueduct from Pardee Reservoir. With the addition of a third barrel in 1965, the aqueduct's capacity increased from 224,000 af (0.3 km^3) to 364,000 af (0.4 km^3) per year. In 1934, the city of San Francisco completed the Hetch Hetchy project, which diverts water from the Tuolumne River. The capacity of the Hetch Hetchy Aqueduct is about 330,000 af (0.4 km^3) per year.

In the 1930s, construction began on two major conveyance facilities diverting water from the Colorado River. The All-American Canal System has the capacity to divert over 3 maf (3.7 km^3) annually for use in the Imperial and Coachella Valleys. The Colorado River Aqueduct has the capacity to divert as much as 1.3 maf (1.6 km^3) to southern California.

The federal government began construction of the Central Valley Project (CVP), the largest water storage and delivery system in California, in the 1930s. The keystone of the CVP is the 4.6 maf (5.6 km^3) Lake Shasta, the largest reservoir in California. The CVP delivers about 7 maf (8.6 km^3) annually for agricultural, urban, and wildlife refuge use. Most CVP water goes to agricultural water users in the Central Valley.

Planning for the multipurpose State Water Project (SWP) began soon after World War II when it became evident that local and federal water development could not keep pace with the state's rapidly growing population. Construction of the project began in the 1960s. The major reservoir of the SWP is Lake Oroville, the second largest reservoir in California after Lake Shasta. State contracts with twenty-nine long-term water contractors obligate the department of water resources ultimately to deliver 4.2 maf (5.2 km^3) of water per year from the SWP. Of this amount, about 2.5 maf (3.1 km^3) is to serve southern California and about 1.3 maf (1.6 km^3) is to serve the San Joaquin Valley.

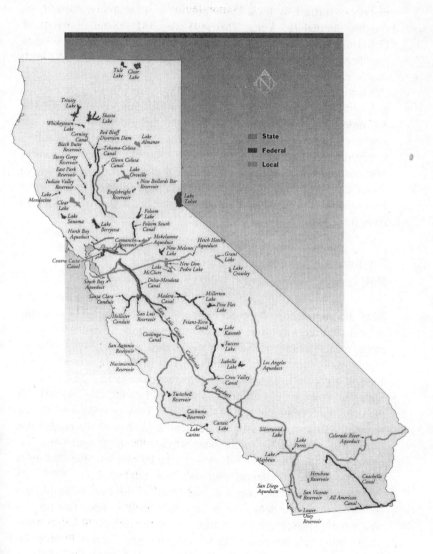

Figure 10.3: California's major water projects

Source: California Department of Water Resources.

The remaining 0.4 maf (0.5 km³) annual entitlement is to serve the San Francisco Bay, Central Coast, and Feather River areas.

Today, there are about 1,200 non-federal dams in California, and the reservoirs formed by these dams provide a gross storage capacity of roughly 20 maf (25 km³). There are also 181 federal reservoirs in California, with a combined capacity of nearly 22 maf (27 km³). Taken together these 1,400 or so reservoirs can hold about 42 maf (52 km³) of water (CDWR, 1994).

Of the state's 42 maf (51.8 km³) surface reservoir storage, over 65 per cent is in the Sacramento and San Joaquin river basins. The Sacramento and San Joaquin rivers are the most developed river systems in California, providing water for three-quarters of the state's irrigated agriculture and over twenty million people. Extensive networks of dams, diversion facilities, and pumping plants on these rivers have affected their ecology and ecosystem. It is impossible to discuss water issues in California without discussing the Sacramento–San Joaquin river delta estuary.

Environmental Considerations in Water Resources Development

Near the confluence of the Sacramento and San Joaquin Rivers, the system of waterways and islands comprises a delta and a series of embayments leading to San Francisco Bay and the Pacific Ocean. Figure 10.4 shows the San Francisco Bay and Sacramento and San Joaquin river delta estuary. This estuarine system has long been an important resource to California. The rivers flowing into and through the delta play a multiple role in the estuary. They provide conduits for migratory fish, such as salmon, to move to and from the ocean. For other fish species, the delta waterways provide a spawning and nursery habitat. The river inflow contributes much of the dissolved nutrients needed to support estuarine food chains.

The major state and federal water export facilities are located in the southern part of the delta. These facilities consist of large fish screens and pumping plants, together with numerous dams built on most of the major tributaries upstream. These facilities play important roles in determining the abundance and distribution of fish, wildlife, and plants in the estuary. There are many unscreened agricultural diversions in the delta and on tributaries to the delta that cause fish loss. Upstream dams

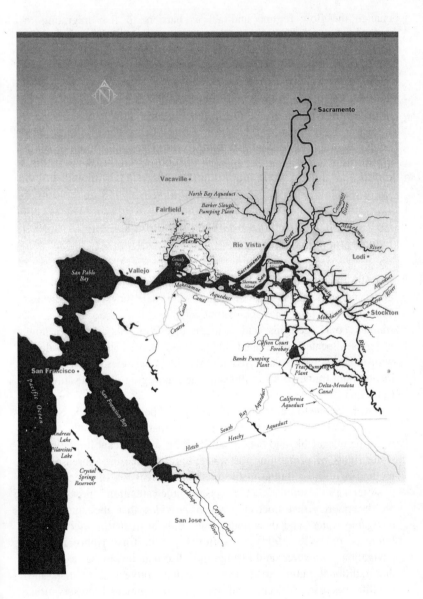

Figure 10.4: The Sacramento and San Joaquin rivers converge to form the delta, the hub of California's surface water delivery system

Source: California Department of Water Resources.

change the flow regime and act as barriers to fish migrating to spawning habitats. Diversions from the estuary reduce the flow through it. In short, water development is one of the major factors adversely affecting fish and wildlife resources in this ecosystem.

Since the mid-1950s, attitudes towards and methods for managing the state's natural resources have gone through many changes. People in the US in general, and in California particularly, have become more environmentally sensitive. This sensitivity is reflected in statutes such as the National Environmental Policy Act, California Environmental Quality Act, the federal and state Endangered Species Acts, the Clean Water Act, and the federal and state Wild and Scenic Rivers Acts.

Institutional Framework for Water Resources Development in California

In California, the use and supply of water are controlled and managed under an intricate system of federal and state laws and agencies. Common law principles, constitutional provisions, state and federal statutes, court decisions, and contracts or agreements all govern how water is allocated, developed, or used. All these components, along with the responsible state, federal, and local agencies, constitute the institutional framework for allocating and managing water resources in California.

The keystone to California's water law and policy, Article X, Section 2 of the California Constitution, requires that all uses of the state's water be reasonable and beneficial. It places a significant limitation on water rights by prohibiting the waste, unreasonable use, unreasonable method of use, or unreasonable method of diversion of water.

After a 1983 landmark court decision, much attention is now focused on the public trust doctrine, which provides that the state holds navigable waters and their underlying lands in trust to protect public interests (CDWR, 1994). Traditional public trust rights include navigation, commerce, and fishing. In California, the law has expanded the traditional public trust uses to include protection of fish and wildlife, recreation, scenic values, and environmental preservation. Consequently, in issuing or reconsidering any rights to appropriate and divert water, the state must balance public trust needs with the need for other beneficial uses of water.

The heightened environmental awareness that flourished in the 1960s and 1970s led to enactment of state and federal laws that protect free-flowing rivers under a 'wild and scenic' designation. The National Wild and Scenic Rivers Act was passed in 1968, while the California Wild and Scenic Rivers Act was passed in 1972. Both acts are intended to preserve rivers with extraordinary scenic, recreational, fishery, or wildlife values, in their natural state for the benefit of the public. The acts generally prohibit construction of any dam, reservoir, diversion, or other water impoundment on a designated river. Diversions needed to supply domestic water to residents of counties through which the river flows may be authorized if it is determined that the diversion will not adversely affect the river's free-flowing character. Most of the rivers in the north coast region are designated as wild and scenic, prohibiting water development and export of water supply from these rivers.

The National Environmental Policy Act (NEPA) directs federal agencies to prepare an environmental impact statement for all major federal actions that may have a significant effect on the human environment. It is a procedural law requiring all federal agencies to consider the environmental impacts of their proposed actions during the planning and decision-making processes. The California Environmental Quality Act, modeled after the NEPA, requires public agencies in California to document and consider the environmental impacts of their actions. It requires an agency to identify ways to avoid or reduce environmental damage and to implement those measures where feasible.

The Federal Endangered Species Act is designed to preserve endangered and threatened species by protecting individuals of the species and their habitat and by implementing measures that promote their recovery. Over 650 species have been listed as threatened or endangered in the US. Of these, 110 are native to California—the largest number in any state (CDWR, 1998). Once a species is listed, the act requires that state and federal agencies ensure that their actions do not jeopardize the continued existence of the species or habitat critical for the survival of that species. The Endangered Species Act has been interpreted to apply not just to new projects, but also to ongoing project operations and maintenance.

Section 404 of the federal Clean Water Act regulates the discharge of dredged and fill materials into the waters of the US, including wetlands. This section of the act has been defined broadly to include the construction of any structure involving rock, soil, or other construction material in US waters. This section also requires the project proponent to evaluate all alternatives that meet the project purpose and select the least environmentally damaging practicable alternative. This section of the act has significant implications in water development, as it requires all alternatives, including water conservation, recycling, and land fallowing, to be evaluated along with the proposed water projects.

Section 10 of the 1899 Rivers and Harbours Act requires a permit for obstructions to navigable water. Most water development projects must comply with Section 404, Section 10, or both.

Increased environmental awareness, public interest in protecting and preserving the state's natural heritage, and the enactment of various laws to protect environmental resources have changed water resources development planning in California. Water resource planners are learning the new tools of planning as they develop a new planning vision that balances urban and agricultural water needs with the need to protect the environment. This evolving process, as well as the higher costs of development, has slowed down water resources development in California. Today, competing water needs and water users must find acceptable and financially feasible solutions to complex water management issues.

As evident from Table 10.1, which shows the historical development of reservoirs in California, the majority of large dams (impounding reservoirs with capacities of 50,000 maf [61.7 million m^3 or more]) were constructed prior to the 1970s. In the past two decades, only a few large reservoirs have been built. The current planning process focuses on increasing water supply reliability by various means, including non-structural alternatives such as demand management (that is water

Table 10.1: Historic development of reservoirs in California

	Number of reservoirs built	Capacity	
		Maf	Km^3
Pre-1940	29	5.7	7.0
1940–49	7	5.7	7.0
1950–59	21	6.5	8.0
1960–69	32	12.6	15.5
1970–79	12	7.0	8.6
1980–89	2	0.6	0.7
1990–99	3	1.0	1.2
Total	106	39.1	48.2

conservation, water recycling, and water transfers). The most recent major reservoirs planned and constructed in California have been off-stream storage projects. Off-stream storage reservoirs are constructed on a small stream that does not significantly contribute to the water supply of the reservoir. Most of the water supply for an off-stream reservoir is conveyed from other sources. Off-stream storage has an inherent environmental advantage because the reservoirs tend to be on minor tributaries, which reduces impacts on live stream habitat and does not prevent fish from migrating upstream to their spawning habitat.

Most large reservoirs in California are multipurpose impoundments designed to provide water supply storage, hydroelectric power, flood control, recreation, water quality, and downstream fishery needs. Often, large reservoirs would not be economically feasible as single purpose projects. Multipurpose designs maximize the beneficial uses of large reservoir sites and provide regional water supply benefits.

Regional Impact of Water Resources Development in California

In California, urban and agricultural water use is currently about 35 maf (43 km^3) annually. This use depends on both locally developed water (surface and groundwater) and imported surface water—in some cases, imported over great distances by large water projects. About 14 maf (17 km^3) annually, or about 40 per cent of the state's total urban and agricultural use, is served by the major water projects—state, federal, and local.

About 70 per cent of California's economy depends on manufacturing, financial, insurance, real estate, and service sectors, a majority of which are centred in the major urban areas of the state, in coastal and semi-coastal areas that rely upon imported water supplies. Three-quarters of California's businesses are located in the San Francisco Bay and the south coast regions (California Trade and Commerce Agency, 2000). These regions originally developed because of their proximity to harbours and the Pacific rim trade.

Water imported to urban areas is used not only for drinking and sanitation, but also for power plant cooling and as industrial process water, critical to maintaining the economy in these regions. Imported water is also very important for providing a desirable environment for people who live and work in these areas.

The urban environment supported by the imported water serves as an 'air conditioner' both in terms of air quality and ambient temperature. It also provides a habitat for birds and other urban mammals. Parks, golf courses, gardens, and industrial and residential landscaping are important to the attractiveness of an area, which help lure prospective industries and employees. Terminal and other regulation reservoirs provide for camping, boating, fishing, and other recreation opportunities, adding additional attractiveness to nearby cities.

All these water-related factors have been instrumental in building a vibrant economy in California. Even industries having a relatively low need for process and cooling water benefit immensely from the amenities that water provides in an urban setting.

Consider the economic benefits of water development in California. Its human and natural resources support a gross state product of over $1 trillion (TCA, 2000). Based on 2000 California (UCLA 2000) and national values (World Bank Atlas, 2000), a country producing as much value as California would be placed fifth among the nations in the world. California is the leading agricultural export state in the US and the nation's number one dairy state. It also produces 55 per cent of the nation's fruits, nuts, and vegetables and accounts for almost 17 per cent of the nation's exports (TCA, 2000). One of the most important natural resources making this level of agricultural and industrial production possible is water.

In the San Joaquin Valley, where over half of the $27 billion of statewide gross agricultural production value was generated in 1999 (California Department of Food and Agriculture, 2000), developed surface water projects serve about 80 per cent of agricultural and urban water use. About 35 per cent of this surface supply is imported. Groundwater serves the remaining 20 per cent of use.

In the San Francisco Bay region, over 70 per cent of the water is imported. The south coast region imports over 60 per cent of its supplies. Figure 10.5 shows regional imports and exports. Over three-quarters of the gross state product is generated in these two regions, evidence of the crucial link between water availability and economic strength. However, this economic strength does not come without a price. The impacts of water development on the quality of life and the environment are evident in these regions. A clear example is in the south coast region.

In 1900, of the 1.5 million people living in the state, only 15 per cent lived in the south coast region of California, which covers about 7 per

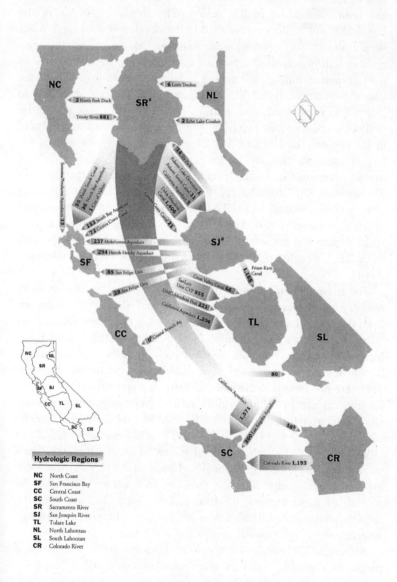

Figure 10.5: Regional imports and exports of water

Source: California Department of Water Resources.

cent of the state's land area. Since then, extensive water development has been carried out in this region. The steady expansion of the population and of the economy led to the development of large water supply projects for importing water to the region. As a result, the south coast is California's most urbanized region. With a population now of over seventeen million, the region is home to more than half of the state's population. By 2020, the region's population is expected to increase to over twenty-four million. Urban growth has led to traffic gridlocks, noise and air pollution, and the loss of natural environment in the region. The cities of San Diego and Los Angeles, two of the largest cities on the south coast, rank one and three respectively in roadway congestion in the country (United States Department of Transportation, 1999). The congestion cost in the Los Angeles area alone amounts to over $12 billion per year (United States Department of Transportation, 1999). The greater Los Angeles metropolitan area ranks among the highest in the nation in the number of days with hazardous air quality (United States Environmental Protection Agency, 2000).

The environmental impacts of water development are also evident, not only within this region, but also statewide. For example, diversions from the Mono Lake basin to the Los Angeles area have lowered the lake's water levels dramatically, altering its ecosystem. Mono Lake, located east of the Yosemite National Park at the base of the eastern Sierra Nevada is recognized as a valuable environmental resource. There are two islands in the lake that provide a protected breeding area for large colonies of California gulls and a haven for migrating waterfowl. Diversions from its tributaries have lowered Mono Lake's level from an elevation of 6,417 feet (1,956 m) in 1941 to a historic low of 6,372 feet (1,942 m) in 1981. With decreased inflow of fresh water, the lake's salinity has increased dramatically. Lower water levels have also exposed a land bridge, allowing predators to reach gull rookeries. As a result of these impacts, the lake and its tributaries have been the subject of extensive litigation between the city administration of Los Angeles and environmental groups since the late 1970s.

Although California's vast system of dams and canals has benefited people by providing water for drinking, agriculture, and industry, the impacts on many fish and other aquatic species have been great. California has more endangered or threatened aquatic species than any other state in the nation.

Since the construction of diversion facilities in the Sacramento and San Joaquin valleys and diversions from the delta for export, populations of native anadromous fishes (chinook salmon, steelhead, white sturgeon, and green sturgeon) have declined dramatically. Declines have been so dramatic that several species may be in danger of extinction.

Habitat degradation is the primary cause of these declines. Habitat quantity and quality have declined due to the construction of barriers to migration and levees, modification of natural hydrologic regimes by dams and water diversions, elevated water temperatures, and water pollution. Rates of decline for most anadromous fish species have increased following the completion of major water project facilities.

As the state's population expands, greater attention will be directed to preserving and restoring California's ecosystems and to maintaining the natural resources which have attracted so many people to the state. A theme dominating much of water management planning at the state level is ecosystem restoration. Voter approval of Proposition 204 in 1996 provided $450 million for state restoration actions directly associated with the delta, and another $93 million in state matching funds for restoration actions of the US Bureau of Reclamation's Central Valley Project Improvement Act. In 2000, California voters approved a $2-billion bond; over half of the money is dedicated to ecosystem restoration.

Current Water Development Planning in California

Today, water resource management in California is at a critical point as evolving policies, financial realities, and physical limits of the state's water supply infrastructure collide. Three major interest groups— urban, agricultural, and environmental—must work their way through California's institutional framework towards solutions that benefit all Californians and their environment. The challenge for water resources planners and policy makers is to create a planning strategy that can balance the diverse and often conflicting interests.

Water resources development planning in California today must adhere to some fundamental principles to succeed. These principles include:

a) Selecting the least environmentally damaging practicable alternative. This principle requires a comprehensive environmental evaluation of all alternatives. This evaluation must include non-structural alternatives such as water conservation and land fallowing, and the no-action alternative.

b) Using an integrated resources planning process. This principle recognizes that meeting future water needs is often best served by implementing a variety of water management actions in an integrated form. Critical to the success of integrated resources planning is that its goals are best accomplished in an open planning process with all the affected parties actively involved. Integrated resources planning ensures that all resources are evaluated and considered in a water management plan, and all the concerns of the interested parties are addressed fully and without bias.

c) Using a least-cost planning process. This principle gives all available alternatives an equal chance in the selection process. In this process, the water manager's objective becomes one of meeting all water-related needs of customers. For example, if a growing service area's need for additional water and sewer treatment can be met by an ultra-low-flush toilet retrofit programme rather than by additional water supplies and treatment facilities, the retrofit programme should be considered on its merits and compared with all other options when developing a water management plan. With the least-cost planning process, the cost of implementing new water management programme must be weighed against the benefits resulting from implementation of a project or the economic losses, including unattained benefits, associated with not developing additional supplies.

d) Protecting ecological and ecosystem health. This principle is a primary consideration in any water development planning process in California. Such protection is needed to preserve the state's natural heritage for future generations.

Because the Sacramento–San Joaquin river delta estuary is the hub of California's water plumbing system, most of the water resources development projects should take into account the estuary's complex issues. To address these issues and to formulate a long-term strategy for water management in California, the CALFED programme was established by the Governor of California and the US Secretary of Interior.

The CALFED Bay–Delta Programme, initiated in 1995, is a collaboration of state and federal agencies and the state's leading urban, agricultural, and environmental interests to address and resolve the environmental and water management problems associated with the San Francisco Bay and the Sacramento and San Joaquin river delta estuary system. The mission of the programme is to develop a long-term comprehensive plan that will restore ecological health and improve water resources management for beneficial uses of the bay–delta system.

The stakeholders within CALFED have been evaluating various water management strategies and negotiating an acceptable solution for meeting future water needs, while protecting the ecosystem of the Sacramento–San Joaquin River delta estuary. The result of these efforts is a record of decision (ROD) on the CALFED Bay Delta Final Programmatic Environmental Impact Statement and Report issued in August 2000. The ROD outlines a long-term plan titled Preferred Programme Alternative.

The Preferred Programme Alternative consists of a set of broadly described actions which set the long-term, overall direction of the thirty-year CALFED programme. The description is programmatic in nature, and is intended to help agencies and the public make decisions on broad methods to meet programme purposes. It is made up of a number of programmes that will meet CALFED's objectives for ecosystem restoration, water quality improvements, delta levee system integrity, and water supply reliability (CALFED Bay–Delta Programme, 2000).

The Integrated Storage Investigations Programme was developed by CALFED to meet CALFED's objectives including improving the state's water supply reliability. This programme is evaluating the feasibility of conjunctive management of surface and groundwater resources in major groundwater basins throughout the state and five major surface storage projects. Surface storage projects include an off-stream surface storage project north of the delta, surface storage on delta islands, enlargement of two existing reservoirs, and additional storage in the San Joaquin River watershed. This is the most comprehensive storage investigation in California in recent years and could significantly contribute to the state's water supply reliability. This programme was formulated after years of negotiations and assurances that the Sacramento–San Joaquin river delta estuary ecosystem will be restored and protected, and that all reasonable water use efficiency programmes will be in place throughout California.

References

CALFED Bay–Delta Programme, *Programmatic Record of Decision*, CALFED Bay-Delta Programme, Sacramento, CA, 2000.

California Department of Food and Agriculture, *Resource Directory*, Sacramento, California Department of Food and Agriculture, CA, 2000.

California Department of Water Resources, *Water Resources of California*, Bulletin 1, Sacramento, California Department of Water Resources, CA, 1951.

—— *California Water Plan Update*, Bulletin 160–98, California Sacramento, Department of Water Resources, CA, 1998.

—— *California Water Plan Update*, Bulletin 160–93, California Sacramento, Department of Water Resources, CA, 1994.

California Technology, Trade, and Commerce Agency Office of Economic Research, *California, an Economic Profile*, Sacramento, California Technology, Trade, and Commerce Agency Office of Economic Research, CA, 2000.

Hundley, N. Jr., *The Great Thirst, Californians and Water 1770s–1990s*, Berkeley, University of California Press, CA, 1992.

University of California, Los Angeles, 'Anderson Forecast Project', *www.uclaforecast.com*, 2000.

United States Department of Transportation, 'Bureau of Transportation Statistics', *www.bts.gov.*, 1999.

United States Environmental Protection Agency, 'Office of Air and Radiation', *www.epa.gov/oar*, 2000.

World Bank, 'World Bank Atlas', *www.worldbank.org/data/wdi2000/atlas.htm*, 2000.

11

Contribution of Water Resources Development to Regional Development: Case Studies from Latin America

Luis E. Garcia • *Diego J. Rodríguez* • *Felipe B. Albertani*

Introduction

Discussing cases where water resources projects have contributed to development would have been very easy and straightforward in the 1960s and 1970s. To do so now calls for some background studies.

The role that water resources have played in Latin America's development, both positively and negatively, cannot be analysed independent of the overall political, economic, and cultural conditions prevailing at a given time. This poses a tremendous challenge to anyone attempting such a task, since Latin America, although sharing some traits, is a heterogeneous region in many aspects, including political, economic, and cultural. To exacerbate matters, there is a vast array of concepts and definitions of development. This is a quicksand that this chapter intends to detour. Let us just use, as a working reference, the instinctive notion that one country or region is more 'developed' than other if more of its inhabitants' income is enough, not only to satisfy their basic needs of food, health, shelter, education, and security, but also to enjoy some of the more fuzzy 'needs' that every society uses to define 'quality of life' at a given point in time.

There have been, and still are, great differences in the socio-political canvas of the region. Being dynamic, these have evolved differently in different countries. Nevertheless, some tendencies that can be used to characterize a specific point in time can be identified. This 'point', of course, usually spans many years, even decades. It should be kept in mind, however, that the length of these spans were much longer in the past and have, progressively and rapidly, shrunk during the last one or two decades.

Pre-Columbian Phase

Let us first attempt to paint a very broad picture of some of the circumstances that have influenced present-day Latin America, starting with pre-Columbian times. Although all pre-Columbian native Latin Americans have made important cultural contributions to the region, the relationship with water of some of them, like the Mayans in Mesoamerica and the Incas in the northern Andes highlands, stand out. In general, it can be said that water has always been an important factor in their well-being and, later, for their agriculture and commerce, exerting a big influence on the location of their dwellings. Latin American rivers have always played an important role as means of communication and transportation of goods and people. The Amazon River and Rio de la Plata are obvious examples that last to present times. But even rivers like the Usumacinta, shared by Guatemala and Mexico, were once important routes for commerce and transportation of minerals, agricultural products, and even for cultural and other invasions of the lesser kind in the land of the Mayas.

But what distinguished both the Mayans and the Incas was the existence of irrigated agriculture. Archaeological and remote sensing findings point to irrigated agriculture and the existence of reservoirs from which Tikal inhabitants obtained their water supply. These are strong arguments in favour of the positive role that water plays in cultural and economic development when used in large quantities. In this sense, some clues for the demise of some of the city-states of the golden era of the Mayan civilization, if the proponents of such theories are to be believed, may also have been related to water: the drying of the reservoirs in Tikal due to prolonged drought, and/or the use of the river by invading cultural and/or military invaders.

In any case, water was so important for these ancient cultures that frequently it was, in one form or another, or in a phase of the hydrological cycle, a deity in their pantheon. So much so that, even now, the concept that water is 'a gift from God' and, therefore, free, is strongly engrained in many groups of Latin American societies. The importance of this for present-day water resources management should not go unnoticed.

Colonial Phase

Although orographic or any other physical barriers posed no obstacle for the fifteenth-and sixteenth-century Spanish and Portuguese 'Conquistadores', rivers played a major role in the cultural/military conquest, both at their origin such as the Guadalquivir River in Fernando and Isabel's Spain, as well as at their destiny; but also later on in the commerce of colonial America with the European metropolises. Many of the major cities in the new world were located in the coastal areas and along major rivers.

Some colonial cities were built on top of Mayan, Aztec, and Inca cities, as a show of dominance. Since the locations of these and other colonial cities were mostly in the highlands due to strategic security reasons, they confronted problems with water supply not faced by their riverside counterparts. Moorish familiarity with water problems, aided by Roman engineering and know-how, came to Latin America through Iberia to build aqueducts and cisterns, thus providing a temporary solution to the problem and allowing many new Latin American urban centres to grow, 'prosper', and thrive. Water was a contributing factor in the development of these new cities.

Surely, nobody at the time imagined the mega centres that some of these cities would turn into centuries later (Mexico City, São Paulo, La Paz, etc.), stressing the limits of their water supply sources and/or degrading the quality of the nearby rivers that once originated and supported their growth and prosperity (or misery). It is tempting and not entirely unreasonable to speculate that this initial water availability may have been one of the factors attracting large numbers of rural population into the cities, contributing to their present-day problems. But in the minds of the authors, the lure of jobs and other opportunities that the city offers, casts a longer shadow.

Project-by-Project Phase

During the period from independence to the 1960s, water did not seem to be an important factor for the region's development. It achieved prominence in the last three decades. Before then, water was available almost everywhere and was, probably, not a limiting factor for the social and economic development of the mostly agricultural and commerce-based economies. Until the beginning of the twentieth century, water requirements were manageable and water projects were mostly small, local, and privately financed. Water-based services such as electricity generation and drinking water supply, were provided by the private sector. It was not until the 1920s that these services were given to the states or municipalities and not until the 1940s (and in some countries even the early 1950s) that agencies of the Central government undertook the task of providing water supply and sanitation services (Lee, 1990).

From 1930–1940 (its end marked by the beginning of the post-World War II era) and later during the 1950s and 1960s (its end marked by the beginning of the economic planning era and the creation of many regional instruments, such as the Inter-American Development Bank to support economic development), water-related projects started to attract renewed attention from administrators and politicians. First in the larger countries and then everywhere, they were conceptualized as part of the much needed infrastructure that would support the 'take-off' of the Latin American economies. The era of 'economic development' and the big infrastructure projects started. Every infrastructure project would axiomatically contribute to development. Therefore, there was no need to develop indicators to measure this contribution.

Since these projects were capital intensive and took a long time (usually more than ten years) from inception to construction and operation, strong public sector involvement and planning effort was needed. But water needs, although much larger than before, were still for the most part localized; water use conflicts were not in sight. Water was still very much regarded as 'God's gift' and its availability unlimited.

Regional planning was unnecessary since a project-by-project approach was adequate under those circumstances. The institutional framework in the Latin American countries was set very much in line with the needs of this approach to water resources development. This framework remained consistent with the needs of the sectored approach phase described in the next section which still remains—and its consequences still persist—in many countries of Latin America.

Sectored Approach Phase

The last three decades of the twentieth century started with a strong emphasis on sectored water resources planning. As the water requirements of different sectors of the economy increased, it became evident that the project-by-project approach was no longer adequate and that some planning was needed within these sectors. Sectored 'master plans' came in vogue for water supply, irrigation and drainage, energy, etc. (Garcia, 1998).

Attention to river basins also started in various Latin American countries in the late 1960s and early 1970s, encouraged by the experiences of the Tennessee Valley Authority (TVA) in the United States, in an attempt to replicate them. The TVA model was looked at as a way to achieve much needed regionally planned decentralization (Cordeiro, 1994). Regional development corporations were thus created, one example being the Cauca Valley in Colombia.

Evidently, this was a technocratic and top-down approach, not very different from other regional planning attempts of those days. The difference was that the regions concerned were river basins and more interest and hope was placed on hydraulic structures. Direct governmental intervention and strong centralized planning were its main characteristics (Cordeiro, 1994). This approach and its consequences persisted, with some exceptions, practically until the end of the century.

Towards a New Paradigm

After what has come to be known as the 'lost decade' of the 1980s (Paiva, 2000), Latin America faced a socio-economic reality that presented both challenges and opportunities as the region's economy played a more active role in the global economy. Sizable investments in water and other 'development projects' have been made in the region. Table 11.1 depicts, as an example, the Inter-American Development Bank's[1] investments in water-related projects from 1961 to 2001.

Despite these investments and important gains, there still is much to be accomplished in the region. About seventy-eight million people still lack piped water supply in their households and some 120 million do not have access to regular sanitation services, mostly in rural areas (PAHO, 1997; WHO/UNICEF, 2000).

---

[1] The Inter-American Development Bank (IADB) is a multilateral development financing institution governed by regional and extra-regional donor countries and regional borrower countries. As a regional development bank, it is the principal development financing institution for the region.

Table 11.1: Water-related projects approved by the Inter-American
Development Bank, 1961–2001 (in 1995 US$ million)

Period	Potable Water and Sanitation	Irrigation and drainage	Hydroelectric	Other*	Total	% of bank loans
1961–65	1390	409	283	0	2082	26
1966–70	873	961	728	0	2562	23
1971–75	1036	1009	3293	54	5392	35
1976–80	1532	1862	3262	110	6766	31
1981–85	1806	816	3541	0	6163	27
1986–90	2058	147	1893	123	4221	25
1991–95	3191	352	1298	243	5084	16
1996–01	2407	0	166	87	2660	6
Total	14,293	5556	14,464	617	34,930	20

*   Includes flood control and watershed management projects
Source: IDB, *Strategy for Integrated Water Resources Management*, Strategy
Paper No. ENV-125, 1998 (updated by the authors).

At the beginning of the new century, Latin America is still
experiencing serious economic and social problems (Paiva, 2000). The
most urgent question to be addressed relates to economic growth.
Given the size of the population and current income levels, Latin
America needs to grow at least twice as fast as the rates observed in
the recent past to reduce the high levels of unemployment and
underemployment. Additionally, income distribution remains unequal.
Income inequality has remained high relative to other regions of the
world. Thus, the region has two major challenges: to accelerate
economic growth and, at the same time, to promote and improve access
to economic opportunities to the large share of its population living in
poverty (Paiva, 2000).

Several important trends and tendencies that appeared timidly in the
late 1960s came of age in the 1990s. This happened after local political
agreements were reached within Latin American societies and a global
consensus on important water issues was reached through a series of
international events, the most notable of which is the 1992 Dublin
Conference.

Elected governments are now the rule rather than the exception in
Latin America, and important spaces for discussion and participation
have been opened to local communities and civil society in general. User
groups are now empowered to make crucial decisions without being

regarded as 'subversive'. Women, minorities, and other groups previously regarded as unprivileged have a new voice and have made important inroads into the official decision-making mechanisms in many countries.

External pressures from bi-national donors, multilateral financing institutions, and non-governmental organizations (NGOs) have been added to the internal forces (economy, population, declining resources, political pressures) to prompt change in the Latin American water scene (Rogers, 2002). The environmental movement championed by NGOs, the neo-liberal economic school of thought being pushed by North America, the relevance and importance of public participation and meaningful decentralization, the integrated water resources management (IWRM) approach favoured by many Nordic European donors, and the emphasis on the watershed or river basin approach for water resources planning and management, favoured mostly by France and Spain, now dominate the Latin American water resources scene. A summary of the changes from the project-based approach to the IWRM approach is presented in Table 11.2.

The case for the environment was less evident in the 1950s and early 1960s, since the emphasis was on development, which considered it more important, for example, to provide water supply to a community than to treat its sewage, or to generate much needed hydropower than to conserve a highlands lake of unique limnological characteristics. Besides, the environmental impacts were less known, studied, and understood than now.

Also, the case for good governance is well made by Rogers (2002) when he states that 'in many countries, the best laws and regulations and the best institutional framework are in place, but the actual performance of the sector is very poor'. He goes on to say that 'important areas for governance which are often neglected are those...that control financial decisions...'. The relevance for this in present-day Latin America cannot be overemphasized. Water resources management, once guided by engineering and development economics, is now dominated by issues of environment and governance. For some of these players, although well intentioned, water resources projects and the 1960s and 1970s notion of development are actually regarded as an anathema.

Case Study Challenge

Many of the environmental and governance considerations that are now considered noteworthy of attention and financing, would have

Table 11.2: Paradigm shift in water resources approach

Project-oriented water resources development	Sub-sectored water resources development	Sub-sectored water resources management	Integrated water resources management
Isolated projects for water supply, irrigation and drainage, hydroelectric generation, navigation, recreation, etc.	Projects for similar beneficial uses, but conceived within a sub-sectored framework.	Similar approach as before, but tries to solve water use problems derived from scarcity, public interest, externality or open access, through infrastructure projects and/or institutional innovation.	Similar approach as before, but individual projects and/or actions result from consideration of all uses, including the environment. Tries to solve conflicts between users and uses through increasing the supply, but also through institutional innovation and demand management.
Each project tries to maximize the benefits for that particular project. An implicit assumption is that a given source of water exists exclusively for that project.	Benefits for the sub-sector are maximized. An implicit assumption is that the sources of water exist solely for the purpose of that sub-sector, for example, irrigation, hydropower, etc.	These projects and/or actions evolve from sub-sectored restructuring or modernization of the state programmes (such as for the water supply and sanitation sub-sector, the agricultural energy sector, etc.), where an attempt is made to individually maximize the benefits for given sub-sectors or sectors. For example, the unilateral assignment of water use permits by the energy sector.	It usually responds better to the adjectives of 'comprehensive', 'environmentally conscientious', 'incentive oriented' and 'participatory', that water resources activities need to be associated with in order to be sustainable.
What happens with water use return flows has lesser importance.	Projects are generally derived from sub-sectored master plans, such as irrigation and drainage, and sanitation, tourism, energy, water supply, etc.		
Emphasis is on solving individual water use problems derived from scarcity or public interest by increasing the supply.	Emphasis in solving problems by supply augmentation remains, but generally regarding the needs of a particular sub-sector.		

(Contd.)

316

Table 11.2 Contd.

Project-oriented water resources development	Sub-sectored water resources development	Sub-sectored water resources management	Integrated water resources management
May create serious conflicts between users and uses, but may be adequate if water is abundant and user requirements can be easily satisfied.	May solve conflicts between users, but may still create conflicts between uses. May be adequate under similar conditions as in the previous case and when only a few uses are predominant.	It is a more efficient way to solve problems, especially when important conflicts exist between users or the scarcity is a consequence of the inefficiency of the providers.	
May create serious environmental problems.	May still create serious environmental problems.	May still cause conflicts between uses.	
		May still create serious environmental problems.	

Source: García, L.E., *Integrated Water Resources Management in Latin America and the Carribean,* Technical Study ENV-123, Inter-American Development Bank, Washington, DC, 1998.

317

been regarded as irrelevant and a waste of time and resources just a few decades ago, and in fact many of them were. A project that in the 1950s or 1970s was considered 'good' and contributing greatly to the development efforts of a country, may now be considered a villain (many water projects of that era are now vilified). There is no doubt that not all projects contributed the same and that many were even detrimental for the specific development goals of a given country (Garcia, 1971), but a serious effort ought to be made to avoid falling into a case of 'judging Columbus by the twenty-first century moral codes and standards'.

Given the previous background and that many projects might have been conceived in one era, built in another, and judged in yet another (that is why, among other reasons, it is so difficult to define development impact indicators for these projects), the question for this chapter is: how to present case studies of water projects, at the dawn of the new millennium, which illustrate their contribution to development?

In societies where the belief that water is a 'gift from God' has been engrained for millennia, do projects that charge for water contribute to development? Do those that do not charge, contribute? Do projects that bring water to heavy urban concentrations contribute to development? Do those that are not built (the 'do nothing' alternative) contribute? Do projects conceived in the project-by-project or sectored approaches contribute to development? What is the contribution of these vis-à-vis the contribution of projects conceived within an IWRM approach? Do projects built in the 1960s or 1970s contribute to development, even if no environmental or governance considerations were made at the time? If not, does this mean that all previous projects have a negative contribution to development because no integrated, environmental, and/ or governance issues were incorporated in their design? If so, how can we explain the growth experienced in the developed countries?

Facing these dilemmas, the authors decided that the best course of action was to follow the simplest and most pragmatic manner. Paraphrasing Rogers's (2002) statement about the goals of governance, it seems, no matter what tendencies are in vogue at a given point in time, the major driver for water resources projects within a country has been the distribution of the economic benefits and costs and how they are spread over society. In other words, who pays for what and who benefits? A caveat is in order here: Rogers (2002) recalls that 'a major…political…dilemma faced in the development of water resources in Latin America, according to Garcia and Valdes (2000), is the tendency to privatize the benefits and socialize the costs'.

On the other hand, the authors pose another caveat: although water resources projects are needed for development, in any way this may be defined, a higher degree of development or poverty alleviation in a country or region is not going to be achieved just by water resources development. In other words, water projects are a *necessary*, but not a *sufficient* condition for development and poverty alleviation.

With that in mind, the following illustrative case studies are discussed in this chapter:

a) The El Cajón hydroelectric project in Honduras and the management of renewable natural resources in its watershed.

This case illustrates how a major energy project, initially aimed at the macro national strategic level, initially had little or no benefits for the local population, and how this was later amended.

b) The Segredo hydroelectric project in Brazil.

This case illustrates how a much needed energy development project successfully took into consideration the needs and concerns of the local stakeholders from the very beginning.

c) Urban development and sanitation in Bolivia.

This case illustrates how a water and sanitation programme helped to stimulate much needed decentralized investments in the sub-sector, following a decade of crisis.

Case Studies

El Cajón Hydroelectric Project in Honduras and the Management of Renewable Natural Resources in its Watershed

Background

The construction of the El Cajón (General Francisco Morazán) Hydroelectric Project was finished in 1984 at a cost of US$ 709 million. It received financing from the Inter-American Development Bank, the El Cajón International Consortium, the Honduran Government, and the Empresa Nacional de Energía Eléctrica (ENEE). The dam is 226 m high, with the reservoir covering a 94 km^2 area. The plant's installed capacity is 292 MW. It generated 501 MWh in 1999.

The project is located in central Honduras, catching the upper basin of the Ulua River, which flows towards the Caribbean Sea, with a contributing area of 8,630 km^2.

The project generates 69 per cent of the country's electric power. Electric power is also exported to neighbouring Nicaragua. It is estimated that between 1985 and 1990, 1.36 billion kilowatts (KW) were exported, representing an annual income of US$ 8 million. In addition, the reservoir served an important flood regulation function during the 1998 floods generated by hurricane Mitch, which in Honduras alone caused US$ 58 million in damage. Although no data was available to quantify it, the figures mentioned here give a qualitative idea of the importance and impact of the project in the development of the downstream Sula Valley region (Honduras's most important agricultural production region) and in the country in general.

These intuitive positive impacts at a macro scale, unfortunately, were not accompanied by positive effects in the population and economy of the project region itself. In fact, some suggested (without quantification) that the project was detrimental, reducing the level of material well-being of the majority of the population in its zone of influence. In addition, loss of agricultural land and many negative environmental impacts in the watershed were, rightly or wrongly, attributed to the project.

Description of the Watershed Management Programme

In November 1993, the Inter-American Development Bank (IADB) approved a US$ 17-million loan to improve natural resources management of the El Cajón reservoir watershed in Honduras. This improvement would be achieved through an array of different activities included in four major components: (i) agro-forestry management; (ii) forest development; (iii) management of protected areas; and (iv) research studies.

Under the agro-forestry management component, the programme included soil conservation measures and activities that would increase productivity. The forest development component included the preparation of management plans for mature forests and the protection of forested areas. The component for management of protected areas was aimed at the preparation and implementation of management plans for up to 65 per cent of the existing protected areas in the watershed. These areas, although declared for protection, lacked any management plans. Research studies in geomorphology, climatology, hydrology, and water quality were to supplement the implementation of activities in the other components.

Issues

Mounting concerns grew near the end of the execution period of the watershed management programme due to the lack of any sustainability mechanism that would ensure the continuity and ownership of the programme by more than 10,000 families and numerous NGOs, and municipalities participating in the programme.

The programme included technical assistance on soil conservation techniques but this was provided on a grant basis, with no cost-recovery instrument attached to it. Forestry management plans were fully financed by the programme, and participating municipalities lacked the capacity to implement them. The fate of the protected areas component was similar to the others; the management plans were prepared with financial resources from the programme but there was no financing beyond that stipulated in the programme for their continuity. There is only one case where the responsibility for the implementation of the plan was transferred to an NGO. All related expenses were covered by the programme and there were no foreseen activities that would allow the municipality to generate revenues that could be used to cover the recurrent costs incurred by the NGO to manage the protected area.

Realizing this, a study was commissioned to evaluate the results of the programme and the resulting socio-economic impact on different stakeholders, mostly small producers. The study was prepared by the International Centre for Tropical Agriculture (CIAT). In general, the report concluded that: (i) producers participating in the programme adopted soil conservation techniques promoted by the Programme. It seems, inconclusively, that there are some spillover effects; (ii) productivity levels improved due to an increase in the area for coffee production; (iii) producers participating in the programme increased their income by implementing better soil conservation practices, including decreasing the use of agro-chemicals; and (iv) producers who did not participate in the programme experienced a decrease in their income levels.

The study concluded that, if the programme ceased operations, some producers might continue to apply soil conservation measures adopted with the assistance of the programme. Experience says that, since these techniques are much more labour intensive, in the long run producers may not have necessary incentives to continue adopting them. Unless they see a rapid increase in their income levels, there would be no incentives to continue with the promoted activities. At the municipal level, the main hurdle with the forestry management plans was the lack of revenue-generating activities that would allow the appropriate implementation and management of the resources in the short term. Municipalities, which lack enough financial resources, cannot be expected to manage a forest for years until it matures (for production) and begins to generate a positive cash flow.

In conclusion it can be said that the El Cajón programme fell short of providing any mechanism or instrument to allow for cost recovery. This made it necessary to extend the programme in order to identify and apply mechanisms to guarantee its financial sustainability. This was done in consultation with the beneficiaries. Therefore, after expiration of the loan in early 2002, only if the additional sustainable financing mechanisms that are devised are implemented and become successful, there will be a relative guarantee that the efforts to promote soil conservation measures, forestry management plans, management of protected areas, etc., will continue.

Lessons Learned from the Water–Development Point of View

There are four important lessons to be learned from this case:

a) *The impact of a project on economic growth has many aspects, macro and micro, which should be evaluated.* In the present case, only the impact on the macro scale was taken into account to go ahead with the hydroelectric project. It seems reasonable to expect that this was accomplished because of the relative weight of the project in the overall energy picture in the country and in the generation of income from exporting energy. Even taking into account any agricultural productivity that might have been lost because of the flooded area, this appreciation would probably not change, given that the contribution of agriculture to the gross domestic product (GDP) of Honduras is hardly over 20 per cent. The reasons are the relative small size of the sector and the fact that the reservoir is located in an area characterized by low agricultural productivity. The flood peak attenuation effect during the hurricane Mitch event, although circumstantial because the project was not designed for that, also seems important. This would not have increased development, but most probably attenuated the setback suffered by the country in this respect, since it diminished the effect of the flood in the most productive area of the country (downstream). In balance, the effect on development (economic growth) at a macro scale of the El Cajón hydroelectric project most probably has been positive. However, on a micro scale, in the adjacent project area there are indications that the effect was partially negative, at least initially, only benefiting a small proportion of the population (only those that participated in the project). Unfortunately, no quantitative data was at hand to attempt these balances and the net effect on a more objective basis.

b) *In low-income areas, the effect of a project on equity and poverty alleviation should also be evaluated locally.* In this case, this analysis was done only after the project had been completed, which prompted the initiation of a supplemental programme for the management of the natural resources of the contributing watershed. Although the main objectives of this supplemental programme were environmentally related, it included activities aimed at increasing the income and welfare of local communities. In this case, some quantitative data was available to evaluate the effect on the local population, as stated previously. The result of this evaluation gave indications of an apparent positive contribution to development at the local and regional level by increasing the income of the population participating in the programme. Unfortunately, this data also indicated a low likelihood for these activities to survive beyond the execution period of the programme. If this happens, the positive contribution would have been no more than a mirage.

c) *To have a positive effect on development at a local and regional level, the local income-generating activities should be sustainable.* Unfortunately in this case, this was only evaluated near the completion of the programme. Fortunately, it was done just in time to include additional activities in yet another supplemental programme, to design financial sustainability mechanisms. As this new programme has just started, the long-term positive or negative contribution to development at the local and regional level still hinges in the balance.

d) *The importance of defining impact indicators cannot be ignored.* In the present case, some impact indicators that attempted to measure the effect of the supplemental programme on local income generation, gave clues about the sustainability of the activities and, therefore, the likelihood of the project making a positive contribution to development at the local and regional level. Without impact indicators, all considerations about a project having a positive or negative effect on development, however it may be defined, is merely speculative. It is very important, when designing a programme such as the one depicted, to be able to invest enough resources and efforts to design a series of impact indicators that could provide the necessary information from the beginning (baseline) to allow the proper monitoring and evaluation of the project.

Information for this case study was obtained from the following sources: IDB Project Information Centre (2002); García, et al. (2002); Rossi (1999); San Martin (2002).

Segredo Hydroelectric Project in Brazil

Background

Latin America and the Caribbean have the highest electricity coverage (84 per cent) of any region in the developing world. This percentage, however, as usually happens with coverage percentages related to other basic services, disguises the fact that approximately seventy-five million people still lack electricity, mostly in rural areas. Driven by economic development and demographic growth, it is expected that the demand for energy in the region will continue to increase over the next decade.

But hydropower is not what it used to be during the 1970s when the whole region favoured this source of energy. Most of the best sites have already been developed and new sites face severe environmental restrictions, causing the long-run marginal cost of hydropower to increase relative to other alternatives. A key feature of investments in hydroelectric power generation projects is that they involve long-term loans with long grace periods, lengthy construction phases, and significant risks. Consequently, the new trend of using private financing for energy projects favours thermal plants.

There are, however, many countries in Latin America in which hydro alternatives still offer the most economic alternative. One of these is Brazil, where the energy sector has been one of the most important sectors for the economic development of the country, especially since the 1950s when the process of industrialization began.

In the mid 1970s and into the 1990s, Brazilian authorities aimed to achieve specific goals in the energy field, namely: (i) the economic use of available domestic resources; (ii) replacing foreign sources of energy with domestic energy sources; (iii) accelerating the exploration of new sources; and (iv) conservation and rational use of energy. As a result of policy initiatives, investments to increase hydroelectric generation boosted capacity from 16,200 MW in 1975 to 45,600 MW in 1988, or 86 per cent of the total installed capacity.

During the same period, total energy use increased at an 8.3 per cent annual rate, fuelled by, among other factors, a rapid growth of the GDP, and state support for electricity-intensive industries. Industry accounted for 54 per cent and household for 21 per cent of total use by 1988.

From 1978 to 1988, in the state of Paraná, the Companhia Paranaense de Energia (COPEL), increased its gross system generating capacity from 1,500 GWh to 6,200 GWh. Most of this increase was produced in hydroelectric plants. Over the same period, energy use increased at an annual rate of 9 per cent and the number of consumers rose from 850,000 to 1,800,000. Per capita energy use in Paraná in 1988 was estimated at 1,135 kWh, below the 1,400 kWh for the whole country. The Segredo Project, once delayed because of slow demand, now fitted within the Brazilian and Paraná programmes for expanding and strengthening the electric energy sector.

Description of the Project

The Segredo Project included a hydroelectric power plant on the Iguaçu River with a first-stage capacity of 945 MW along with its dam, spillway, powerhouse, associated structures, and extension of the existing transmission lines. It was estimated to be the power facility with the lowest per-megawatt cost in Brazil. The project also included studies on potential environmental impacts and the implementation of measures to alleviate any adverse effects. Financing from the IADB for US$ 135 million, out of a total cost of US$ 745 million, was obtained in 1990.

Issues

The creation of the reservoir, however, affected some 550 families totalling approximately 2,750 persons, 40 per cent of which were property owners or long-term occupants. Fifty per cent of the households earned less than US$ 110 per month.

A resettlement programme was designed to assure land tenure, housing, and the means to increase family income levels. Resettlement was geared to provide not only improved living conditions for the rural population but also to establish new homesteads on much better land for farming activities along with agricultural services. The company carrying out the project, COPEL, was actually the first hydroelectric company in Brazil to carry out, on its own, environmental plans and programmes to preserve the natural surroundings and improve living conditions of the families in the affected region.

Already in 1986, five years before the start of disbursements, COPEL had established a group to propose solutions for the resettlement of families to be displaced by the project. In addition to representatives from various government departments, members from rural and city workers' associations and unions, and the mayor's office of affected municipalities contributed. Together they approved criteria, basic principles, and procedures for the appropriation of land and relocation of the population.

Each resettled family received operating funds to cultivate 12 ha of soybean. Community wide, the electric company has provided 11,000 fence posts, 385 rolls of wire, lime for soils, the construction of bridges over two rivers, a health post with equipment as well as legal assistance. In addition, COPEL has constructed churches, a community centre, a state-run elementary school, and provided telephone services. COPEL also agreed with local associations (representing the resettled communities) to establish a revolving credit fund. The credit fund is a common pool of resources to which all association families have access. Typical investments include: mulching machines, milk cows, beef cattle, horses, beehive boxes, sewing machines, refrigerators, tractors, and fruit trees. Loans could be paid back in sacks of corn.

Lessons Learned from the Water–Development Point of View

This case illustrates how an important investment in infrastructure, despite facing important environmental and social issues, was able to contribute to development not only at a macro strategic level, but also at the local level. This contribution came at a crucial moment when Paraná and Brazil needed it most. Not only it was able to contribute significantly to ease supply restrictions to an increasing demand for the industrial expansion, but may also have contributed to stimulate this demand because of the low cost of the energy generated. In this case, although the project by itself did not create development, it contributed to it by making it possible. An important necessary condition was met.

On the other hand, it had the vision, from the very beginning, to include activities aimed at the local scene, minimizing the social cost involved and supporting income-generating activities and projects of the local communities.

Unfortunately, impact indicators were not at hand to prove these intuitive speculations. The information for this case study was obtained from the following sources: Fortin (1999); San Martin (2002); Millán (1999); IDB Project Information Center (2002).

Urban Development and Sanitation in Bolivia

Background

The generalized economic and social crisis that affected Latin America in the 1980s did not exclude Bolivia, one amongst the countries in the region with the lowest per capita income. Successive governments worked to overcome the results of that crisis, stabilizing the economy and embarking on a decentralization process. This process began in the mid-1980s with strong emphasis on municipalities and encouraging participation of the civil society in decision making and the regulation of local investments. Municipalities played a key role in this decentralization process, as they were made responsible for basic sanitation and other infrastructure. However, public investments had not recuperated by the end of the decade, and the conditions in the cities were still acute due to rural migration and increasing urban poverty.

Description of the Project

To support Bolivia in this effort, financing for the Global Programme for Urban Development and Sanitation (PRODURSA-I) in the amount of US$ 60 million, out of a total cost of US$ 80 million, was approved by IDB in 1990. The programme included three components: (i) credit; (ii) institutional strengthening; and (iii) infrastructure. It included water supply, sewerage, solid wastes, drainage, urban and inter-urban roads, bridges, markets, energy, and bus terminals, with a demand-driven approach.

In fact, PRODURSA was conceived as an operative instrument of the social and economic policy of Bolivia, whose short-term objectives included among others, to recover the public investment levels that had been severely depressed in the 1980s.

Thus, the objectives of the programme were stated as 'supporting the country in its regional and municipal strategy, in order to enhance the quality of life of the beneficiary population through the allocation of resources for the financing of investment projects that meet the pressing needs of the population at the local and municipal level, and are compatible with national and regional priorities and financial resources available at the national and regional levels'.

Issues

A more closely and explicitly related project with stated development goals and objectives of a country would be difficult to find. At its completion, it was determined that the general objective had been accomplished since it channelled investment resources responding to priorities set by the population at local and municipal levels, which were compatible with national priorities and the availability of financial resources.

However, some interesting results were also noted. First, being a demand-driven programme, only two sectors received financing: roads and sanitation. In financing amount, roads scored higher, while in number of projects, sanitation took the lead. Second, investments were concentrated in the La Paz, Santa Cruz, and Cochabamba major urban axis. One of the reasons behind this rather surprising outcome, despite regional incentives made available to less developed communities, was the limited institutional capacity of potential borrowers.

Also, lack of quantitative information precluded determining the degree to which the programme had reached the poor, as well as its significance in terms of benefits received by women and children or how it contributed to improve health conditions. The results of PRODURSA-I, therefore, were described as 'mixed'.

In 1993, US$ 64-million financing for a total of US$ 80-million PRODURSA-II programme was approved, targeted mainly at overcoming the shortcomings of PRODURSA-I, resulting in a more even geographic distribution of investment projects, although there was criticism about the lack of promotion of decentralization. Also, sanitation and transportation investments continued to dominate, although other types were also demanded. Again, lack of numerical indicators precluded the determination of the low-income population effectively reached.

Lessons Learned from the Water–Development Point of View

Independent of how a project such as this performed, several interesting conclusions can be made.

a) There is no doubt about the intentions of the project to specifically link the programme with development objectives, although no numeric specific goals were given.

b) The fact that water supply and sanitation projects were included, along with other types of infrastructure, only underscores the fact that water projects, along with the others, are a necessary but not sufficient condition for achieving the stated development goals. But interestingly enough, sanitation, along with roads, were deemed 'more necessary' than others by the population.

c) The availability of funds and existing demand notwithstanding, the actual investments were limited. This also underscores another interesting fact: institutional capacity is needed to meet resources with needs. The mere existence of funds is not enough and 'software' is as important as the 'hardware'.

d) Finally, the lack of numerical indicators to monitor and measure the impact is, again, a major hindrance to evaluate the relationship between water and development.

The information for this case study was obtained from the following sources: IDB Project Information Centre (2002); IDB (1997); FNDR (1992).

Conclusion

The discussion presented in this chapter and the lessons learned from the three cases leads us to the following reflection: They might be pointing in the direction of an approach that received much attention in the 1970s but was later abandoned—that of promoting the analysis of water–related projects from a development impact perspective. A methodology that would enable the analyst to 'isolate' the impact of water resources projects from the rest and thus allow for the evaluation of their contribution to development goals was needed then and is still needed now.

References

Cordeiro, Newton V., 'El Manejo de Cuencas Hidrográficas Internacionales', Paper presented at the II Watershed Management Latin American Congress, Mérida, Venezuela, 1994.

Fondo Nacional de Desarrollo Regional (FNDR), *Evaluación Técnico Operativa del Programa Global de Crédito Regional para el Desarrollo Urbano y Saneamiento*, La Paz, FNDR, Bolivia, 1992.

Fortín, Charles, *Segredo Hydroelectric Project*, EVO, IADB, Washington DC, 1999.

García, Luis E., *The Effect of Data Limitations on the Application of Systems Analysis to Water Resources Planning in Developing Countries*,

CUSUSWASH Water Management Technical Report No. 7, Fort Collins, Colorado State University, Colorado, 1971.

—— *Integrated Water Resources Management in Latin America and the Caribbean*, Technical Study ENV-123, Inter-American Development Bank, Washington DC, 1998.

García, L.E. and J. Valdés, 'Water Resources Sustainability for the Next Millennium, the Latin American Case', in H. Maione, B. Maione and R. Monti (ed.), *New Trends in Water and Environmental Engineering for Safety and Life*, A.A. Balkema, Rotterdam, 2000.

García, Luis E., Diego J. Rodríguez, and Mark Wenner, 'The Financial Conundrum of Watershed Management Programs: Experiences from the Inter-American Development Bank', Paper presented at the Watersheds 2002 Conference, Ft. Lauderdale, Florida, 2002.

IADB, *Ex-Post Evaluation, Prodursa I: The Global Program for Urban Development and Sanitation, Bolivia*, EVO, Washington DC, 1997.

—— *Strategy for Integrated Water Resources Management*, ENV-125, Inter-American Development Bank, Washington DC, 1998.

IDB Project Information Centre (2002), *http://www.iadb.org/exr/pic/pictx.htm*.

Lee, Terence, *Water Resources Management in Latin America and the Caribbean*, Studies in Water Policy and Management, No. 16, Westview Press, Boulden, Colorado, 1990.

Millán, Jaime, *The Future of Large Dams in Latin America and the Caribbean: IDB's Energy Strategy for the Region*, Inter-American Development Bank, Washington DC, 1999.

Paiva, P., Presentation made at the meeting *Multilateral Institutions: A New Approach for a New Millennium*, Madrid (Unpublished), 2000.

Pan American Health Organization (PAHO), *Mid-Decade Evaluation of Water Supply and Sanitation in Latin America and the Caribbean*, PAHO, Washington DC, 1997.

Rogers, Peter, *Water Governance*, Working Paper for Panel 2, presented at the Seminar on Strategic Water Resources Issues in Latin America and the Caribbean: An Action Agenda, Fortaleza, Inter-American Development Bank Annual Meeting, Brazil, 2002.

Rossi, Lorena, 'IDB's Dam-Related Projects', Inter-American Development Bank, (Unpublished), Washington DC, 1999.

San Martin, Orlando, *Water Resources Problems and Challenges in Latin America and the Caribbean and Alternatives for Solution*, Working Paper for Panel 1, presented at the seminar on Strategic Water Resources Issues in Latin America and the Caribbean: An Action Agenda, Fortaleza, Inter-American Development Bank Annual Meeting, Brazil, 2002.

WHO/UNICEF, *Global Water Supply and Sanitation Assessment Report 2000*, WHO/UNICEF Joint Monitoring Program for Water Supply and Sanitation, New York: WHO/UNICEF, 2000.

12

Salto Grande: A Binational Dam on the Uruguay River

Lilian del Castillo de Laborde[1]

Origin of the Binational Project

Description of the Uruguay River and the Stretch between Argentina and Uruguay

The Uruguay River is a long and large waterway that has its origin in the Federative Republic of Brazil in Serra do Mar. After running for 1,800 kilometres it joins the Paraná River to form the River Plate. It is one of the most important rivers of the River Plate basin covering some 3,200,000 km². The Uruguay River basin has a surface area of approximately 368,000 km². In its first section, of approximately 500 kilometres, in Brazilian territory, it has a winding course running from east to west on undulated land. When the Pepirí-Guazú River coming from the right margin joins it, it becomes the border between Brazil and Argentina and changes its course towards the south-east running some 800 kilometres until it is joined by the Cuareim River (Queray) on its left margin. In this three-border area (Brazil, Uruguay, and Argentina), it changes course again, this time towards the south, and becomes the border between Uruguay and Argentina, and finally flows into the River Plate. In this last section, from Santo Tomé onwards, which is

[1] The author wishes to thank the officials of the Joint Technical Commission of Salto Grande for their invaluable assistance in the preparation of the present paper, which would not have been possible without their collaboration and concrete support. Nevertheless, the text and any involuntary mistake or omission that it might have are of her own.

approximately 500 kilometres long, the undulating ground changes, and the river flows evenly, with less turns (Figure 12.1). Its course slowly widens as it flows through flat land. Nevertheless, the basin is on evenly distributed ground with areas that do not exceed 200 m above sea level. The whole basin is non-seismic. Only the after effects of faraway earthquakes have been recorded here.

The climate of the basin varies at different heights, lengths, and latitudes. The Uruguay River has its course at 30° south, which has a subtropical climate. As it runs south, the climate changes to a Mediterranean one. Hence, rain and vegetation also change. While at its origin 2,400 mm of rain are recorded annually, in the south, in the

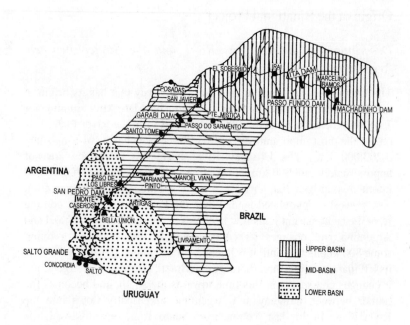

Figure 12.1: The Uruguay River Basin

Salto Grande dam area, 1,200 mm of rain are recorded annually. The rains are not constant due to meteorological phenomena, which may change its volume from one year to another. Therefore, the quoted figures are only an average between the years of rain and drought. The temperature in the basin is quite regular. The only variations are the result of different heights and latitudes and the influence of the atmospheric mass of the tropical Atlantic Ocean all along its course. In the high part of the basin, this influence moderates the wind speed, which blows at 7 km/h, while in its last section winds are much more violent and may even become small cyclones.

The River Plate was discovered at the beginning of the sixteenth century by Spanish explorers. While exploring the South American hinterland, these explorers sailed along the Paraná River and founded the cities of Santa Fe and Corrientes on its margin, and the city of Asunción, the capital of the Republic of Paraguay, on one of its biggest tributaries, the Paraguay River. From the seventeenth century onwards, the Uruguay River was also explored, and it became the only access to the subtropical regions of its high basin, a part of the Guaraní nation. Sailing along the Uruguay River was easy, from the River Plate to Barra Concepción, in Misiones, today an Argentine province, except in its central section where a drop of 35 metres breaks a slope along 175 kilometres, from the Bella Unión–Monte Caseros line to the so-called Hervidero area, 25 kilometres downstream from the Concordia–Salto line. In this section, there is a reef, called Ytú by the native inhabitants and renamed Salto Grande by the Spanish discoverers, another waterfall called Salto Chico and many smaller rapids. Only when water levels were high could ships with a shallow draft sail over these reefs. This is why the Jesuits, colonizers of the subtropical region and founders of missions in the area, which are still called 'Misiones', sailed to and from Buenos Aires in a small fleet of rafts built with tree trunks that had to be dismantled piece by piece to be able to cross the Salto Grande area.

After the Spanish Crown expelled the Jesuits in 1768, Francisco de Paula Bucarelli, the Spanish governor, promoted a regular flow of goods and travellers from that area. Mate, tobacco, cotton cloth and cow hides, horses and otters were the goods most frequently traded. Where sailing was not possible, goods were carried in carts. Between 1777 and 1789, roads were improved in the Yapeyú area and new towns were founded. One of them was Federación, a city-port in the current province of Entre Ríos, Argentina. In 1820, a navigation route was established between Sao Borja (Federative Republic of Brazil) and Concordia (Argentina)–Salto (Uruguay) despite the difficulties found in the mid-section of the Uruguay River. When the railroad was built in 1874, the cities of Concordia and Federación were interconnected and the hurdles of Salto Grande were avoided. Later, the railroad was extended to Monte Caseros and it was possible to cross the Uruguay River. In the same period, 1874–87, the Republic of Uruguay extended its railroad to Salto and Bella Union. In this way, the tripartite area of the upper Uruguay basin, partially extended into the Republic of Paraguay, continued to be the preferential north–south water and railroad connection that ended in Concordia and Salto.

By the end of the nineteenth century and the beginning of the twentieth century, these cities became important regional overseas ports. Local shipping companies linked ports from the lower section of the Uruguay River (Paysandú, Fray Bentos, Mercedes, Carmelo, Colón, Concepción del Uruguay, Gualeguaychú) to the capital cities of Buenos Aires and Montevideo on the River Plate. Small vessels were used in the upper Uruguay and bigger ones in the lower section to favour trade and urban development using river transportation. Later, the development of railroads and newly built roads caused a substantial drop in river transport and the rate of growth of the city-ports such as Monte Caseros, Bella Unión, Concordia, Salto and others was interrupted. Navigation also received a setback because larger vessels could not be used in the mid-section of the Uruguay River. During the last third of the nineteenth century, electricity brought about a qualitative change in economic activities. Electric power, appearing first in the United States and the European countries, started to be used in the region and the demand for power production increased. Although requiring an important investment of capital at the beginning, hydroelectric power has low operational costs and production is continuous and inexhaustible, resulting in a low price for the electricity obtained.

The region we are particularly interested in is the lower section of the river running between the territories of Argentina and Uruguay. While Argentina has a federal political system, the Republic of Uruguay has a unitarian one. The Argentine provinces on the banks of the Uruguay River are, from north to south, a small part of Corrientes and all of Entre Ríos. The departments of the Republic of Uruguay on the Uruguay River are, from north to south, Artigas, Salto, Paysandú, Rio Negro, Soriano, and Colonia. On both banks, agricultural production prevails: grains, flax, citrus, and rice are the most important crops, followed by cattle breeding, and industrial production in third place. Industry is mostly centred on the transformation of farming raw materials in the Federative Republic of Brazil. The area of influence of the Uruguay River is the state of Rio Grande do Sul, south of Brazil, a state with borders with Argentina and the Republic of Uruguay. The economy of this area is essentially based on agriculture and cattle; some areas have important industries as well, supplemented with forestry.

The abundant flow of the Uruguay River, a determining factor for production of electric power, together with its slope, has a historic average of 4,622 m^3/sec. The maximum flow measured, since 1898, was 37,714 m^3/sec, recorded on 9 June 1992, and the minimum flow as recorded on 3 February 1949 was 109 m^3/sec. The total outlet capacity of the plant is 64,000 m^3/sec.

First Projects

At the end of the nineteenth century, a businessman from Concordia, in the province of Entre Ríos, Gregorio Soler, prepared a modest first draft project to take advantage of the Salto Grande falls to benefit the two nations, and in 1890 requested a licence for the production of electric power. Soler's project was improved over two decades and finally he submitted it to the Argentine Senate on 21 September 1911, asking for a ninety-year licence 'to use the waters of Salto Grande to produce power'. He submitted it again in 1912 and 1913, 'on his own behalf and for a mostly Argentine and Uruguayan capitalist group'. In his last presentation he asked for the approval of his project and the rejection of one proposed by Mollard and Company, as it was a copy of his own. The existence of two projects brought about confusion even in the Argentine Senate. Gregorio Soler had contracted an engineer, Maurice Mollard, in Paris for technical advice, but then Mollard, acting on his own with European financial assistance, proposed a much more important project than Soler's. A peninsula of the existing lake in Salto Grande, used for recreation and tourism purposes, has been named after the pioneer of this bi-national project.

In 1907, Juan T. Smith, an engineer, submitted another project to the Government of the Republic of Uruguay. This project included electric power production as well as the improvement of navigation in the waters upstream the Salto area on the Uruguay River. The governments of Argentina and Brazil agreed to study Smith's project, which was done applying European methodologies in hydroelectric production. But when, in 1911, Gregorio Soler submitted his project, the negotiations stopped, and the project remained just one of several projects that were submitted at the end of the nineteenth century and the beginning of the twentieth century. One of them was the project proposed to the Argentine Congress by one of the representatives of the Chamber of Deputies, C. Otaño, from the province of Entre Ríos in 1894. He proposed to build a canal running on the side of the falls. The Salto Grande hydroelectric complex has been built, but this part of the project is still pending.

Maurice Mollard submitted his request for a licence to the Argentine Senate on 23 August 1912, on behalf of European capitalists, to 'be given the right to use and exploit the flow of the Uruguay River'. He proposed the construction of a dam, on the Salto Grande, approximately 17.50 m high and 2,200 m long, crossing the Salto; a navigation canal with a lock on the river bed; a diversion-canal on each margin; and hydroelectric power stations of equal power at the end of each of the canals. At the same time, he proposed a dam with wickets, a waterway, and locks in Hervidero; and eventually a diversion canal and a hydroelectric power station on each margin. The project would have the following benefits: (i) it would guarantee navigation of vessels of ten-feet (3.05 m) draught, which could sail to Santo Tomé, on the Uruguay River between Paso el Hervidero and Concordia and between Concordia and Santa Rosa; (ii) it would reduce the risk of flooding; (iii) it would produce electric power; and (iv) it would allow irrigation, either by directly using the water in the reservoir or by raising the water of streams and lakes with the electric power provided by the Salto Grande power stations. The power station on the Argentine territory would have a minimum 75,000 horsepower capacity. The license was requested for ninety years. Similar requests were to be made to the governments of Uruguay and Brazil. The project requested a total subsidy of 14,500,000 pesos in strong currency, of which 8,350,000 was to be paid by the Argentine government, 4,500,000 by the Brazilian government, and 1,300,000 by the Uruguayan government, all in the same currency. The total cost of the project was estimated at 28,500,000 pesos: 16,225,000 for navigation and 12,275,000 for electric power production. As compensation the consortium offered to produce and sell a minimum of 200,000,000 kilowatt-hour (kWh) annually, which would be sold at one cent per kWh to benefit agriculture, and which it would be free to sell at a price directly agreed upon for industries, the municipalities, and the provinces. Of that minimum production, the Argentine government would benefit with a thousandth of a peso per kWh and with an additional half thousandth per kWh of the two hundred millions fixed as a basis.

The French technician had come to the country accompanied by other French and German experts. Engineers and politicians from Uruguay and Brazil helped him in the study and support of his draft project, the most complete project presented at that time. The request for a licence was discussed during the sessions of the Argentine Senate held on 11 and 13 September 1913, with a favourable report of the Commission of Public Works. The reporting representative, engineer Virasoro, stressed the advantages of the project for the navigation of the Upper Uruguay River and the growth of the Corrientes and Misiones economies in Argentina. He underlined the contradiction set by this navigable river, 1,800 kilometres long, with an obstacle 200 kilometres from its estuary that had not yet been solved. He joined the Commission in support of this last project since it doubled the amount of power that Soler's project would be able to produce, which proposed a lower dam and did not include navigation facilities, although it did not request a subsidy from the government. Senator Del Valle Ibarlucea maintained that such a large project should be built by the public sector, so that such a resource would not be given to an individual. A timetable was proposed: the future licensee, had to obtain the approval of the governments of Uruguay and Brazil before 1 January 1915. The Brazilian government approved the project and included the amount in its 1914 budget to eventually take part in the project. Although the Argentine Senate approved this draft project, it was not discussed at the Chamber of Deputies the following year, when the First World War started. Mollard did not drop his project. Once the international conflict ended, he returned to the region in 1921 and started new studies on the geological and hydrological characteristics of the basin related to his project. Nevertheless, the desire of sponsoring a project proposed by the national administrations had been born, and in 1919 the Argentine government ordered the General Office of Navigation and Ports to study the hydroelectric potential of the Salto Grande falls in the Uruguay River, as well as the Iguazú Falls, in the tributary of the Paraná River and the Apipé rapids in the Paraná River. All of these have currently been transformed into the Yacyretá bi-national dam (Argentina-Paraguay). The final report published by engineers Humberto Gamberale and Francisco Mermoz in 1928, included the three areas.

With respect to the Uruguay River, the project on Salto Grande was largely based on Mollard's project, with the addition of a detailed technical analysis of the variable water flows and general and specific estimates concerning the transmission of electric power to the consumer areas, specially the city of Buenos Aires. Moreover, in 1930, the Uruguayan government asked Adolfo Ludin, an engineer to contribute new ideas to the project. To this, the Italian engineer Angel Forte added his project in 1941. While the technical and economic aspects of these projects were being improved, the Congress began to show interest in the project and began to discuss the corresponding legal and financial matters. On 27 August 1936, a representative of the Chamber of Deputies from the province of Entre Ríos, Bernardino Home, submitted a bill before the Argentine Congress, requesting that the executive be allowed to build, directly or through contracts with private companies, after a public tender, the necessary works to improve navigation on the Uruguay River, in order to have a depth of 'a minimum of nine feet north of Concordia and of eighteen feet south of Concordia'. In relation to hydroelectric production, it was decided that the Office of Navigation and Ports would prepare the final project of the works that would be carried out on Salto Grande and Hervidero 'based on the projects and studies of Engineers Maurice Mollard, Humberto Gamberale and Francisco Mermoz'. It was also decided that the Argentine government would negotiate with the Republic of Uruguay and the Republic of Brazil all necessary agreements on the construction of the projects, navigation, electric power production, and irrigation. Also, those agreements would establish the contributions of each party with regard to the payment for the design and construction work. The bill also considered the expropriation of the land that would be covered by water as a consequence of building the reservoir, as well as the indemnity to be paid to the affected landowners, and the projects of the ports to be built, planning an overseas port for Concordia. Finally, it provided for the issue of government bonds and other treasury papers to finance the future works. This far-sighted bill, foresaw the regular navigation along the upper Uruguay River by vessels of nine-feet draught, and by overseas vessels up to the city of Concordia. The benefits of the hydroelectric power from the Salto Grande and Hervidero falls, the irrigation of a large agricultural region, the regular preservation of the river bed, and the elimination of the effects of periodical flooding, were not immediately discussed. Nevertheless, it was under its influence that the Salto Grande project and the creation of the Technical Joint Commission (JTC) were included in the 1938 Agreement.

Bilateral Understanding Between Argentina and Uruguay

The agreement signed on 13 January 1938 was not the result of negotiations aimed at the shared use of the Uruguay River. Its main purpose was to find solutions to the conflict about the islands on the Uruguay River that had arisen between Argentine and Uruguayan authorities and settlers since no clear borders had been established. The lack of definition of the sovereignty of the many islands on the Uruguay River in the bilateral Argentine–Uruguayan section led to an exhaustive negotiation, which, in 1960, ended in a treaty on borders[2] where, in some sections, due to the difficult task of setting a single line to serve all purposes, different demarcation borders were established to divide waters and islands.[3] However, under the circumstances prevailing in 1938, setting borders was still a faraway aspiration; at that moment the main concern was to avoid conflict between neighbouring states and reach an agreement on hydroelectricity and any other possible uses of this fast-flowing river. No reference was made to specific districts, although the area of the rapids known as Salto Grande and Salto Chico, with its natural falls, was the most appropriate place for such a project. Since the Uruguay River was the border, the projects could only be carried out with the consent of both countries. Therefore, it was agreed to establish a 'status quo' on the situation of the islands on the Uruguay River and go ahead with the use of its waters. To that end, the aforementioned Commission would be appointed to 'jointly study the hydrographic features of the Uruguay[4] River'.[5] Although the reasons to negotiate were different, they led to the agreement. It should be stressed that, in this case, the conflict on the border did not bring about a dispute but became the starting point for

[2] Treaty on Borders of the Uruguay River, between Argentina and Uruguay, signed on 7 April 1961 in Montevideo, Republic of Uruguay, which was enforced after the exchange of the ratification instruments on 28 December 1965.

[3] Treaty, Article 1, B, II.

[4] Argentine-Uruguayan Agreement, 13 January 1938, Article 1.

[5] Agreement mentioned in footnote 2, Article 2.

a joint project and the basis for a bi-national agency, which planned, engaged, controlled, and later managed the hydroelectric complex called Salto Grande.

In 1946, the Argentine–Uruguayan Agreement was finally concluded with the signing of the Convention on the Utilization of the Uruguay River Rapids in the Salto Grande Area and of an Additional Protocol on 30 December 1946. One of the main points of the Convention is the determination to ensure that waters of the Uruguay River are 'jointly utilised in equal shares' (Article 1). Accordingly, it provides that the costs related to surveys and projects such as 'common works and facilities, mainly the dam, the mechanic and electric installation for generation' (Article 4, first paragraph) shall be borne by both parties in equal shares. On the other hand, costs related to the access works, complementary works, transmission lines, as well as indemnities and expropriations in the territory of each country, shall be borne by the respective governments (Article 4, second paragraph). With respect to the ownership of the undertaking, it is clearly set forth that 'Whatever should be the proportion contributed by each of the Contracting Parties, common works and facilities shall be jointly owned in equal shares by the signatory States at maturity of the amortisation period' (Article 4, last paragraph). As regards water management, the agreement provides that 'The various manners of water utilization shall respond to the following order of priorities and no hindering or restrictive application thereof shall be permitted: 1) Utilization for domestic and sanitary purposes; 2) Utilization for navigation; 3) Utilization for power production; and 4) Utilization for irrigation' (Article 3a).

Two years later, the Argentine Congress[6] passed the Convention but the Uruguayan Congress delayed it for more than a decade. In fact, some of the Uruguayan legislators were opposed to the Convention because they felt that it was necessary to set the borders between Argentina and Uruguay in the Uruguay River before the bi-national project was carried out. Although the opposition continued, the commission pushed forward the project until the negotiations on the borders along the Uruguay River began. The agreement on the border was finally signed on 7 April 1961. The Uruguayan Congress did not pass the Convention until 1958.[7] This opposition was not meant to delay the project; so on 19 January 1947 the exchange of notes took

[6] Law 13.213, 2 July 1948 (Official Bulletin 16 July 1948).
[7] Law 12.517, 13 August 1958 (Official Newspaper 27 August 1958).

place and the Protocol was signed, together with the Convention, on 30 December 1946. The Protocol foresaw that while the ratification of the Convention was still pending, the Joint Technical Commission, established under Article 5 of the 1938 Agreement, would continue working. However, despite this, the bi-national agency took no measures to complete the project and implement it.

It was under the initiative of the people from the cities on the margins of the Uruguay River in Argentina, Uruguay, and Brazil, some of whom still did not have electric power such as the citizens of Monte Caseros in the province of Corrientes, that the project was reactivated. In fact, in 1956, the 'Executive Commission on Behalf of the Salto Grande Works' was established in the city of Concordia in the province of Entre Ríos. Other similar commissions were created in other Argentine and Uruguayan cities on the riverside, as well as in some Brazilian cities, such as Quaraí and Sao Borja. In the latter city, a meeting was promoted to join efforts,[8] and this resulted in the creation, in 1957, of the 'International Committee on Behalf of the Salto Grande Dam'. This was to become the origin of joint actions by the cities on the margins of the Uruguay River. The committee got in touch with the media and the authorities of each country in order to implement the project. Between 1956 and 1973, people in support of the organized implementation of the Uruguay River project finally influenced the political sectors. Salto Grande is, therefore, a good example of the joint concern of people and government towards the implementation of a hydroelectric complex. The conflict was generated because the demands made by the people of the region to implement the project had been left unanswered.[9] Its most evident result was when on 25 October 1964, people marched from the city of Salto, on the margins of the Uruguay River, to the city of Montevideo, the capital of the Republic of Uruguay, demanding the implementation of the project. Later, negotiations were started with officials of the governments of both countries to complete the draft project stage and begin the implementation.

The approval of the JTC Technical-Administrative Regulations was a significant step forward on 20 October 1972, and the Argentine and Uruguayan governments signed the Convention on 30 December

[8] II International Congress on support of the Salto Grande works, Concordia, province of Entre Ríos, 28 October 1956.

[9] *When People and Governments want...*, Technical Joint Commission of Salto Grande, 1992, 71 pages.

1946.[10] The Convention defines the joint and non-joint works, the coordination needed for the construction of the projects, in order to avoid causing more delays, the financial resources needed, the ownership of the project once completed, the rules for expropriation and indemnity and the rights on electric power production, as well as the customs-tax exemptions for the materials, equipment, engines, and other articles to be used in the building of the complex, and the tax exemptions awarded to the JTC, its assets and contracts in the territory of both parties. The bi-national agreement, which was first started in 1938 and then in 1946 and in 1973, was completed after the approval of all ancillary documents, setting the conditions to build the projects. In fact, construction started a few months later, on 1 April 1974.

International Legal Framework of the Rio de la Plata Basin

The Uruguay River is one of the large waterways of the Rio de la Plata basin. It flows through the territories of five South American countries: Argentina, Bolivia, Brazil, Paraguay, and Uruguay. The Plata basin follows an institutional system established by the Plata Basin Treaty, adopted by the neighbouring countries on 23 April 1969 in the city of Brasilia, Brazil. The treaty establishes general principles for the development of the region, especially on infrastructure and use of the natural resources of the basin. With respect to water resources, the treaty provides that the participating countries shall 'rationally use the water resources, mainly by controlling the preservation of the waterways and their multiple and fair utilisation' (Article 1).[11] In the institutional framework of the Plata basin, the country members to the agreement adopted the Declaration of Asunción on the use of international rivers, where principles on the use of international adjacent and successive rivers were set up. With respect to the adjacent rivers, it was decided that the projects shall be agreed to between the river-edge countries, and with respect to the successive rivers, 'each State may use the waters according to its needs, without causing significant damage to any State of the basin'.[12] Consequently, the principles of general

---

[10] Passed by Decree No. 789-M74, 20 December 1973, Argentine Republic and Decree No. 1035/973, 5 December 1973, Republic of Uruguay.

[11] *U.N.T.S.*

[12] 'International Rivers and Lakes', document from *OAS/Ser. I / VI, CIJ-75 rev. 2, page 187.*

international law, which acknowledge an equitable and reasonable use of the water resources as well as the obligation to avoid causing significant damage to the other countries of the basin, are incorporated into the Treaty. At the same time, these documents acknowledge the practice followed by the riverside states in prior understandings, as, for example, when the project of Salto Grande was being planned.

The 1946 Convention, as discussed earlier, proposed a conference where the Government of Brazil would be invited to 'consider the modifications that may arise as a result of this Convention in the navigation of the Uruguay River and in the river system under the provisions of Conventions currently in force' (Article 11). The conference met on three occasions and the respective minutes were drafted. During the last meeting, the participating countries, Argentina, Brazil, and Uruguay, agreed to a joint declaration. It is recorded in the declaration[13] that the Government of Brazil 'is pleased to see the joint execution by the Governments of Argentina and Uruguay of the Salto Grande works,' and that 'the Argentine and Uruguayan Governments welcome this statement with satisfaction' (first paragraph). At the same time, 'according to international doctrine and practice, and as recognised by the Argentine and Uruguayan Governments, the Brazilian Government reserves its right to: a) Attempt and obtain, at any time, a fair indemnity derived from any damages that might be caused in Brazilian territory, either during the construction or the operation of the works; b) Be listened, in the case that the participating countries may wish to introduce during the development of the studies any changes that modify the currently scheduled conditions' (second paragraph). Mention is also made to the part that 'In accordance with the international instruments in force and other rules of International Law, the Governments of Argentina and Uruguay recognise the Government of Brazil the right to the free construction of hydraulic works of any nature, in the Brazilian section of the Uruguay River and its tributaries' (fourth paragraph, first part). It was also declared that if the Brazilian government decides to implement works in the river 'according to international doctrine and practice, the Brazilian Government shall previously consult the other border States in the case of constructing hydraulic works that may bring about the alteration of the present

---

[13] Technical Joint Commission of Salto Grande. *Documents and Background. 1938–June 1998,* Montevideo, Republic of Uruguay, 1998, pages 89–90 (Spanish).

condition of the Uruguay River' (fourth paragraph, second part). The declaration of the three countries of the Uruguay River sub-basin agrees specially to incorporate the principles concerning the right of use and prior knowledge of the other countries, to be applied in case of future works in its waterways, which later would be supplemented with equitable and reasonable conditions of use incorporated into the documents of the Plata basin.

Joint Technical Commission

Background

The agreement signed on 13 January 1938 by Argentina and Uruguay created a Joint Technical Commission[14] to carry out surveys to determine the possibility of hydroelectric production on the Uruguay River. Although both governments designated delegates to the commission, and work was started, with the beginning of World War II, a generalized concern for the international community stopped the regular progress of works. Certain studies were carried out, such as the aerophotogrammetric survey in 1945, taking advantage of the river's exceptionally low tide; geologic profiles and perforations were made mainly by Argentine government bodies, transmission lines and access roads were designed, and other aspects that would help to define the project were analysed.

For multiple reasons, this enterprise took decades before it started, and all along this process the Commission went through periods of uncertainty and delays, and other times of enthusiasm and progress. Eventually, the Commission was formally established in 1946 and started to draft the bi-national legal instruments without delay, defining the characteristics of the project and the tasks to be developed. This joint work resulted in the signing of the convention and additional protocol for the utilization of the Uruguay River rapids in the Salto Grande area by Argentina and Uruguay on 30 December 1946.

The 1946 Convention reiterated the need to appoint and maintain a Joint Technical Commission (JTC). Even though this JTC was not the same Commission established under the 1938 Agreement, the studies and previous works already carried out by the later were to be used by the JTC established in 1946. So, according to the provisions of

[14] Agreement mentioned in note 1, Article 5.

Article 6 of the Convention, the JTC was to carry out 'any studies still pending at the time the Commission took office', an example of the continuity that was to be confirmed as time went by.

The Convention of 1946 had to be ratified by each party which could lead to possible delays in Parliament. In order to prevent this drawback, an additional protocol to the Convention of 1946 was signed, which became effective at the time of its signing, and stated that the Commission founded by Article 5 of the Agreement of 1938 (Article 1) would remain in office for the period necessary to ratify the agreement, ratified in 1958. The original JTC would end its tenure after the designation of a new commission by handing over all the documentation it possessed. During this first stage, all the JTC's expenditures, salaries, and personal expenses of the delegates were to be paid by the respective governments.

As regards the Commission's composition, the 1946 Convention establishes that it would have the same number of delegates for each country, without naming any specific national agency in charge of the designation or integration of the respective delegations. The additional protocol, on its part, required each delegation to include an engineer, in order to advise the JTC on the development of studies, the collection and harmonization of the data gathered, and the preparation of graphs on the progress made.

The Commission's headquarters was to be in the city of Buenos Aires, where monthly meetings would be held, though it could meet in other places as deemed necessary. Apart from ordinary sessions, it could hold extraordinary sessions, drafting the corresponding minutes.

Purpose

The function of the JTC, according to the 1946 Convention, was to be in charge of 'all matters concerning the utilisation, the dam, and the change of the course of the Uruguay River waters'. The JTC, during its first stage, made topographic, geologic, and hydraulic studies to define the river's characteristics in the Ayuí area. To that end, two meteorology stations were installed, one on each margin of the river. The cost of these studies and facilities were to be paid by the respective governments, according to the distribution of tasks between the parties. However, these contributions had to be added to the cost of the works and facilities that were to be built jointly.

The JTC was in charge of the design of the project and facilities, and the preparation of estimates, bidding conditions, provisions applicable to the general labour system, and financing programmes, which would be presented for approval to the respective governments. With the approval of the project, the JTC would start the execution, and would be in charge of the acceptance of all partial and final stages.

As a result of the engagement provided by the bi-national understanding reflected in the 1946 Convention, the JTC went into a more detailed definition of the utilization and the characteristics of the studies and design. Simultaneously, financial support for the project was negotiated with international bodies. Negotiations were carried out in 1966 and 1967 with the World Bank. It became necessary to give the JTC the necessary institutional framework to perform its tasks, and this aim was reached with the approval of the Technical-Administrative Regulations, on 20 October 1972. These regulations are still in force with certain modifications.

Legal Status

The JTC has legal status in the internal legal system of both countries, as well as internationally. The 1946 Convention decided to appoint and maintain a joint technical commission. Its functions had to be carried out by the joint technical commission created by the 1938 Agreement, until the convention became effective. The 1946 Convention saw to it that at the end of the construction process, the parties would establish a 'competent inter-state body' for the operation and management of the works and facilities. It further stated that in order to avoid an institutional gap until the creation of that body, 'the Joint Technical Commission shall be competent' for all operations and management tasks.

The Convention also empowered the JTC to adopt its own regulations and a working plan (Article 3). However, the technical-administrative regulations were only agreed upon on 20 October 1972, when both governments decided, through decrees passed that very same year, to allocate funds to continue the JTC's works and to conclude the project's final definitions, especially with regard to its funding, call for bids, and date of initiation. The regulations reiterate that the JTC, defined as the 'Joint Technical Commission established by Article 2 of the Convention',[15] would be 'responsible for all matters concerning the utilisation, the dam and the change of the course of the Uruguay River

---

[15] Article 1c, Technical-Administrative Regulations.

waters', according to the provisions of Article 2 of the 1946 Convention. Nevertheless, it was also to be in charge of the tasks stated in Article 7 of the Convention.[16] Consequently, the JTC becomes the inter-state body referred to in Article 7 of the Convention following the provisions of the regulations that said that the JTC would be the body in charge of paying for the execution of the project, with the funds provided by government sources, with 'those obtained from the use of credit and those resulting from the exploitation of the Salto Grande project, under the terms established in Article 7 of the Convention'.[17]

From the institutional standpoint, the adoption of the agreement in December 1973 was decisive, regulating the 1946 Convention, with the incorporation of some provisions from the Technical-Administrative Regulations that refered to the JTC's legal status. Therefore, the Convention of 1973 reaffirmed the JTC's legal capacity 'to perform public and private actions for the fulfillment of its purpose' (Article 1, Convention of 1973).

Other additional instruments were approved recognizing the JTC's international legal status. The Headquarters Agreement signed on 15 April 1977 between the Argentine Government and the Commission, acknowledged that the Commission would 'have legal status in the territory of Argentina and the capacity to hire, purchase and dispose of goods' (Article 2). Consequently, the Agreement granted the JTC executive, administrative, judicial, and legislative immunity regarding the inviolability of its facilities, files, goods, and possessions (Article 3); it also had 'immunity against any judicial or administrative proceedings' (Article 4). The JTC's goods and possessions were exempt from any kind of taxes, although no exemption could be claimed on contributions or duties for public services (Article 5); the JTC was also exempted from paying taxes. This definition did not include public utilities, that exported or imported goods for the Commission's official use were to be exempt from customs duties (Article 6).

Although the JTC's headquarters are located in the Argentine Republic, it also has facilities, goods, and personnel within the territory of Uruguay. For this reason, an agreement on privileges and immunities under the jurisdiction granted to the Joint Technical Commission of Salto Grande was signed between the JTC and the Republic of Uruguay, on 6 March 1979. This agreement acknowledged that the

[16] Article 2, Technical-Administrative Regulations.
[17] Article 4, Technical-Administrative Regulations.

'Commission is an international organisation with the necessary legal status for the fulfillment of its specific purposes', and added that 'The Commission has legal status in the territory of Uruguay to hire, purchase and dispose in any manner goods and chattels and real estate ...'.

International Court of Arbitration of Salto Grande

The JTC's international legal status and its jurisdiction immunity has been recognized by the contracting parties in the respective agreements, and has led to certain relevant cases at the national courts.[18] The Commission's officials decided to establish an International Court of Arbitration, after having asked the advice of outstanding jurists of both countries,[19] in order to overcome the lack of competent jurisdiction. The absence of jurisdiction would lead internal courts to regulate the granted immunity and to exercise their own jurisdiction according to domestic law. Among the various possible solutions adopted by international organizations, one is to establish arbitration proceedings, which might be ad hoc or permanent. The JTC chose to establish a permanent court[20] with evident advantages, since these previously existing jurisdiction and procedures would not lead to any expenses, and judges are proposed from a list belonging to the Commission when a matter to be solved arises.

The Commission drafted and approved the by-laws and the rules of procedure of the International Court of Arbitration,[21] and nominated the first group of ten judges, five for each country.[22] The judges remain

---

[18] Cases: *Saier S.R.L. vs. Joint Technical Commission of Salto Grande,* National Federal Court of Appeals related to Administrative Litigation, 10 February 1979; *Cabrera J.E. vs. Joint Technical Commission of Salto Grande, on dismissal,* Supreme Court (Argentina), 12 May 1953.

[19] The jurists consulted were Eduardo Jiménez de Aréchaga (Uruguay), former member of the International Court of Justice, and Isidoro Ruiz Moreno (Argentina).

[20] Resolution No. 718/79 of 9 December 1979.

[21] Resolution No. 520/80 of 3 October 1980 and resolution No. 339/81 of 16 September 1981.

[22] Resolution No. 363/81 of 25 September 1981. The Commission designates the judges, selecting five out of a list of ten candidates proposed by the governments, basing themselves on the candidates' background, Article 12, by-laws of the International Court of Arbitration of Salto Grande.

in office for four years, and they cannot be removed, but they may be re-elected indefinitely for equal periods; some judges have been there right since the creation of the court. The Court would have three members for each specific case.[23] The Commission decided that the national majority of two judges would correspond to Argentine judges in even years, and to Uruguayan judges in odd years.[24] The Court decides its own jurisdiction in case of dispute, whereas in case of subject matter jurisdiction it shall intervene in labour issues of the JTC agents, in matters of contractual origin, in the absence of any contractual specifications, and in those of extra-contractual responsibility. Criminal jurisdiction is excluded. The Court shall also be competent to settle all those matters that the JTC decides to put forward,[25] as well as in every other case where its jurisdiction has been agreed upon by the parties, or has been offered by the JTC and accepted by the other party.[26]

The Court of Arbitration, in order to support its decision, shall apply all contractual rules available, the Joint Technical Commission applicable rules and the conventions binding the contracting parties.[27] Should the enumerated rules be insufficient in a specific case, the application of complementary rules is not specifically foreseen and remains a judicial prerogative. The Court is the only jurisdiction, its sentences are then final and unappealable.[28] Moreover, this has been confirmed by the Supreme Court case law of the Argentine Republic.[29]

Structure

The commission's legal representatives are the chairman and the secretary, elected among the members of the delegation. The chairman shall be chosen for six-month periods, in an alternate and rotating basis

[23] Article 11, by-laws of the International Court of Arbitration of Salto Grande.

[24] Resolution No. 365/81 of 25 September 1981.

[25] Article 1, by-laws of the International Court of Arbitration of Salto Grande.

[26] Article 2, by-laws of the International Court of Arbitration of Salto Grande.

[27] Article 6, by-laws of the International Court of Arbitration of Salto Grande.

[28] Article 5, *Ibidem.*

[29] *Fibraca Constructora S.C.A vs. Joint Technical Commission of Salto Grande, on Appeal of Action,* CS, 17 September 1993; *Ghiorzo.*

by each country. The parties shall alternate the chair and the secretariat and at no time shall both belong to the same delegation.[30]

The system for decision making was already agreed upon—the 1946 Convention had established that all resolutions should be adopted by the majority of its members. However, in case of a tie, that is, in the event of it being impossible to reach a decision with an affirmative vote of the majority of all its members, the minutes with the issues pending are to be submitted to the respective governments. The governments would try to solve the problems in direct negotiations, but in case a consensus is not reached, 'disputes shall be submitted to arbitration'. The system to solve disputes determines, therefore, a first stage of negotiations, and if unsuccessful, the creation of a court of arbitration to solve them. The convention also states that all communications between the JTC and the respective governments shall be addressed to the ministries of foreign affairs of both countries, due to the commission's international status and because the topics under its jurisdiction regarding the shared utilization of an international river, belong to the international policy of the countries along the riverside.

The JTC estimates the expenses and resources for each fiscal year, and presents the budget with the annual statement and the action plan to both governments.[31] The budget includes salaries, maintenance expenses for the whole project, the purchase of equipment, and an overall insurance.

In human resources, whenever possible, the principle of having the same number of nationals from both countries is followed.[32]

According to the JTC's international legal status, the delegations from both countries shall have the same immunities in the other party's territory regarding their duties. It is thus stated in the Headquarters Agreement with the Argentine Republic[33] and in the Agreement on Privileges and Immunities with the Republic of Uruguay, granting the members of the Argentine delegation the same immunities and privileges corresponding to foreign diplomatic officers authorized before the Government of Uruguay, and to the administrative staff the same system applicable to the administrative staff of foreign diplomatic

---

[30] Article 16, Technical-Administrative Regulations.

[31] Article 8, Technical-Administrative Regulations.

[32] Article 3(d) and Article 6.4, 1946 Convention, and Article 36, Technical-Administrative Regulations.

[33] Article 9.

missions.[34] The JTC may consider as international officers, those staff members who perform duties at the Argentine headquarters and do not have Argentine nationality.[35] The delegation members, the international officers, and all JTC personnel, are exempt from paying any kind of taxes and contributions on their salaries, and are allowed to import and export, tax free, their furniture and belongings.[36]

Operation

The JTC has been operating and maintaining the hydroelectric system since 1979, making all efforts necessary to provide the highest safety standards to the riverside populations and to the dam. Informatics has been incorporated to optimize management with high technology, which was non-existent when the project was designed. As of 1998, the Commission has introduced and is applying a quality management system (QMS), according to the ISO 9000 (ISO 9002) standards, to all areas of electric power generation and transmission.

More than 100,000 people visit the project annually showing the interest this complex has created. This has contributed to boosting tourism in this region. In addition, as part of its relationship with the community, the JTC carries out ongoing cooperation programmes through the national delegations. The delegation from Uruguay has developed an important plan for rural electrification, improving dramatically the lifestyle of the surrounding population. Simultaneously, the Argentine delegation has sponsored the overall repair of rural schools in the province of Entre Ríos. Both delegations cooperate in the preparation of classes and conferences regarding the use of electric power and in the organization of school visits to the dam. Publication and research activities are also jointly carried out, such as the International Seminar on Operative Hydrology, held in May 2000. This seminar drew experts from the field of water resources management from several South American countries and Canada, and representatives of the bi-national bodies of Yacyretá and Itaipú.

Power generation by the project is efficient. The levels forecasted in the design of the project have been exceeded. This operative effectiveness must, however, not be at the cost of inconveniences

---

[34] Articles 11 and 12, Agreement on Privileges and Immunities.

[35] Articles 9 and 10, Headquarters Agreement.

[36] Articles 9 and 10, Headquarters Agreement, and Articles 13 and 15, Agreement on Privileges and Immunities.

caused to the populations located around the lake and downstream area of the dam. The demand of electric power by the consumption centres requires an operation that maximizes the electric power generation up to the limit of its best operational level. This high yield point is determined not only by natural conditions but also by the multiplicity of uses and environmental preservation. Therefore, it is the JTC's task to achieve the delicate balance of responsible management.

The operational effectiveness of the complex, from an environmental standpoint, is further conditioned by environmental policies applied by the communities located on the riverside. In fact, the JTC's jurisdiction is limited to engineering work, dam waters and surrounding areas, and a park area of around 500 ha under its administration. Every municipality on the riverside is in charge of environmental management, setting the quality standards in their respective districts. The JTC promotes the relationship with the communities along the river to work jointly and for their mutual benefit.

Salto Grande Hydroelectric Complex

Building the Dam

As previously mentioned, the Uruguay River flows through a rather flat and even land, except in some areas. Most of its course runs across the South American fertile plain called the Humid Pampa (Pampa Húmeda), a region which comprises the south of Brazil, Uruguay, and the central region of Argentina. The geography of the region defined the Salto Grande project and the people of this region chose the site, the technology, and the system of operation.

The Salto Grande complex has a reservoir 144 kilometres long, a surface of 78,300 ha (783 km^2), and a volume of 5,500 hm^3. Its maximum width is nine kilometres. It is the only power link between Argentina and Uruguay and is connected to the Yacyretá dam (the Argentine-Paraguayan bi-national dam). Works on the project started in 1974 with the building of the lateral landfill and the engine rooms, one on each margin. The civil works made significant progress between 1947 and 1979; 1,500,000 m^3 of concrete were used. The river bed was dredged to a depth of 30.50 metres and the total volume excavated was 2,500,000 m^3. At the beginning of 1979, the civil works had made good progress, allowing the water gates to be placed to build the reservoir. The filling began in March 1979. It was a slow operation due to the characteristics

Figure 12.2: The Salto Grande Dam site

of the dam which was built partially from concrete with earth and stone laterals. The filling was finished in October of that same year reaching a storage level of 33 m, as agreed, although this is not the maximum capacity of the complex so far as electric power production is concerned. It must be taken into account that in all the documents the storage level is based on the hydrographic zero level of the Riachuelo (Argentina)[37] (Figure 12.2 and 12.3).

On 21 June 1979, the turbines were tested for the first time and on 12 July of the same year commercial production began increasing its

---

[37] According to the Additional Protocol to the Convention of 30 December 1946, Article 2a.

Figure 12.3: The Salto Grande lake

capacity gradually until its full operation in 1982 with the start up of
the last of the fourteen production units, seven in each jurisdiction.
The dam is 69 m high measured from its foundation and 39 m over the
river level. The railroad bridge over the dam is 39.50 m above sea level.
The two engine rooms and the two control buildings situated
downstream are part of the complex. The central spillway has nineteen
hydraulic radial water gates. The total capacity of the dam is 64,000 m³/
sec and it has two outlet tunnels, one on each margin, used in case
of an extraordinary flood.

Resettlement of the Population

In 1990, the filling of the lake of the dam flooded productive land, and
the people and the city of Federación, Argentina, had to be moved—the

first experience in the country due to the building of a hydroelectric complex. The long waiting time after the first projects were submitted— the Convention was signed in 1946 but work on the project started only in 1947—brought about a decade of uncertainty for the people living in the Uruguay River region. The placement and characteristics of the project would have an effect on the areas to be flooded. This was discouraging for the producers and inhabitants on both margins. The 1946 Convention stated that the 'indemnity and expropriation to be carried out in the territory of each of the countries shall be made by the respective Governments'.[38] Consequently, the properties on both margins, which would be affected by the building of the dam, were declared to be of public interest and, therefore, subject to expropriation. The area of expropriation and the conditions in which the land was going to be acquired were determined.[39] Only part of the urban area of the city of Federación was included. On 12 October 1974, the city dwellers had to vote in a plebiscite with five possible alternatives. Most of them chose a place called La Virgen, on the banks of the future lake of the dam, to build the new city of Federación.[40] In 1976, while the dam was being built, the construction of the new city was decided upon.

Federación was founded in 1810 in the province of Entre Ríos and in 1847 was moved to the banks of the Uruguay River. In the 1870s, European immigrant farmers settled there. In the 1970 census, the municipality of Federación had 6,162 inhabitants, in both rural and urban areas. In March 1979, although the new city was not yet finished, the inhabitants were moved from the old Federación to the new Federación, officially inaugurated on 25 March 1979. According to its inhabitants, adapting to the new place took a decade. As seen today, the results have been positive in the mid-term, as the local economy has been reactivated by tourism, not only because of the dam but because of the thermal waters discovered in the region. In 2000, the population was over 14,000 people, twice the population of the old Federación. This is above the average index of urban growth of the country. The inhabitants, who have already adapted to their new city, are worried about plans to expand the complex and oppose the rise of

---

[38] Article 4, second paragraph, 1946 Convention.

[39] Acts No. 19,210 of 1971 and No. 20,139 of 2 June 1973, Argentine Republic.

[40] Consequently, Act No. 21,125 of 1975 of the Argentine Republic enlarged the expropriation to all the city.

the dam's water level. They believe that part of the city would be affected by it and it would result in more than just economic problems for them. Although they are aware that most of the new dwellings are better and more comfortable than those now under the water, they stress that the social effects produced by the forced change were significant. It must be clarified that the JTC will not take any decision without the consensus of the population of the dam area. It expects complete participation of the people in the decision-making process.

At the same time, the law on expropriation of the lands linked to the complex was passed in the Republic of Uruguay and the people had to be moved.[41] The dam led to the expropriation of real estate in the towns of Belén and Constitución and the city of Bella Unión in Uruguay.[42] On both margins, the rural zones were considered case by case.

Energy Production, Transmission, and Marketing

The hydroelectrical power station has fourteen generators with Kaplan-type turbines of Russian origin. The fourteen turbines, seven on each end of the dam, and their respective engine rooms are part of one single power station.

Both engine rooms are interconnected with a total installed capacity of 1,890,000 kW and an annual production average of 6,700 GWh. In the year 2001, its annual average production reached 8,486 GWh. The flow, since the time of the filling of the dam in 1979 until 2001, has been 31 per cent over the average value increasing from 4,288 m³/sec to 5,611 m³/sec.

The electric system of Salto Grande is operated from the control centre that coordinates the activities of the complex between the two countries to which it supplies energy. This is done through the national agencies of each country in charge of load transmission (CAMMESSA in Argentina and UTE in Uruguay). The centre controls, in real time, the operation of the entire system of the Salto Grande plant and is in charge of the surveillance of the system through the main parameters. It also operates by remote control the main functions of the power sub-stations.

The two countries agreed to begin the energy interconnection simultaneously with the initiation of the works,[43] and finished when the

[41] Act No. 13,958 of the Republic of Uruguay, 10 May 1971.
[42] Act No. 14,627 of the Republic of Uruguay, 28 December 1976.
[43] Agreement on Energy Interconnection, 12 February 1974.

dam was already in full operation.[44] It is a permanent and constant system to exchange energy and electric power in each national interconnected system.[45]

The JTC has never traded energy directly. In the first stage this task belonged to the two state companies, Agua y Energía Eléctrica (A y EE) (Argentina) and Administración Nacional de Usinas y Transmisiones Eléctricas (UTE) (Uruguay) which were responsible to the JTC.[46] While the same company still operates in Uruguay, in Argentina it was replaced by Emprendimientos Energéticos Binacionales S.A (EBISA).[47] They are in charge of energy marketing activities for the Salto Grande complex, the Yacyretá bi-national entity and the nuclear stations.

Tariffs were discussed since the first bi-national understandings. In the 1946 Convention it was provided that the 'value to be assigned to joint works and facilities shall allow power production at a lower cost than the one that could be obtained at a thermal plant of equal power installed in the work areas'.[48] Tariffs are revised periodically and when permanent modifications are needed specific agreements are concluded.[49] The income collected through tariffs is devoted to services and the payment of the debt contracted to build the project, which was partially paid for with contributions from the governments and external finances.[50] The tariffs of the Salto Grande complex are, owing to its

---

[44] Agreement on the Implementation of the Agreement on Energy Interconnection, 27 May 1983.

[45] Agreement on the Implementation of the Agreement on Energy Interconnection. Article 2.

[46] The Argentine Republic and the Republic of Uruguay guarantee the commitments made by the companies A y EE and UTE respectively or its successors or licensees before the JTC, Article 19 of the 1973 Ruling Convention.

[47] Created by Decree 616/97 (Argentina), 7 July 1997.

[48] Article 4, fourth paragraph, 1946 Convention.

[49] Different agreements were signed after an exchange of notes. The last one was on the System of Tariffs for the Power Produced by Salto Grande signed on 31 July 1996.

[50] Each government would contribute, to the common projects, US$ 40 millions, of which US$ 5 millions would be paid per year and per country as of the year 1973, according to the implementation schedule. The rest would be financed with private credits or with credits from international agencies. The non-common projects would be the responsibility of each party. Article 7 of the 1973 Ruling Convention.

operating efficiency, very competitive with respect to the other production sources of interconnected systems in both countries.

Between 1977 and 1986, the transmission network was built to transport the energy produced by the hydroelectric station to the consumer centres of Argentina and Uruguay. Six sub-stations linked through 1,300 kilometres of aerial lines form the transmission network, of 500 kW. The transmission tension used allows a great deal of energy to be conveyed in a technically and economically profitable way through long distances since a line of 500 kW has a transmission capacity of approximately fourteen lines of 132 kW. Two of the sub-stations, which are part of the joint project, are situated in Argentina and two in Uruguay, and they transmit energy to the inner area; two additional ones interconnect Argentina and Uruguay. Transmission lines, which have not been built in common but by each of the nations, stretch out from the sub-stations in the territory of each state. The Commission operates the common sub-stations, connecting the lines and the hydroelectric power station, the so-called 'common ring' of the whole 500 kW, twelve lines, switching sub-stations and interconnecting lines (Figure 12.4).

The passage of the high tension lines are fastened to a right of way. The aerial lines of high tension travel across rural as well as urban zones, restricting the use of the land. The right of way has a maximum width of 77 m, following the unevenness of the ground, in rural zones. Homes may not be built within 69 m of this strip of land. Buildings of one floor, without terraces or balconies may be built on the rest of the strip. Forestry is limited as the trees near the buildings should never be very tall so that if they fall, they should be at least 5 m away from the conductors. Plantations under the transmission line should not exceed a height of 3.40 m. There must always be a road of free access to the line which the owners of the land are obliged to provide.

The 1946 Convention did not establish the water level of the reservoir for the future, an aspect that could be of interest for Brazil as it is the country with longer river borders. In the convention it was decided to invite Brazil to a conference to discuss the modifications that could take place in the navigation and the river system, after the convention had been concluded.[51]

On 8 August 1967, the Joint Technical Commission informed the two governments that the studies were finished. The time schedule proposed

---

[51] Article 11, 1946 Convention.

Figure 12.4: The Salto Grande power network

indicated that the complex would start operations in 1979. Work was begun in the first semester of 1974. A hired international consultant, Acres, proposed to leave the old project of two twin power stations aside and build one single power station in the middle of the dam with the spillway on the left margin of the river (Uruguay). Finally, the idea of the two power stations prevailed but this delayed the project in 1970 and 1971, circumstance that also delayed the call for tender and the negotiation of financial resources.[52] The company chosen to implement the final project, Main y Asociados, developed, since 13 April 1973, the

---

[52] The Yacyretá project, an Argentine-Paraguayan project on the Paraná River that was on the design stage at the same time, adopted the design having only one engine room which was set on the right margin (Paraguay) and only one navigation lock on the left margin (Argentina).

project including two power stations and added the railroad bridge. This company was in charge of controlling construction during the whole implementation stage.

The electromechanical equipment was commissioned to the Russian company Energomachexport associated to the Argentine company Ingeniería Tauro S.A. In December 1973, the first loan was received from the Inter-American Development Bank (IADB) and the bidding started for the civil engineering works, which would cover the construction of lateral dikes, a spillway, nineteen openings, outlet tunnels, two engine rooms, an international railroad bridge, and a navigation lock. This work was to be done by the consortium Empresa Constructora Salto Grande S.A. formed by several international and national companies. Companies of different countries were hired for ancillary buildings. A list of their names appears at the entrance of both pavilions. The dates included in the 1967 schedule were respected since work began on 1 April 1974 and the first turbine started to produce electric power on 21 June 1979. In the year 2001, Salto Grande was producing approximately 50 per cent of the electric power of the Republic of Uruguay system and 7 per cent of the Argentine Republic system.

Other Uses

The lake of the Salto Grande dam has favoured sports and tourism activities. Apart from the facilities for water sports, 500 ha around the lake have been assigned for recreational activities. The discovery of hot springs in the area have added to its attraction. Hotels, which provide thermal waters and facilities for water sports and vessels, have been built on both margins. The Commission has prepared a plan for the whole area, which has been incorporated into the domestic law of the countries, and in which the urban, rural, and industrial uses of the area around the lake have been established.

Navigation along the Uruguay River has been interrupted by the Salto Grande dam, upstream the cities of Concordia (Argentina) and Salto (Uruguay). The Ayuí falls, covered by the lake, has always been a hurdle for navigation and one of the purposes of the hydroelectric complex, pointed out as second priority in the Convention of 20 December 1946, is to use that waterway for navigation. Priorities changed during the implementation stage of the project. During the nearly three decades that went by between the signing of the convention and its implementation, bridges were built over the Uruguay River to facilitate the movement of people and goods. The design for the navigation system of the Salto Grande, on the Argentine margin, has been fully finished and the system partially built. The site chosen, after analysing sixteen alternatives which combined different possibilities for the site and the dimensions of the navigation canal and locks, does not modify the bi-national character of the project.[53] The navigation canal is formed by two locks, the Ayui lock and the Salto Chico lock and the navigation canal connecting both, approximately 13 kilometres long, and the bridge lock. Of the complete project, the Ayuí lock, which is 135 metres long and 24 metres wide, has been built; but its hydraullic equipment is not operational. The Salto Chico lock shall have similar dimensions and both locks shall have a hydraulic system to let the water in and out. Once the works are finished it shall be possible to sail downstream through the Uruguay River to the border with Brazil without any obstacles.

Above the dam, at a height of an 2,486 m, is an international road and railroad bridge. Thus, the expanded road system of the region is interconnected and the railroad systems of Argentina, Paraguay, and Uruguay are linked for the first time. The bridge, toll free, is part of the hydroelectric complex. There is only one customs and sanitary migration control on one end of the bridge carried out by the authorities of both countries. Traffic of vehicles is very active throughout the year.[54] There is a neighbourhood system that allows the inhabitants of the cities of Concordia (Argentina) and Salto (Uruguay) and of other nearby towns to move about freely. The cargo trains transport grains, cotton, wood, and other commodities every day. The importance of the permanent links between Argentina and Uruguay have to be underlined owing to the infrastructure deficit of the region.

Operations at the Power Station

Environmental Management

The environmental factor has been taken into account in the Salto Grande project, right from the time of the feasibility stages, to the

---

[53] December 1973 Agreement to Rule the Convention of 10 December 1946, Article 3.7.

[54] During the year 2001, 86,570 vehicles, cars, trucks, and buses passed over it.

construction of the dam's start up. The JTC carries out a systematic control programme in order to meet its environment quality goals, and performs the necessary surveillance to detect any perilous situation that might be caused by the dam's operations. An environment master plan has been designed, including the environmental impact assessment programme and the permanent environmental surveillance programme. The first one is based on level fluctuations and flow management for operations at the station. The environmental surveillance programme is used for emissions, flow and waste control of all activities performed at the station, as well as the supplies used. The evaluations of eutrophication, of toxic substances, recreation areas, fish, and wild fauna and flora is carried out. Different areas are monitored. To this end the complex has been divided into the reservoir and its immediate area of influence, the station and its facilities, the transmission system, the dam and all sub-stations. Simultaneously, an environmental system has been set up, based on the ISO 14000 model for all JTC activities. The regional integration assistant-management and its ecology department, a part of the general management, has taken care of all environmental aspects, with increasing resources. During the construction period the quality of the waters was improved, effluents were controlled, and forest species and local vertebrate animals were protected. In addition, an important number of species, threatened by changes in their natural environment and the filling of the reservoir, were rescued and relocated. Simultaneously, the control of environmentally deleterious agents was carried out, specially important in the transmission of diseases, and a basic cleaning infrastructure was installed, which included the regular elimination of industrial wastes. The JTC developed other activities to protect water conditions, such as deforestation and removal of vegetation and its waste from the area of the artificial lake bed, in order to minimize the eutrophication of the waters of the dam and to contribute to recreational navigation and fishing. Keeping in mind the new conditions of the ecosystem, specific legislation was passed with regard to the need to reforest the lake margins, to increase their strength and renew its vegetation, to build up piers and marinas in order to have access to the lake, to start programmes to carry out human and industrial settlements, and space distribution.

Despite these efforts, during certain periods, algae have grown disproportionally in the lake, and research on this issue has been done jointly with the University of the Republic of Uruguay. Another topic which is being studied in the dam is the presence of invasive molluscs (*corvícula flumínea* and *corvícula largillierti*), as well as in other reservoirs of the region. Ichthyc resources which are abundant in the Uruguay River and its tributaries, have been affected considerably by the dam. Two scales of fish were formed, one on each margin. The area affected by the dam has been modified by the reservoir and the productive development of the upper basin. So, representative environments of ecosystems with natural evolution are scarce. However, there are some areas which have their own characteristics, such as the dam islands and other environments, with important flora and fauna demanding a better knowledge to insure their preservation. The semi-natural environments where a spontaneous. colonization of wild fauna was formed, deserve special consideration. Surveys carried out are mainly inventories of the region's representative species, management of the wild fauna found in the dam area, re-breeding of endangered species, and planting of indigenous tree species.

Operation and Maintenance

Electrical power generated by the Salto Grande station is destined to the interconnected systems of Argentina and Uruguay. The figures of power production for January 2002, are as follows: monthly production was 197,131 MWh, with a historical average of 370,391 MWh for January; the average flow was 823 m^3/s, with a historical figure of 2,400 m^3/s for January. These relative figures show the sharp drop to a low-water mark in the present season, reducing the flow to one-third, and power production to one-half. Out of the total amount of power produced, 86,181 MWh (45.97 per cent) was for Argentina, and 101,307 MWh (54.03 per cent) for Uruguay. Power available was 1,575 MW. As regards its worth, the amount of electric energy and the power supplied reached a total amount of US$ 4,856,998, with an average sale price of US$ 27.32 /MWh. These figures clearly reflect the effectiveness and competitiveness of power production by the Salto Grande station.

In order to optimize its operation, the station has a very important hydro-meteorology network, formed by 158 conventional stations and fifty-one automatic stations. The latter report almost immediately, the amount of rain and the height variation of the tributaries of the Uruguay River. The reservoir capacity allows for irrigation of an area of 100,000–130,000 ha.

During the 1990s, the Argentine electric model went through a qualitative transformation that led to big investments in electric power production, improved the station, and brought tariffs down. However, distribution and, especially, transportation of power did not meet the initial expectations, so surplus electrical power did not increase further. Demand stagnation in Argentina, caused by recession, and the Brazilian energy crisis, derived in the export of power to Brazil, which took place in two stages of 1,000 MW each. Thus, the power surplus was reduced to 5 per cent of demand. It must be considered, however, that should this situation be reversed, an energy crisis in the sub-region can be expected within two or three years. So far, the equipment's effectiveness has been maintained after twenty-two years of operation. The Commission plans to create a fund to pay back assets, following a programme based on the facilities' useful life, in a twenty-five year period. Periodically, an inspection of the dam's soil and concrete structures is carried out. The adequate instruments of general and particular resistance have been placed in the soil and concrete dams. The results observed meet the project's defined standards.

Moreover, the repowering of turbo-generators has been planned, with no environmental impact, in order to incorporate high technology not only in equipment but also in evaluation and planning tools. Technological progress will increase power generation and production at a better price, without any environmental impact.

New Multiple Use Projects

The Commission has carried out surveys to increase the dam's yield. More investment could improve its operations and the environmental cost–benefit ratio. The first stage will be to study the environmental impact of a fixed increase of the water level of the reservoir to 36 m. This will also increase the volume of the lake by 27 per cent. It will improve the station use and will lessen flooding. The affected lands shall be expropriated up to a 36.5 m water level. The estimated investment to pay for expropriations and some ancillary works would be US$19 million, which could be paid back in two years. The water level will be adapted to the present characteristics of the Uruguay River basin, although the river level upstream will be increased on the dam up to the city of Monte Caseros. The 4 per cent increase in the falls will produce an important power surplus.

As already mentioned, navigation is the hydroelectric complex's pending stage. The Convention of 1946 states that 'all measures to be adopted by the Salto Grande project shall not affect any rights of the

Contracting Parties regarding navigation on the Uruguay River'.[55] The construction of a compensatory dike, located 100 kilometres downstream from the present dam would allow navigation at nine feet in the rocky and most difficult section. The project would not only allow navigation on the Uruguay River, but would also increase power production. In fact, reduction in power generation because of the new works would be compensated by the installation of additional generation. The dam's water level would be set at +8 m, thus preventing current variations on the river bed under normal conditions of operation, with a greater capacity to reduce overflows and floods. Work will be carried out through a public works license, and the investment will be paid back with surplus power after an estimated thirteen-year period. The set of projects for technological improvement of the higher water level of the complex, and the construction of a compensatory dike, might produce, according to modest estimates, an increase of 400 MW in power generation.

Experience Achieved

After more than twenty years of power generation, various advantages and disadvantages have been detected. For a dam built on a plain, variable volumes cause changes on the river and erosion on the riverside. These damages are more serious in the cities of Salto (Uruguay) and Concordia (Argentina), located twenty kilometres downstream from the dam. These cities go through periodical floods. To lessen the damages the station has had to operate above the nominal water level during extraordinary overflows. It must be noticed that Salto Grande was designed with a medium flow forecasted at 4,288 m^3/s, according to the available historical information (for the period 1898–1969) during the project stage.

Later, considerable modifications of the river's hydraulic characteristics were detected that will probably be permanent, produced by alterations in the use of soil, deforestation, and climate changes. Consequently, the dam sometimes exceeded the project's chosen expropriation water level, causing damages to nearby population. The mean record for the 1983–2000 period, as mentioned, was 5,611 m^3/s, that is, 31 per cent higher than that forecasted in the project. Eight out of the ten biggest overflows of the last hundred years took place during the period of

---

[55] Article 10, Convention of 1946.

operation of the station. In 1979, 1982, 1983, and 1986 extraordinary overflows occurred, and in the decade of the 1990s, floods took place in 1992, 1997, and 1998. Indemnity for damages caused by this increase in water level have already been paid to the injured parties, or have started to be paid. That is why the decision to increase the original number of twelve generator units to fourteen was made during the execution stage. The maximum flow through the turbines was thus raised from 7,200 m^3/s to 8,400 m^3/s, an approximate increase of 17 per cent. However, this change was not taken into account for the modification of the river edge and for the water levels to be considered for expropriation purposes of plots affected, maintaining the original flow of 7,200 m^3/s. Consequently, damages to lands and farms increased and worsened during the extraordinary floods, thus causing uncertainty in the neighbouring populations even though they had been indemnified. On the other hand, the dam's functioning made it possible to successfully regulate the significant variations in the flow regime. In 1992, during the largest flood of the twentieth century, with a flow of 37,700 m^3/s, the station kept downstream water levels three metres below the height they would have reached, had the river not been regulated. This reduced flood damages.

Planning and building projects for the utilization of water resources is part of any region's development demands. In fact, the best use of water is an essential element for economic growth, meeting the population's growing demand for drinking water, water for agriculture, river transportation, and electric power, which can be obtained from hydroelectric production, while protecting the land from floods, using the reservoirs for irrigation, and many other needs of the population. In order to get a higher efficiency between demand and available water quantity and quality, planning and hydraulic engineering works are needed.

Hydraulic works are essentially an environmental management activity, which will bring about greater benefit to the region and its inhabitants, whenever the negative effects of these projects are reduced, whether at the building or the operational stage. At the River Plate basin, also formed by the Uruguay River, a large number of dams, presses, floodgates, and other forms of flow regulation projects have been carried out mainly during the second half of the twentieth century, modifying the natural systems. The reservoir areas have had sustained growth, especially at the Paraná River and some of its tributaries, such as the Iguazú River, though such a trend has not been equally shown on the Uruguay River.

So far, the Plata basin countries have acquired great expertise in design, construction, and operation of large hydraulic projects. Now, as an essential twenty-first century vision, all this must be complemented with the adequate management of natural resources within the framework of environmentally sound activities. The need to protect the lifestyle of the population living in the region demands the integration of technical-hydraulic engineering with environmental needs and goals. The extensive dimensions of these projects increase the importance of good environmental management. Society is still demanding, though unsuccessfully, that common goals be found to effectively solve the relations between the utilization of water, environmental protection— mainly the human environment from a humanity-centred view—and society's requirements. This equation responds to the elusive concept of sustainable development. The operation of the bi-national hydroelectric power station of Salto Grande must be headed towards this goal. The main priority is the generation of electrical power,[56] although this does not mean that electricity is an end in itself. The riverside countries underlined this purpose when they decided to start this project to take advantage of the great benefits of the natural conditions the Uruguay River rapids offer in the Salto Grande area, for the economic, industrial, and social development of both the countries.[57]

The project was designed under the concept of multiple utilization, to favour regional development through the generation of competitive power, but also building an international bridge, with the creation of areas for sports and tourism, with the control of water, and consideration of flora and fauna. It included improved navigation on the Uruguay River. It is still pending because of low economic demands. It is insufficient to finance a profitable project and operation. Summing up, the Salto Grande is not only a large bi-national dam but a significant contribution to regional development.

[56] Article 5, Act of 13 January 1938 signed by Argentina and Uruguay, which states: 'Considering that the utilization of the water power of the Uruguay river is of common interest for both countries, they agree to promote the designation of a Joint Argentine–Uruguayan Technical Commission that shall start the corresponding study and report accordingly to both Governments as soon as possible'.

[57] Preamble, first paragraph of the Convention and Additional Protocol between Argentina and Uruguay, for the utilization of the Uruguay River rapids in the Salto Grande area, on 30 December 1946.

Index

Adıyaman 156, 159, 160, 163, 164, 166, 167, 172, 175, 186, 190, 200, 225, 242
Africa 65
  Southern 251–289
Agenda 21, 84, 277
Agricultural residues 7, 9, 98
Agricultural subsidies 67
Agroforestry 320
Agroindustries 51, 174, 207, 220, 221, 222
Agrochemicals 91, 130, 321
Ahmad, Z.U. 83–113
Akpinar, 240, 241-242
Albertani, F.B. 309–330
Algeria 258
Altinbilek, D. 190, 211, 224, 228, 240, 248
Angola 254, 259, 261, 264, 270, 275, 276, 277, 278
Aquaculture 34, 209, 211
Argentina 331–367
Arsenic contamination 91, 108
Asia 14–53, 89, 91, 144
Asunción Declaration 342
Asymmetrical development 114
Atatürk, M.K. 1

Bangkok 16, 17, 18–21
Bangladesh 9, 83–113
  National Water Plan 93
Barrage 83
  Farakka 106
  Mahananda 106

Nagara 148–151
Pangsha 107–108
Teesta 9, 106, 107
Basic human needs 41, 42, 43, 44, 51, 108, 161
Bateni, N.J. 290–308
Batman 156, 157, 159, 160, 163, 164, 166, 167, 172, 186, 190
Bay of Bengal 86, 90, 92
Benefit-cost analysis 1, 29, 85, 98, 364
Bhutan 10–12, 84, 85, 87, 98, 102
Bilateralism 85, 105, 112
Biodiversity 10, 59, 95, 139, 273
Biomass 97
Biswas, A.K. 1–13, 191, 248
Bolivia 319, 327–329, 342
Botswana 252, 254, 259, 261, 262, 263, 264, 267, 270, 272, 275, 276, 277
Bottom-up 24, 43, 44, 52, 93, 180
Brahmaputra Master Plan 94
Brazil 6, 25, 319, 324–326, 331–367
Build-own-operate 94
Build-own-transfer 191
Budget constraints 29
Burundi 252

Calabarzon 24, 25
California 290–308
Cambodia 25
Canada 104, 351
Canal lining 79

Caribbean 324
Capacity building 12, 41, 165, 176, 178–179, 200, 278, 321
Central Valley 290, 292
China 6, 16, 22, 23, 24, 25, 48, 84, 102, 103, 104, 114–136
Climate
    Change 91–92, 365
    Sub-tropical 24, 26, 45, 52
    Tropical 24, 26, 45, 52, 91
Colombia 313
Coastal submergence 92
Commission, Orange-Senqu 270–271, 279, 280
Communication 8, 21, 41, 43, 200, 201, 221, 240, 244, 246
Comparative advantages 42
Compensation 4, 95, 151, 153, 224, 228, 243, 245, 246
Conflicts 41, 83, 84, 114, 130, 139, 261, 268, 312, 317, 339
    Southern Africa 254–260
    Upstream-downstream 116, 121, 125, 133, 142, 152
Congo 252
Corruption 95
Cost allocation 27
Cost effectiveness 1, 47, 52, 77, 93
Cost recovery 94, 123, 134, 321, 322
Credit 67, 68, 224, 326, 327
Crop cycle 33
Crop diversification 51, 220
Crop rotation 79
Crop yield 5, 50, 80, 200
Cropping intensity 7, 38, 180
Cropping pattern 38, 39, 111, 172, 180, 221
Cuba 258, 264, 265
Cultural heritage 198

Dam
    Aswan High 7, 9
    Atatürk 8, 174, 190–250
    Bireçik 191, 194
    Calueque 276
    Check 54, 58, 78
    Driekoppies 224
    Earthquake-resistant 96
    Gariep 256
    Ilisu 195
    Kapti 98
    Karakaya 192, 197, 202, 209
    Keban 202, 209
    Letsibogo 271
    Maguga 274
    Massingir 272
    Molatedi 263, 271
    Pongolapoort 274
    Rockfill 95
    Salto Grande 331–367
    Sanmen Gorge 123
    Sapta Kosi High 106, 109, 111
    Sterkspruit 273
    Sunkosh 106
Decentralization 17, 97, 133, 188, 313, 319, 327, 328
Deforestation 8, 9–10, 11, 59, 72, 73, 98, 100, 362, 365
Del Castillo, L. 331–367
Demand management 54, 300, 316
Desalination 72, 146
Desertification 73, 77, 83
Development objectives 1–2, 27
Devolution 17
Disaster management 105
Diyarbakır 156, 157, 158, 159, 160, 163, 165, 167, 172, 175, 186, 225
Drainage 6, 46, 121, 132, 137, 165, 180, 191, 242, 313, 316, 327
Dredging 106, 108, 148, 276, 299, 352
Drought 29, 34, 55, 57, 58, 59, 65, 67, 91, 114, 117, 123, 125, 126, 127, 129, 138–139, 143, 144, 148, 260, 292, 310, 333
Dublin Conference 314

Earthquakes 59, 96, 147, 151
Economic efficiency 1, 2, 139
Economic zones 23, 124
Ecosystems health 40, 245, 306, 307
Education 8, 14, 59, 65, 69, 86,
    156, 162, 165, 175, 176, 200,
    201, 222, 223, 240, 241, 242,
    244, 246, 309
Egypt 25, 112
Employment 5, 6, 8–9, 12, 68, 69,
    70, 96, 127, 154, 160, 171, 172,
    175, 178, 182, 187, 200, 201–
    207, 221, 222, 243, 244
  Skilled 6, 202, 204, 207, 213,
    222–223, 242, 244
  Unskilled 6, 202, 204, 206, 222,
    227, 241, 242, 244, 246
Endangered species 363
Energy
  Audit 80
  Conservation 324
  Crisis 78, 142, 364
  Price 11
Environmental activists 2–3, 139,
    315
Environmental degradation 73, 80,
    83, 100, 175, 198, 199, 221,
    244, 245, 247, 299, 300, 305
Environmental impacts 9, 11, 46,
    48, 72–76, 98, 148, 149, 150,
    152, 197, 198, 199, 208, 243,
    246, 247, 299, 304, 307, 315,
    320, 362, 364
Environmental issues 2, 12, 91, 93,
    95, 98, 100, 107, 111, 129–130,
    132, 139, 142, 143, 145, 148,
    152, 153, 180, 223, 224, 245,
    247, 274, 296, 298, 299, 302,
    304, 305, 306, 316, 317, 323,
    352, 361–363, 366, 367
Environmental security 84, 165
Equity 22, 85, 133, 247, 323
Equitable use 343
Erosion 89–90, 121, 137, 198

Europe 334, 336
Eutrophication 362
Evaluation 1, 27, 28, 29, 32, 152,
    180, 181, 200, 300, 305, 306,
    307, 323, 329
Evaporation 30, 78, 91, 121, 276
Export processing zone 23

Family planning 69
Farmers, small 64, 67, 171, 220
Fertilizer 70, 133, 198
Fisheries 65, 83, 84, 92, 140, 149,
    150, 174, 187, 207, 208, 209,
    210, 211–220, 241, 242, 296,
    298, 299, 301, 304, 305, 362,
    363
Fish migration 150
Flood 91, 92, 105–106, 114, 117,
    118, 121, 123, 129, 138, 140,
    143, 144, 146–147, 148, 198,
    320, 322, 336, 339, 354, 356,
    365, 366
  Control 27, 29, 47, 85, 87, 89,
    90, 96, 98, 106, 118, 137,
    138, 141, 143, 145, 148,
    151, 292, 301
  Forecasting 85, 105
  Warning 85, 105
Fluoride contamination 74–75
Fluorosis 74–75
Food and Agricultural Organization
    38, 198
Food security 6, 65–68
Foreign investment 105
Forest 11, 14, 30, 60, 65, 95, 137,
    140, 145, 146, 169, 187, 320,
    321, 322, 362
Free trade 26, 52
France 315, 337
Fuelwood 7, 8, 9, 11, 98
Funding agencies 108, 113

Gabon 252
Ganga Action Plan 91

Ganges Treaty 85, 87, 106, 107
Gansu 120, 121, 122, 124, 129, 131
Garcia, L.E. 309–330
Gaziantep 156, 157, 158, 159, 163, 164, 165, 167, 172, 186, 190
Gender issues 5, 8–9, 60, 68, 70, 92, 160, 162, 163, 164, 165, 166, 176, 177–179, 181, 201, 242, 244, 245, 315, 328
Geographic Information System 34, 37, 38
Geology 30, 96
Geomorphology 111, 320
Geopolitics 29
Germany 258, 261
Globalization 42
Global warming 91, 102, 121
Gobi 102
Governance 181, 278, 318
Grassland 30, 132
Grassroots 43, 93
Greenhouse gases 98, 102
Groundwater 26, 29, 30, 31, 33, 34, 36, 45, 46, 58, 72, 77, 80, 81, 91, 108, 125, 132, 144, 147, 169, 195, 276, 301, 302, 307
  Assessment 30, 31
  Pollution 45, 46, 108, 118, 119, 120, 147, 148, 198, 223
  Mining 58, 65, 66, 73, 79, 123, 137
  Recharge 77, 78, 81, 87
Guatemala 310
Gujarat 54–82
Gupta, R. 54–82, 154–189

Harran Plain 172, 173, 181, 197
Hashimoto, T. 14–53
Health issues 5, 8, 44, 59, 60, 65, 68, 69, 70, 74, 86, 87, 91, 92, 165, 175, 176, 179, 180, 198, 200, 207, 223, 227, 240, 242, 244, 245, 309, 328

Helsinki Rules 270, 277
Henan 115, 117, 121, 124, 126, 129, 131
Heyns P.S. 254, 266, 267, 268, 269, 272, 273, 274, 275, 276, 277
Himalayas 10, 89, 92, 94, 95, 96, 98, 102, 111
Honduras 319–320, 322
Hong Kong 138
Human Development Index 59, 60
Human resources development 29, 41, 43
Human rights 76
Hydrogeology 31, 78
Hydrological data 29, 37, 86
Hydropower 6, 7, 10, 11, 12, 27, 39, 48, 50, 77, 81, 85, 95, 97–105, 107, 121, 137, 140, 141, 145, 168, 170, 190, 201–207, 294, 300, 315, 316, 319–320, 324–325, 336, 337, 338, 339, 340, 351, 361, 363, 364–365, 366
Hygiene 45, 59

Immigration 73, 121
Impact indicators 327, 329
Imperial Valley 294
Incentives 221
Income distribution 10, 158–165
Income generation 7, 60, 68, 154, 173, 177, 178, 179, 200, 323, 326
India 7, 8, 10, 12, 16, 54–82, 83–113
Indonesia 22, 23, 24, 25, 45, 47
  BAPENAS 23
  Brantus River Development Office 22
Indus Treaty 112
Industrial effluents 91, 130
Industrialization 5, 12, 48, 70, 72, 89, 90, 100, 138, 141, 151, 181, 207, 221–224, 244, 326, 367

Industry, small-scale 72, 227
Infant mortality 156, 191
Infiltration 147
Information flow 42, 43
Information technology 42, 43
Inner Mongolia 116, 119, 120, 121,
    124, 128, 129, 131–132, 133
    Hetao Irrigation District 131–
        132, 133
Integrated rural development 12,
    23–24, 28, 41, 45, 84, 87, 111,
    190, 243, 248, 312
Integrated water resources
    management 314, 315, 316–317,
    318
Institutional issues 23, 27, 47, 93–
    94, 112, 137, 145, 148, 149,
    181, 190, 247, 252, 254, 266,
    298–301, 312, 328, 342, 346,
    347
Institutional strengthening 133, 327
Inter-American Development Bank
    312, 313, 314, 319, 320, 324,
    325, 327, 329, 330, 360
Interbasin water transfer 14, 15,
    29, 37, 40, 46, 48, 49, 50, 109–
    111, 116, 121, 123, 260, 273,
    275, 278–279, 294
International Labour Organization
    178
Interregional disparities 16–17
Iraq 209
Irrigation 5, 6, 7, 27, 31, 33, 34,
    35, 36, 37, 38, 48, 50, 51, 58,
    59, 60, 65, 66, 67, 77, 80, 89,
    95, 106, 107, 115, 117, 120,
    123, 125, 126, 128, 129, 132,
    141, 144, 149, 160, 165, 168,
    169, 171–175, 176, 179, 180,
    182, 183, 185, 186, 187, 190,
    191, 192, 197, 198, 199, 200,
    207, 220–221, 240, 242, 244,
    292, 296, 310, 313, 316, 336,
    338, 340, 366

Drip 79, 80
Efficiency 79, 198
Expansion 5, 117
Return flow 77, 199
Sprinkler 79, 80
Israel 261

Japan 23, 25
Bank for International
    Cooperation 23
International Cooperation
    Agency 23, 24, 36, 41, 53
Official Development Assistance
    23
Overseas Economic Cooperation
    Fund 23
River Council 143
River Law 137, 139, 142–145,
    151
Special Measures Law 142, 153
Water Resources Development
    Acceleration Law 141, 151
Java 23, 25
Joint Technical Commission 344–
    348, 356, 357, 362, 367
Jordan 25

Kalpasar 78, 81
Kaunda, K. 260
Kenya 25, 252
Kizilcapinar 240, 242–243
Kuri Chu 11
Kutch 55, 58, 59, 67, 68, 69, 72,
    73, 74, 75, 77
Kyoto Convention 102

Labour-intensive 321
Land
    Arable 86
    Arid 10, 12, 13, 31, 257, 275
    Capability 27, 28
    Inundation 191, 198, 200, 326,
        338, 355, 356, 364
    Leveling 79

Reclamation 148
Subsidence 137, 147
Tenure 325
Use 30, 34, 59, 111, 145, 146,
    172, 198, 245
Landless people 2, 5, 8–9, 60, 61,
    62, 64, 67, 191
Lake
    Mono 304
    Oroville 294
    Shasta 294
Laos 25, 50, 112
Laws and regulations 1, 29, 79, 93,
    180, 221
Latin America 309–330
Legal framework 342–344
Lesotho 255, 257, 259, 260, 261,
    262, 264, 265, 267, 268, 270,
    280, 281
Life expectancy 59, 156
Lifestyle 2, 7, 10, 11, 13, 60, 139,
    191, 207, 240, 243, 244, 245,
    246, 247, 351
Literacy 59, 68, 69–70, 86, 177,
    191
Livestock 9, 34, 50, 51, 60, 65,
    132, 187, 221, 224
Loess Plateau 122–123
Low-flow augmentation 27, 85, 95,
    97, 106–108
Luzon 24, 25, 30–31, 34–37

Macmillan, H. 255, 256, 258
Mahakali Treaty 85, 87
Malawi 253, 259
Malaysia 25, 258
Maldives 88
Mandela, N. 263, 264
Manila 24
Manpower 29
Marketing 213, 214
Market-oriented 48
Master Plan 23, 28, 152, 174, 181,
    182, 196, 313, 316

Mauritius 252
McCaffrey, S. 114, 115
Mekong Treaty 112
Mexico 72, 310
Midgley, D. 262
Migration 77, 165, 171, 191, 198,
    201, 202, 222, 241, 243, 247,
    301
Minerals 14, 28, 187, 292
Mini-hydro 11
Model
    Calibration 37
    Global circulation 91
    Rainfall-runoff 37
Mollard, M. 335, 336, 337, 338
Monitoring 42, 91, 93, 96, 108,
    131, 172, 181, 223, 362, 363
Monsoon 67, 78, 87, 91, 92, 108,
    137, 144
Mozambique 25, 257, 258, 259,
    260, 261, 262, 264, 269, 270,
    271, 272, 273, 274, 278
Mugabe, R. 259
Multilateralism 85, 115
Mumbai 16
Myanmar 48

Nagoya 141
Namibia 254, 263, 264, 267, 268,
    270, 275, 276, 277, 280, 281
National security 73
Navigation 12, 26, 83, 85, 87, 92,
    95, 97, 102, 109–111, 298, 316,
    333, 334, 336, 337, 338, 358,
    360–361, 365, 366, 367
Nepal 10, 83–113
    Water and Energy Commission
    99
Netherlands 140
New Samsat 240–241
Nickum, J.E. 114–136
Ningxia 120, 121, 124, 129, 131
Non-governmental Organization 24,
    181, 182, 200, 315, 321

North-Atlantic Treaty Organization 265
Norway 104
Nutrients 132, 209, 296
Nutrition 6, 70, 177, 212

Ocean, Atlantic 333
Osaka 141
Operation and maintenance 47

Pakistan 73, 88, 97, 99, 112
Pampa 352
Paraguay 333, 334, 342, 352, 361
Participatory development 41, 42, 43, 52, 93, 94, 155, 180–181, 188, 321
Participatory irrigation management 79
Pasture 120, 129, 169
Peace 12
Pearson Report 41
Percolation tanks 78
Pesticides 198
Philippines 17, 22, 24, 25, 30–31
  Davao Integrated Development Programme 22
  Metro Manila Development Authority 22
  Subic Bay Metropolitan Authority 22
Plata Basin Treaty 342
Poland 25
Political instability 41
Population density 16, 29, 52, 290
Population growth 16, 34, 38, 51, 67, 156
Poor 2, 3, 4, 5, 7–8, 17, 42, 45, 78, 80, 86, 89, 93, 132, 177, 188, 213, 244, 246, 327
Portugal 258, 262, 265, 271, 272, 273, 311
Poverty alleviation 3, 10, 12, 13, 41, 42–43, 54, 59–65, 72–76, 78, 95, 140, 153, 158–165, 247, 323

Private sector 23, 93, 94, 105, 181, 183, 185, 186, 191, 200, 208, 312, 324
Privatization 47, 52
Project
  Bhima 8–9, 10
  Cabora Bassa 257, 258
  Chukha 10, 11, 100, 101
  El Cajón 319–320, 323
  Gandak 104
  Indira Gandhi Nahar 10
  Kosi 104
  Lesotho Highlands Water 260, 262, 263, 265, 268–269, 270, 271, 280
  Okavango 260
  Orange Vaal Transfer 269
  Pancheswar 104
  Sardar Sarovar 54, 58, 77, 79, 81
  Segredo 319, 324–326
  Southeastern Anatolia 154–183, 190–250
  Tugela-Vaal 259, 260
  Tula 101
Project beneficiaries 3, 4
Protected areas 59, 320, 321
Public awareness 47, 66
Public hearing 143
Public interest 298, 300, 316, 355
Public participation 17, 47, 176, 181, 182, 314
Public-private partnership 79
Public sector reform 186

Qinghai 124, 129, 139
Quality of life 13, 89, 101, 176, 190, 200, 241, 244, 247, 302, 309, 325, 327

Rainwater harvesting 54, 78–79, 81, 146
Recreation 27, 153, 198, 220, 298, 299, 301, 316, 335, 360, 362
Recycling 34, 35, 79, 81, 146, 301

Regional cooperation 85, 107, 112, 113, 255, 260, 265
Regional disparities 51, 54–82, 254
Regional income distribution 2
Regional stability 12, 254, 255, 264
Regional tension 252, 254
Rehabilitation 95, 96, 224–243, 247
Remote sensing 310
Resettlement 65, 96, 98, 139, 200, 224–243, 246, 247, 325, 326, 354–356
Retention basins 78
Rivers
    Amazon 310
    Barak 90
    Bhagirathi 106
    Boro 276
    Boteti 279
    Brahamaputra 15, 83–113, 117
    Brantas 22, 23, 47
    Buriganga 109
    Buzi 253
    Cagayan 35
    Chao Phraya 48, 49
    Chobe 263
    Colorado 130, 290, 294
    Cuareim 331
    Cunene 253, 259, 276, 277
    Cuvelai 253, 272, 273–274
    Danube 119
    Euphrates 168, 191
    Feather 296
    Fen 121, 123
    Gandak 111
    Gandaki 99, 109
    Ganges 15, 83–113, 117, 119
    Gorai 108
    Guadalquivir 311
    Hai 121, 123
    Huai 121, 123
    Huang Ho, see Yellow River
    Iguacu 325, 337, 366
    Incomati 253, 255, 270, 271–272, 273, 275

Indus 112, 280
Irrawaddy 48
Jordan 280
Karnali 99, 109, 111
Kirindi Oya 39–40
Kiso 150
Kornati 274
Kosi 90, 99, 109, 111
Limpopo 253, 255, 268, 277
Mahakali 99
Mahaweli 47
Maputo 253, 255, 264, 270, 272, 274–275
Meghna 83–113
Mekong 15, 50, 52, 112
Menik Ganga 39
Mississippi 119
Moutloutse 272
Nagara 148–151
Nata 252
Narmada 58, 77
Nile 112, 119, 121, 252, 280
Okavango 252–253, 254, 255, 257, 262, 263, 270, 272, 275–278
Olifants 273
Orange 253, 255, 256–257, 267, 271, 275
Pampanga 30–31, 35, 37
Paraná 333, 337, 366
Pepirí-Guazú 331
Plate 310, 331, 333, 334, 344, 366, 367
Pungue 253, 278
Rovuma 253
Sabie 273
Sacramento 290, 292, 293, 296, 297, 306, 307
Salween 15, 48, 49, 52
San Joaquin 290, 292, 296, 297, 306, 307
Sapta Kosi 99
Save 253
Sharda 99

Songwe 253
Sunkosh 11
Teesta 107
Thamalakane 259
Tigris 168
Tone 140
Tvolumne 294
Ulua 319
Uma Oya 39–40
Umbeluzi 253, 269, 273
Uruguay 331–367
Usumacinta 310
Vaal 256, 263, 268
Walawe Ganga 39–40
Wangchu 100
Wei 117, 121, 123
Yang-tze 48, 115, 120
Yellow 48, 114–136
  Delta 115, 119, 121, 123, 125, 128
  Zaire 253
  Zambezi 253, 257, 259, 262, 263
River management 137–153
River restoration 142, 150, 151
River training 27, 105, 109
Rivers, international 15, 26, 44, 48–51, 52, 83–113, 251–289
Rodriguez, D.J. 309–330
Run-off coefficient 29, 30, 32
Run-of-the-river 11, 100, 132

SAARC 88–89, 101
SADCC 259, 260, 261, 264, 279, 281
Saline barrier 140
Salinity 31, 32, 76, 108, 121, 124, 132, 148, 198, 274, 304
Salt water intrusion 46, 73, 91, 92, 148, 149
Sanitation 16, 26, 43, 44, 45, 47, 51, 52, 59, 108, 165, 240, 312, 313, 315, 319, 328
Şanliurfa 156, 157, 159, 160, 163, 164, 165, 167, 172, 175, 186, 190, 191, 197, 200, 225, 244

Scudder, T. 253, 259, 262, 263, 275, 276
Sea, Caribbean 319
Sea level rise 92
Sedimentation 89–90, 91, 108, 111, 114, 117, 121, 123, 129, 137, 198
Self reliance 42, 43, 52, 65, 124
Septic tank 45, 46
Seismicity 95, 198, 332
Service contract 47
Service coverage 45
Sewerage 45, 137, 145, 147, 152
Shaanxi 115, 117, 119, 121, 123, 124, 129, 131
Shandong 115, 116, 121, 124, 125, 126, 127, 128, 129, 131, 133
Shanghai 16
Sharecropper 242
Shengli oilfield 115, 119, 123, 125, 126
Shifting cultivation 72
Sichuan 120, 121, 131
Şırnak 156, 157, 159, 160, 163, 164, 165, 166, 167, 172, 190
Sludge treatment 45
Social activists 2–3
Social concerns 39
Social impacts 3, 244, 245, 246, 323
Social interactions 6, 43
Social issues 1, 2, 47, 93, 95, 98, 154, 179, 180, 197, 243, 245, 248, 274, 309, 326, 367
Social safety nets 43
Social stability 12, 16, 201
Soil conservation 39, 40, 111, 320, 321, 322
Soil moisture 79, 91, 92
Soil loss 320
Solar energy 80
Soler, G. 335, 336, 337
Solid wastes 46, 165, 166, 223, 224, 327
Somalia 252

South Africa 254, 255, 256, 258, 261, 262, 263, 264, 268, 270–271, 272, 273, 274, 280, 281
South African Development Commission 252, 264, 270, 278, 279, 280, 281
South Korea 124, 138
Soviet Union 258, 265, 266
Spain 292, 311, 315, 333
Sri Lanka 17, 22, 24, 25, 37–40, 47, 88
   Southern Development Authority 22
Standard of living 4, 5, 13, 59, 154, 187, 190
Sudan 112
Swaziland 261, 269, 272, 273, 274
Sweden 104
Switzerland 104
Syria 119, 168, 209

Taiwan 261
Takahasi, Y. 137–153
Tanzania 252, 253, 259
Taxation 17
Technology 1, 6, 222
Technological development 34, 174, 245
Technological cooperation 48, 51, 321
Tennessee Valley Authority 313
Terrorism 158–165
Thailand 7, 16, 17, 18–21, 23, 24, 25, 31–34, 47, 48, 49, 50, 51, 112
Third World Centre for Water Management 9
Tibet 102
Tokyo 141, 146, 147, 151
Top-down 22, 24, 43, 44, 52, 93, 313
Tortajada, C. 190–250
Tourism 14, 28, 187, 220, 335, 351, 360, 367
Training 180, 206, 207, 217, 222, 224, 242, 243

Transmigration 15
Transportation 6–7, 8, 12, 21, 89, 187, 200, 221, 240, 243, 327, 328, 329, 333, 334, 361
   Intermodal 12, 89
Trickle down effects 43
Turkey 7, 8, 154–183
Turton, A.R. 251–289
Typhoon 140, 148, 151

Unemployment 1–2, 72, 158–165, 206, 242
United Kingdom 256
United Nations 55, 112, 256
   Children's Fund 159
   Conference on Environment and Development 277
   Convention on Non-Navigable Uses of International Waterways 264, 270
   Convention on the Rights of the Child 75, 76
   Development Programme 41, 164, 190, 278
Ünver, O. 154–189, 190, 191, 197
United States of America 1, 313, 334
   Clean Water Act 298, 299
   Endangered Species Act 298, 299
   Federal Reclamation Act 1
   Flood Control Act 1
   National Environmental Protection Act 298, 299
   Rivers and Harbours Act 301
   Wild and Scenic Rivers Act 298, 299
Urban issues 5, 23, 26, 31, 41, 45, 46, 51, 59, 108, 119, 145, 146, 147, 148, 156, 158, 175, 177, 187, 199, 221, 227, 228, 243, 245, 319, 328, 355
Urbanization 5, 86, 139, 141, 151, 152, 201, 224, 304, 355
Uruguay 331–367

Vietnam 24, 25, 138, 258
Vulnerability 42, 92

Warsaw Pact 265
Wastewater 26, 44, 165, 223–224
   Treatment 26, 44, 46, 223
Waste disposal 45, 52, 91, 130, 145
Water
   Allocation 66, 92, 130–131,
      298, 327
   Assessment 27, 28, 29–30, 31,
      32
   Balance 32–40
   Collectors 92
   Conservation 39, 40, 81, 92,
      133, 176, 301, 306
   Crisis 48, 52, 65, 77–78
   Engine for development 2, 3, 4–
      10, 12, 190, 200, 243
   Market 131, 133, 134, 316
   Pricing 45, 46, 47, 80, 92, 127,
      133, 134, 180, 318
   Quality 5, 16, 26, 27, 44, 45,
      52, 87, 90–91, 93, 97, 107,
      108–109, 114, 115, 123, 131,
      137, 139, 145, 147, 148,
      165, 198, 211, 223, 245,
      301, 311, 320, 366
   Rights 27, 29, 38, 93, 112, 116,
      118, 131, 137, 144, 146, 298
   Scarcity 16, 35, 37, 38, 40, 44,
      54, 55, 60, 61, 62, 65, 68–
      76, 77, 78, 80, 83, 87, 128,
      130, 139, 141, 147, 151,
      260, 275, 316
   Security 260, 270
   Supply-demand relationship 52,
      292
   Tankers 57
   Treatment 45
Use
   Agricultural 7, 27, 31, 35, 38,
      50, 59, 60, 83, 84, 107, 118,
      126, 127, 128, 129, 132,
      137, 138, 146, 149, 198,
      292, 294, 300, 301, 302,
      304, 305
   Commercial 7, 107
   Domestic 5, 26, 27, 31, 32, 34,
      36, 44, 54, 57, 59, 68, 77,
      78, 80, 83, 84, 92, 107, 118,
      140, 144, 145, 147, 165,
      198, 310, 311, 312, 313,
      315, 316, 327, 328, 340, 366
   Efficiency 128, 133, 165, 180
   Environmental 83, 118, 294, 305
   Industrial 7, 27, 32, 35, 59, 72,
      73, 78, 79, 81, 84, 92, 107,
      118, 123, 125, 126, 127,
      128, 129, 137, 138, 149,
      198, 275, 304
   Non-consumptive 7, 275
   Rural 33, 38, 191
   Urban 33, 34, 38, 79, 123, 126,
      129, 137, 141, 191, 199,
      242, 292, 294, 301, 302,
      304, 305
   Unaccounted for 45
Water-borne diseases 198
Waterlogging 79, 198
Watershed Management 97, 111,
   321
Water Users Association 27, 133,
   134, 179, 180
Wetlands 138
Wildlife 34, 294, 296, 298, 299
World Bank 41, 59, 64, 72, 108,
   112, 130, 293, 302, 348
World Food Summit 65
World Health Organization 74
World Trade Organization 80
World Water Vision 65

Zambia 257, 259, 262
Zimbabwe 252, 254, 259, 260, 261,
   264, 270, 272, 278